AF780138

Ultrasonography of Muscles and Tendons

Bruno D. Fornage

Ultrasonography of Muscles and Tendons

Examination Technique and Atlas of Normal Anatomy of the Extremities

With 612 Illustrations

Springer-Verlag
New York Berlin Heidelberg
London Paris Tokyo

Bruno D. Fornage, M.D.

Associate Professor of Radiology
Division of Diagnostic Radiology
Chief, Section of Ultrasound
The University of Texas
M.D. Anderson Cancer Center
Houston, Texas, USA

Associate Professor of Radiology
Institut Jean-Godinot, Reims, France

Translation of *Echographie du système musculo-tendineux des membres/Atlas d'anatomie ultrasonore normale*, Vigot, Paris, 1987.
Translation by the author.

Library of Congress Cataloging-in-Publication Data
Fornage, Bruno.
Ultrasonography of muscles and tendons.
Translation of: Echographie du système musculo-tendineux des membres.
Includes bibliographies and index.
1. Muscles—Imaging. 2. Tendons—Imaging. 3. Ultrasonic imaging. 4. Muscles—Atlases. 5. Tendons—Atlases. I. Title. [DNLM: 1. Muscles—anatomy & histology—atlases. 2. Tendons—anatomy & histology—atlases. 3. Ultrasonic Diagnosis—methods—atlases. WE 17 F727e]
QM151.F6713 1988 611'.73 88-4657

Printed on acid-free paper

© 1989 by Springer-Verlag New York Inc.
Softcover reprint of the hardcover 1st edition 1989
All rights reserved. This work may not be translated or copied in whole or in part without the written permission of the publisher (Springer-Verlag, 175 Fifth Avenue, New York, NY 10010, USA), except for brief excerpts in connection with reviews or scholarly analysis. Use in connection with any form of information storage and retrieval, electronic adaptation, computer software, or by similar or dissimilar methodology now known or hereafter developed is forbidden.
The use of general descriptive names, trade names, trademarks, etc. in this publication, even if the former are not especially identified, is not to be taken as a sign that such names, as understood by the Trade Marks and Merchandise Marks Act, may accordingly be used freely by anyone.
While the advice and information in this book are believed to be true and accurate at the date of going to press, neither the authors nor the editors nor the publisher can accept any legal responsibility for any errors or omissions that may be made. The publisher makes no warranty, express or implied, with respect to the material contained herein.

Typeset by Publishers Service, Bozeman, Montana.

9 8 7 6 5 4 3 2 1

ISBN-13: 978-1-4612-8119-1 e-ISBN-13: 978-1-4612-3482-1
DOI: 10.1007/978-1-4612-3482-1

Introduction

Muscles, tendons, and the other soft tissues of the extremities have historically been neglected by radiology, in the main because of the lack of an effective imaging technique. Yet it is surprising that until recently sonography has not received much attention in this application, since, even with conventional equipment, it readily evaluates superficial tissues. In this setting, there is no obstacle from gas or bone to the propagation of sound. Further, sonography permits both depiction in any orientation and real-time dynamic study.

I was induced to undertake ultrasound examinations of muscles and tendons by physicians specializing in sports medicine. Before this specialty emerged, physicians had limited interest in the traumatic and inflammatory disorders of muscles and tendons, given the questionable severity of the lesions, the paucity of late sequelae impairing normal activity, and the absence of life-threatening complications. Now, however, athletics is a major social activity and sonography is used more and more frequently.

In the domain of muscular sonography, a dual expertise is required. First, precise knowledge of the regional anatomy is a prerequisite to any ultrasound examination. Second, the sonologist must be familiar with the clinical evaluation of patients with muscle trauma, in whom physical examination is often equivocal.

In this book, the technical fundamentals of the sonographic examination of muscles and tendons and the normal ultrasound anatomy of the extremities are presented.

Note to the Reader

In this atlas, sonograms are oriented according to the international standards for imaging techniques. All scans have been performed on right extremities. With few exceptions, which are noted, transverse sonograms are "seen from below," whereas longitudinally oriented scans, whether sagittal or coronal, are shown with the proximal part of the extremity on the left and the distal part on the right.

Acknowledgments. The author is indebted to Véronique Dupuis who prepared the manuscript, and Suzanne Simpson of the Department of Scientific Publications, The University of Texas M.D. Anderson Cancer Center, for her valuable editorial assistance.

Contents

Part I
Technique of Sonographic Examination and General Considerations on Ultrasound Normal Anatomy

1
Anatomic Considerations

Skeletal Muscles

Skeletal muscles consist of juxtaposed bundles of parallel muscle fibers with ensheathing connective tissue. Each fiber is several centimeters in length with a diameter of 10 to 100 μm. The fibers are arranged in fascicles separated by septa of connective tissue, the perimysia, that extend inward from an external sheath of dense connective tissue, called the epimysium, that surrounds the whole muscle. This perimuscular aponeurosis, or fascia, separates the various muscles (Fig. 1.1). In the fascicles, muscle fibers are separated by the endomysium. The general orientation of muscle fibers in relation to the overall axis of the muscle depends on the degree of muscular contraction. The striated muscle contains a significant amount of water and connective tissue.

A rich vasculature enables dramatic variations in blood flow. The blood vessels and the nerve form a neurovascular bundle to enter the muscle. Blood vessels run longitudinally in the perimysium between fascicles. During intense contraction, blood vessels may be squeezed so that the muscle has to function anaerobically.

The nerve fibers of motor neurons enter the muscle at the motor end plate. Each axon innervates several muscle fibers. The motor unit comprises one motor neuron and all the innervated muscle fibers.

Histologically, the muscle fiber is a highly differentiated syncytial cell that contains longitudinally oriented contractile filaments of actin and myosin. The contractile filaments are responsible for muscle contraction; they slide past one another when energy is provided in the form of adenosine triphosphate (ATP). Accordingly, the muscle fiber is rich in mitochondria, which lie in contact with the outer membrane (the sarcolemma), and glycolytic and respiratory enzymes. It also contains large amounts of glycogen, ATP and phosphocreatine, and myoglobin.

The over 650 muscles account for 45% of the total weight of the body. There is significant individual variation in the shape and size of muscles. Athletic or occupational activities may result in muscular hypertrophy. Conversely, disuse may be associated with a relative hypotrophy, whereas total immobilization may lead to atrophy. Such variations in muscle size develop fairly quickly and are reversible. Supernumerary (or accessory) muscles have been described, particularly in the hand (extensor digitorum brevis) and the lower calf (accessorius ad quadratum plantae) [2], and other muscles may be absent (plantaris).

Tendons

Tendons are elongated structures made of dense, regular connective tissue whose role is to attach the striated muscle to bone. They cannot be stretched and are extremely resistant to traction forces (up to 400 kg for the Achilles tendon). McMaster [1] has shown that one half of the fibers of the tendon of the gastrocnemius muscle in the rabbit must be injured before the tendon ruptures under violent tension, while the preservation of as little as one fourth of the fibers can prevent rupture during a muscular contraction of normal intensity.

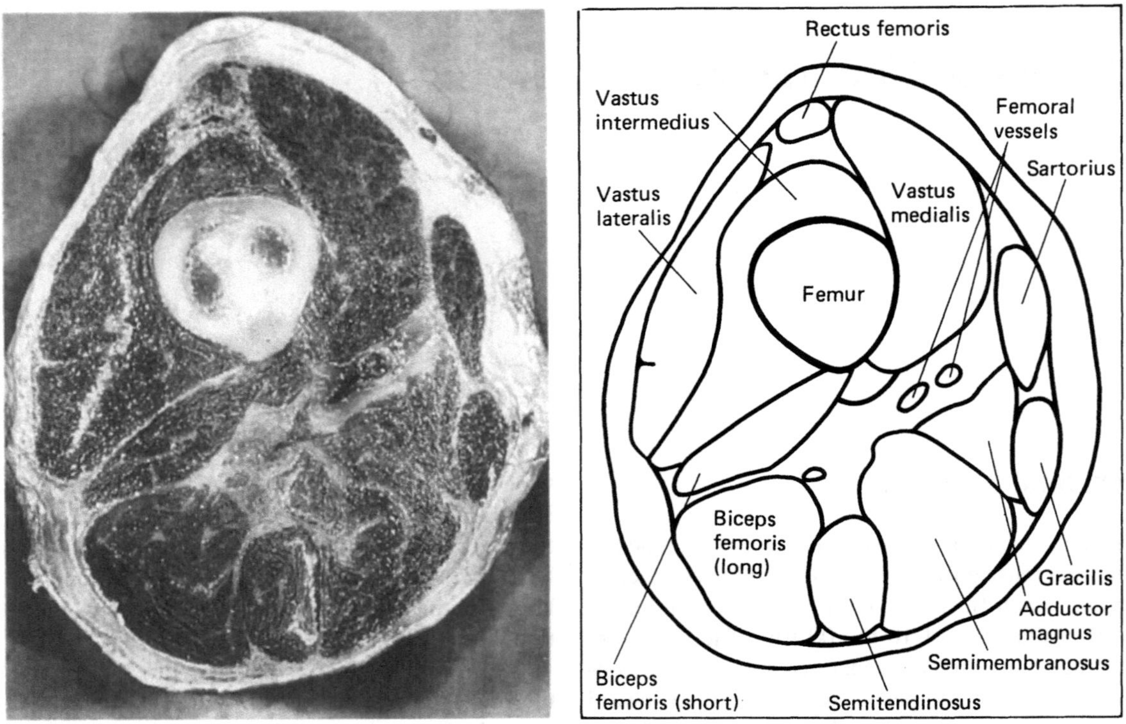

a b

FIGURE 1.1. Transverse anatomic section of the lower third of the thigh shows the various muscles delineated by fasciae.

Tendon fibers are continuous with the muscular connective tissue and are therefore clearly distinct from the muscle fibers. The closely packed, parallel collagen fibers are separated by a small amount of amorphous intercellular ground substance that contains a few fibrocytes. The primary bundles aggregate into secondary, and occasionally tertiary, bundles. The peritenon is a layer of loose connective tissue surrounding the tendon and continuous with the muscular aponeuroses. It originates intratendinous septa, separating bundles of collagen fibers. These septa include a few neurovascular bundles and play a basic role in nutrition. In contrast to muscles, tendon vascularization is remarkably poor, with nutritional exchange occurring mainly in the highly hydrated intercellular substance.

In the absence of mechanical constraint or anatomic peculiarity, the tendons are surrounded by fatty areolar tissue, the paratenon [3]. In areas of mechanical effort or in particular anatomic situations (narrowed passages or curved courses), they are invested with a synovial sheath. This virtual cavity lined with mesenchymal cells includes a minimal amount of synovial fluid (similar to that of joints) that lubricates the tendon and allows it to slide smoothly [4]. Fibrous sheaths maintain some superficial tendons close to segments of bone, particularly in the distal extremities (flexor and extensor retinacula at the wrist, fibrous sheaths of flexor tendons in the fingers, peroneal and flexor retinacula in the foot). Variants of normal anatomy include intratendinous sesamoid bones.

Synovial Bursae

The synovial bursae facilitate the sliding movements of muscles and tendons. Normal bursae have a histologic structure similar to that of synovial sheaths, whereas secondarily acquired bursae have no synovial lining. Synovial bursae are mainly affected by inflammatory disorders (bursitis or hygroma).

Joints

Apposed bone surfaces are in contact through radiolucent articular cartilage, whose thickness varies according to the shape of the articular surface. Some joints include complementary fibrocartilages (menisci). Anatomically, joints are limited by the synovial membrane. Physiologically, there is a small amount of viscous synovial fluid that lubricates the articular surfaces. Periarticular ligaments are made of dense connective tissue similar to that of tendons.

Fatty Tissue

Fat is found everywhere in the body. In the extremities, it is present between the muscle groups, in the vicinity of joints, and in the subcutaneous tissues. The distribution of adipose tissue throughout the body may show dramatic variations depending on anatomic site, age, sex, hormonal status, stress, and other factors. Accordingly, the sonographic appearance of fat may also vary to some extent with such unpredictable factors.

References

1. McMaster PE: Tendon and muscle ruptures. Clinical and experimental studies on the causes and location of subcutaneous ruptures. J Bone Joint Surg [Am] 1933;15:705–722.
2. Nidecker AC, von Hochstetter A, Fredenhagen H: Accessory muscles of the lower calf. Radiology 1984;151:47–48.
3. Razemon JP, Duquennoy A: Tendons (pathologie traumatique et non traumatique). Encycl Méd Chir (Paris), 10-1965, Fasc. 15151 C10, 1–10.
4. Verhaeghe A, Decoulx P, Demaille A, Razemon JP: Affections des gaines synoviales. Encycl Méd Chir (Paris), 7-1959, Fasc. 15153 A10, 1–4.

2
Technique for Sonography of Muscles and Tendons of the Extremities

Type of Sonographic Equipment

As early as 1965, a transverse sonographic scan of a normal leg was published by Howry [13]. The technological development of sonographic equipment since that date is well known. The high quality of today's real-time machines has rendered B-mode static equipment obsolete [4–11]. Real-time examination is faster, more convenient, and, above all, more complete since continuous scanning leaves no blind area. Also, real-time capability allows dynamic examination of muscles during flexion/extension maneuvers or palpation.

Superficial soft tissues are best evaluated using linear-array transducers. The parallel beams provide a high-quality near-field (Fig. 2.1a). In contrast, the near-field of sector scanners is characterized by a reduced size and poor resolution. Furthermore, in the lateral portions of the ultrasound image, the beam is excessively oblique to the skin and the underlying interfaces (Fig. 2.1b) (see also Chapter 5).

For technical reasons, the width of the field of view of a linear-array electronic transducer is related to its frequency. An advantage of B-mode static equipment was the visualization of long anatomic segments. However, modern real-time linear-array units have the capability of displaying two and even three adjacent scans on the video monitor. This permits the visualization and measurement of structures or lesions whose size exceeds the width of the probe (Fig. 2.2).

The current trend is toward miniaturized ultrasound equipment. High-quality but low-cost portable machines are now available with 5.0- and 7.5-MHz probes (Fig. 2.3). These portable units can also be used in veterinary medicine, particularly in the evaluation of tendon diseases of the legs in racehorses (Fig. 2.4).

Probe Frequency

High frequencies are associated with improved spatial resolution, but also with increased attenuation of the beam. Currently, commercially available probes range from 3.5 to 10 MHz, and the operator has to select the appropriate probe according to the depth of the region of interest, keeping in mind the crucial role of the focal zone. Most examinations of muscles are performed using a 5-MHz probe, and the evaluation of superficial tendons usually requires the use of a 7.5-MHz probe [8,10]. The 7.5- and 10-MHz probes are needed for the evaluation of distal extremities and very superficial structures, including dermatologic applications [4,9,11]. It must be borne in mind that the higher the frequency of the probe, the narrower the field of view, which is limited to 3–4 cm for most 7.5-MHz linear-array transducers.

Standoff Systems

The use of a standoff pad allows visualization of the skin and the most superficial structures. Also, by increasing the distance between the probe and the region of interest, the standoff pad places the region of interest in the optimal focal zone of the transducer. Thus, acceptable images of muscles

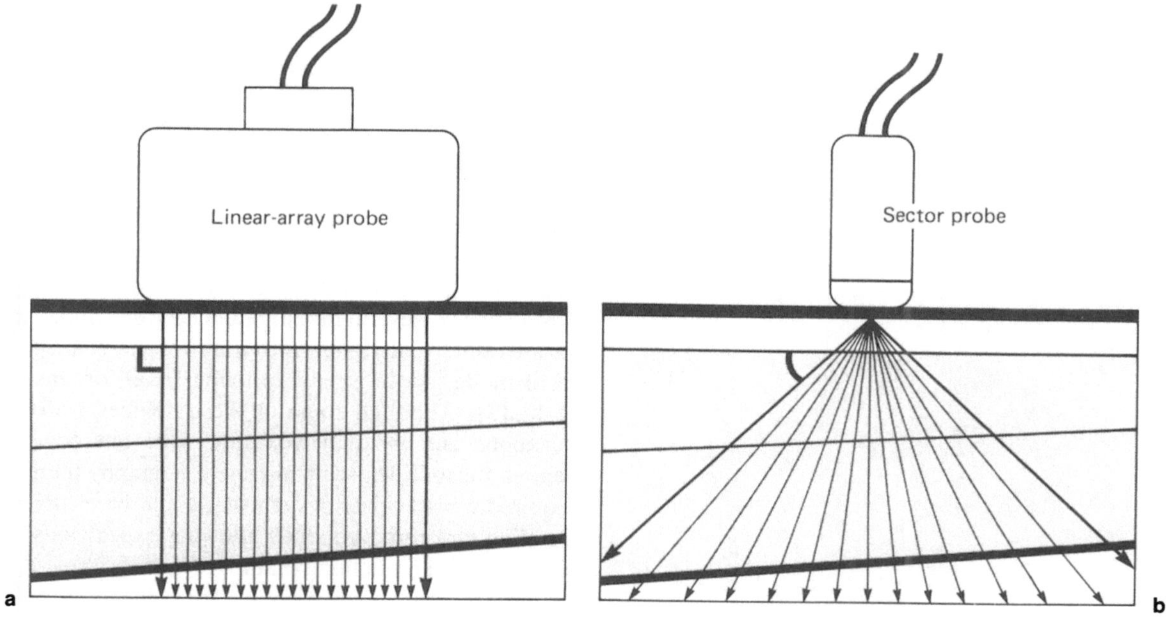

FIGURE 2.1. Diagram comparing linear electronic and sector scanning of superficial soft tissues. Longitudinal scan of the anterior region of the thigh. (a) The parallel beams of the linear-array probe are orthogonal to the skin and the underlying interfaces. (b) In the lateral portions of the sector image, the beam is markedly oblique in relation to the interfaces.

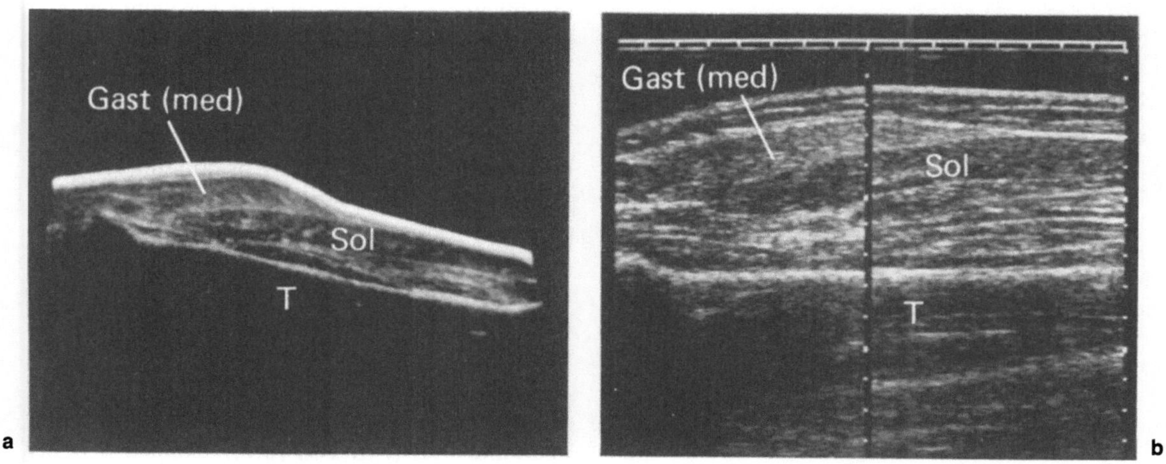

FIGURE 2.2. Longitudinal sonograms of the calf. (a) Scan obtained with B-mode static equipment shows the whole length of the triceps surae. (b) Two contiguous real-time scans are displayed on the video monitor, thus doubling the width of the ultrasonic field of view. *Gast (med)* = gastrocnemius (medial head); *Sol* = soleus; *T* = tibia.

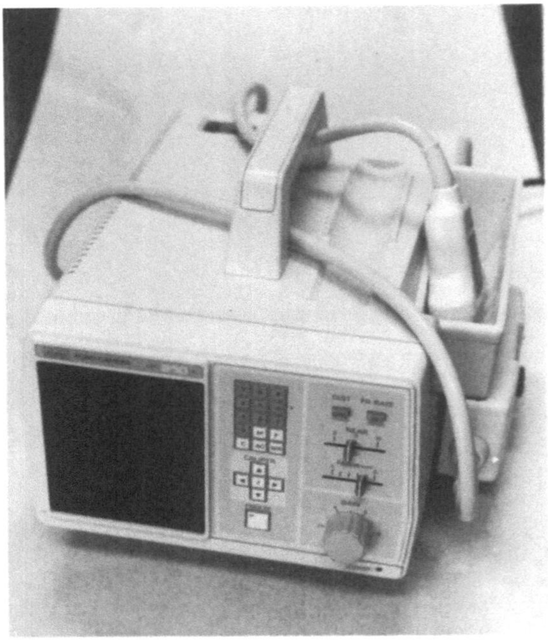

FIGURE 2.3. Portable linear-array sonographic unit that accepts 5.0- and 7.5-MHz transducers.

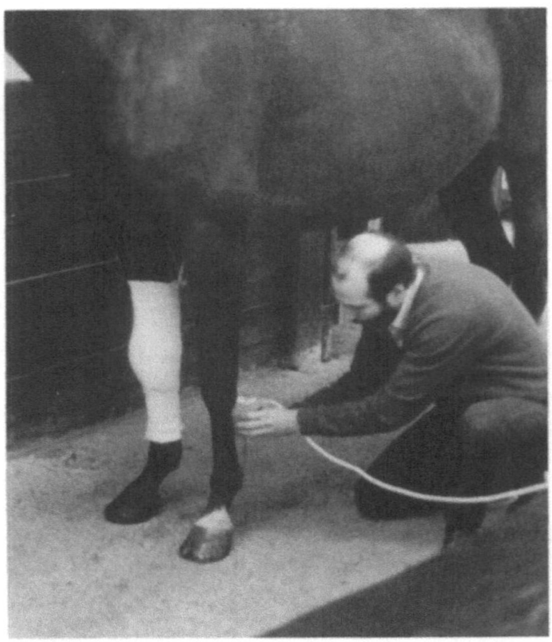

FIGURE 2.4. Ultrasound examination of the tendons of the leg of a racehorse performed using a portable unit at the stud farm.

and tendons can be obtained using the 3.5-MHz linear-array transducers that are routinely employed in abdominal and obstetrical sonography (Fig. 2.5). On the other hand, when only sector scanners are available the use of a standoff pad reduces the obliquity of the beam in the lateral areas of the region of interest (Fig. 2.6). In the same manner, the built-in water-path of very high-frequency (10-MHz) sector probes results in a trapezoidal image format (Fig. 2.7).

Since 1977, several investigators have developed various water-path systems to improve the quality of B-mode static scans of superficial organs [1,14,15]. Although some of these devices were ingenious and inexpensive, they were not convenient for routine use. Real-time sonography then required a water-path system that could be easily installed and removed, thus allowing expeditious changing of the scanning plane.

Our first studies were conducted using disposable latex gloves filled with tap water as water bags (Fig. 2.8). The few air bubbles that remained after the glove was closed were easily moved to the tip of a finger, then trapped by a knot. The disadvantages of such a system were (*a*) strong reflections from the glove walls, which could generate reverberation artifacts, (*b*) a significant acoustic enhancement distal to the water, which could obscure very superficial tissues, (*c*) a cumbersome examination technique due to the instability between the two convex surfaces of the water bag and the extremity, and (*d*) the risk of rupture of the water bag. Some manufacturers have water bags specially adapted to the scanhead.

Since 1982, we have routinely used a standoff pad made of a gelatinous mixture of an oil and a viscoelastic synthetic polymer. This material is employed in the manufacture of flotation rest pads used to prevent decubitus ulceration and its use has also been proposed for the development of tissue-equivalent phantoms in sonography [12]. The material is now commercially available (Kitecko, 3M Co., St. Paul, MN). The standoff pad is placed on the skin after liberal application of coupling medium. Its deformability and compressibility allow optimal contact with even the most irregular anatomic sites (Fig. 2.9) [6,7]. The pad is easily cleaned with soap and water. The ultrasound velocity through this material and its specific gravity (and therefore its acoustic impedance) are close to those of superficial soft tissues, resulting in a

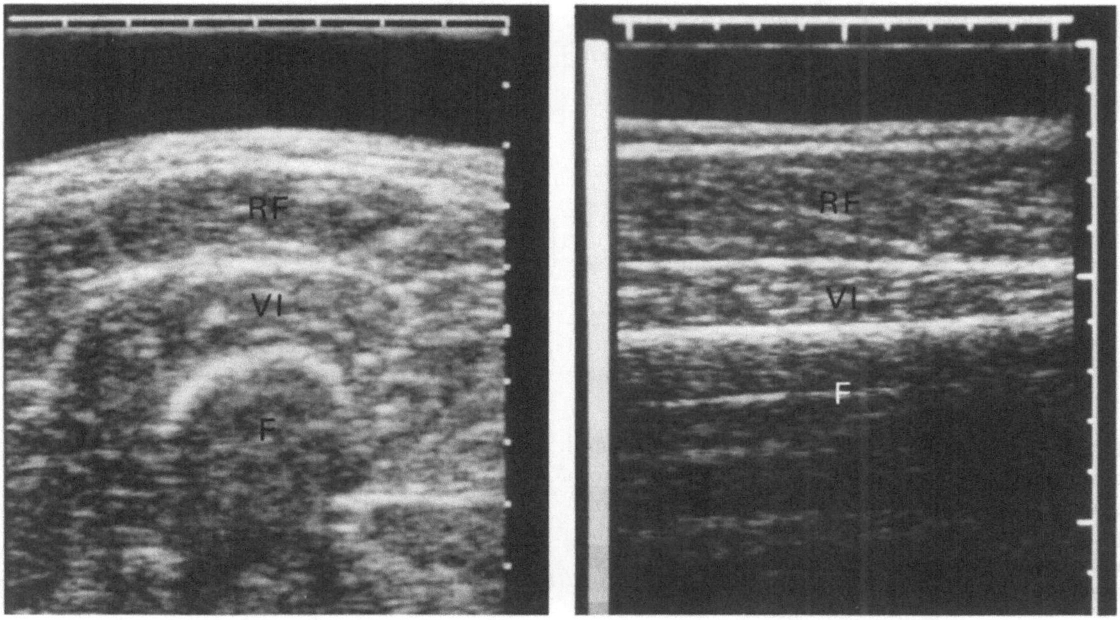

a b

FIGURE 2.5. Sonograms of the thigh performed using a standoff pad and a standard 3.5-MHz linear-array transducer routinely employed in abdominal and obstetrical sonography. (a) Transverse scan of the thigh using an 8-cm-wide probe. (b) Longitudinal scan obtained with an 11-cm-wide probe. F = femur; RF = rectus femoris; VI = vastus intermedius.

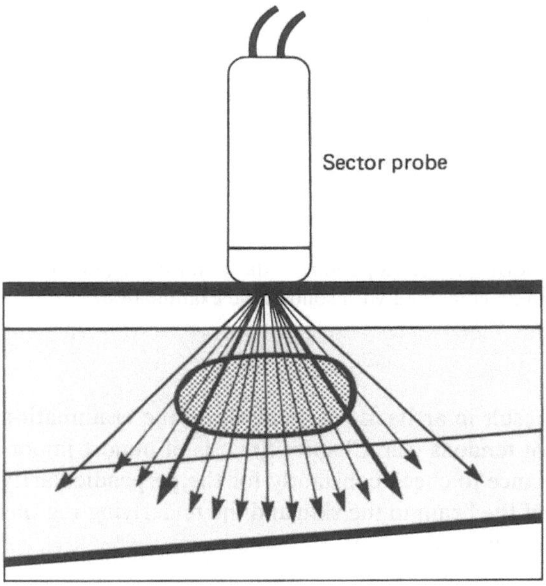

a

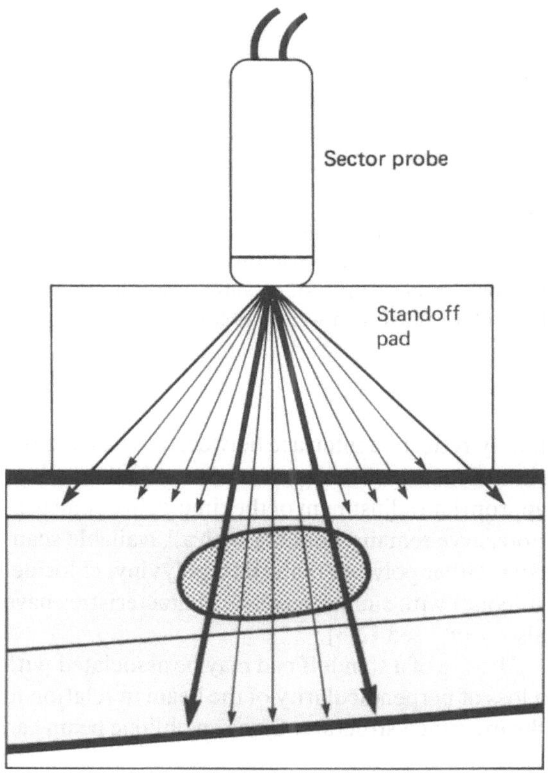

b

FIGURE 2.6. Diagram showing the value of a standoff pad when the examination is performed with a sector scanner. Longitudinal scan of the anterior region of the thigh. (a) The transducer is placed directly on the skin. (b) With a standoff pad, the probe is away from the skin, resulting in a reduced obliquity of the beam in the lateral portions of the region of interest.

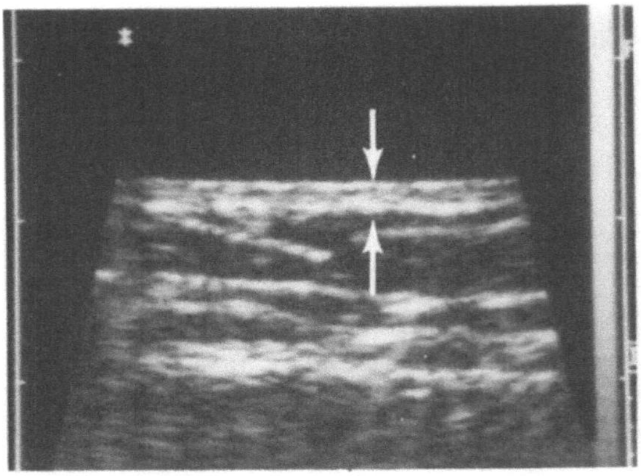

FIGURE 2.7. Sonogram of the skin (*arrows*) and subcutaneous tissues performed using a 10-MHz sector scanner with a built-in water-path and a thin standoff pad. The image format is trapezoidal. The width of the ultrasonic field of view is less than 3.5 cm at 3 cm from the transducer.

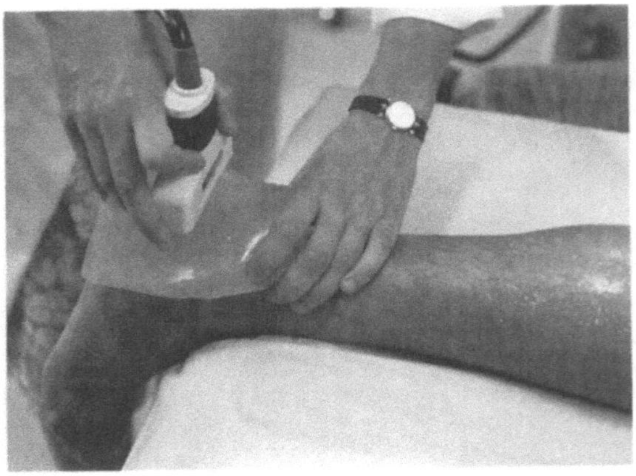

FIGURE 2.9. Real-time sonographic examination of the Achilles tendon. The standoff pad provides satisfactory contact between the linear-array probe and the skin, which is particularly helpful for transverse scans.

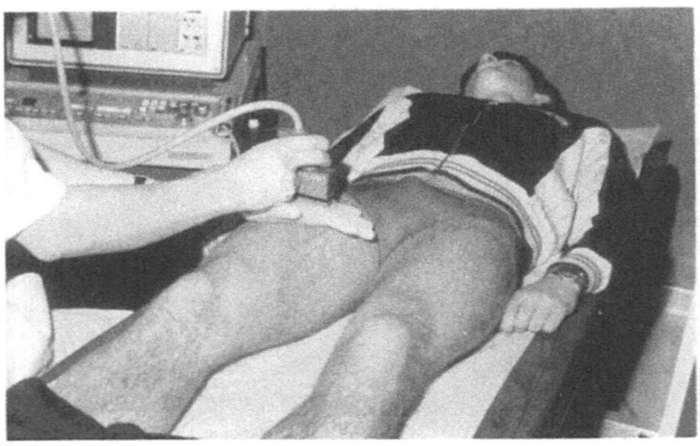

FIGURE 2.8. Sonographic examination of the anterior region of the thigh using real-time equipment and a self-fashioned water-path system consisting of a disposable latex glove filled with water (1980).

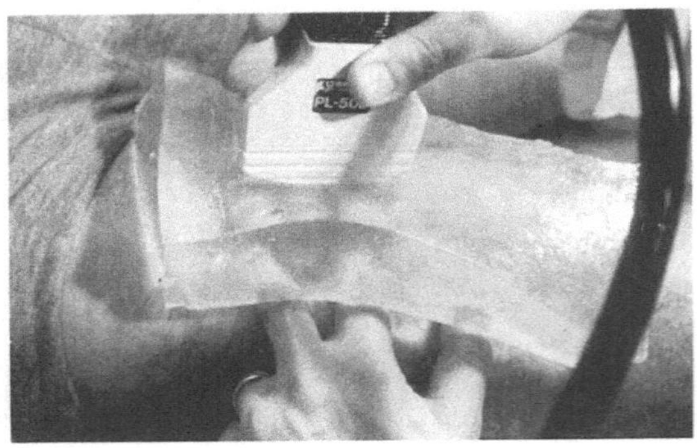

FIGURE 2.10. Palpation of the anterior region of the thigh combined with sonoscopic examination.

mildly reflective interface between the pad and the skin. Because of the moderate sound attenuation, appropriate adjustment of the time-gain compensation curve remains possible with all available scanners. Other polymer materials (polyvinyl chloride, silicone) with similar acoustic characteristics have also been used [2,3].

The use of a standoff pad may be associated with a loss of perpendicularity of the beam in relation to the examined structure. Since an oblique beam can result in artifacts, particularly in the examination of tendons (see Chapter 5), it is of utmost importance to check constantly for the perpendicularity of the beam to the skin and the underlying regions of interest.

Orthogonal Views

The need for orthogonal views in ultrasound examinations applies to superficial soft tissues as well. The combination of longitudinal and transverse scans allows (*a*) more accurate identification of the structures to be evaluated, (*b*) differential diagnosis between a vessel and a minute fluid collection, (*c*) preoperative three-dimensional localization of lesions in particular foreign bodies, and (*d*) determination of the volume of lesions.

Examination of the Contralateral Segment

The contralateral segment represents a crucial reference of normal anatomy for the novice operator. However, it must be kept in mind that some diseases may involve the extremities bilaterally. Examples are inflammatory disorders of large tendons in the lower extremities resulting from certain athletic activities and tendinopathies in patients with systemic diseases (such as gout, lupus erythematosus, and rheumatoid arthritis).

Real-Time Dynamic Study

At present, only sonography allows cross-sectional real-time imaging, which further allows the real-time dynamic study of muscles and tendons. Ultrasound examination is first performed with the patient at rest, then during active and passive flexion/extension maneuvers that involve the muscle or the group of muscles under evaluation. This dynamic examination is particularly useful in the accurate identification of small muscles. Obviously, the operator must know the physiologic function of each muscle, so that the patient can be asked to perform the maneuver that will identify a particular muscle. In pathologic conditions, the examination of the muscle during contraction may better demonstrate the alterations of the normal architecture caused by the lesion. Therefore, dynamic study should be used to enhance the sensitivity of sonography in the demonstration of minute or isoechoic lesions.

Palpation Under Sonoscopy

Prior to any sonographic examination, the radiologist must evaluate the patient clinically by taking a precise history and palpating the involved segment. Palpation under sonoscopic monitoring is a unique advantage of real-time sonography (Fig. 2.10). It represents the meeting point between "live" imaging and physical examination. This technique allows one to focus the sonographic study on palpable abnormalities and, conversely, to concentrate palpation on sonographically equivocal areas. Palpation under sonoscopy also provides information about the dynamic behavior of lesions.

References

1. Berger LA, Lister A, Lilley AC: A water bath for thyroid scanning with ultrasound. Br J Radiol 1981;54:630–631.
2. Chan B, Merton-Gaythorpe JV, Kadaba MP, Zafaranloo S, Bryk D: Acoustic properties of polyvinyl chloride gelatin for use in ultrasonography. Radiology 1984;152:215–216.
3. Claes HP, Reygaerts DO, Boven FA, Leemans J, De Boeck M, Geurts JM: An echo-free silicone elastomer block for ultrasonography. Radiology 1984;150:596.
4. De Flaviis L, Nessi R, Del Bo P, Calori G, Balconi G: High-resolution ultrasonography of wrist ganglia. J Clin Ultrasound 1987;15:17–22.
5. Fornage B: Une nouvelle indication des ultrasons: l'échographie du système musculo-tendineux et des parties molles. Concours Med 1984;106:3833–3838.
6. Fornage B, Touche D: Un matériau d'interposition pour l'échographie des organes superficiels. J Radiol 1984;65:215–216.
7. Fornage BD, Touche DH, Rifkin MD: Small-parts real-time sonography: A new "water-path." J Ultrasound Med 1984;3:355–357.
8. Fornage BD: Achilles tendon: US examination. Radiology 1986;159:759–764.
9. Fornage BD, Deshayes JL: Ultrasound of normal skin. J Clin Ultrasound 1986;14:619–622.
10. Fornage B: L'échographie de la main: technique et anatomie ultrasonore normale. Journal d'Echographie et de Médecine Ultrasonore 1986;7:193–203.
11. Fornage BD, Rifkin MD: Ultrasound examination of the hand and foot. Radiol Clin North Am 1988; 26:109–129.
12. Fyke FE, Greenleaf JF, Johnson SA: Continuous wave measurements of acoustic attenuation in an oil/polymer mixture. Ultrasound Med Biol 1979; 5:87–90.

13. Howry DH: A brief atlas of diagnostic ultrasonic radiologic results. Radiol Clin North Am 1965;3: 433–452.
14. Pochaczevsky R: Simple transducer supported water bath device for ultrasonography. AJR 1979;133: 553–555.
15. Yeh HC, Wolf BS: A simple portable water bath for superficial ultrasonography. AJR 1978;130:275–278.

3
Normal Ultrasound Anatomy of Muscles

The normal ultrasound appearance is similar for all skeletal muscles.

Longitudinal Sonograms

Scans obtained along the long axis of the muscle show a hypoechoic background with fine, oblique, echogenic striae corresponding to the fibroadipose septa that separate the fascicles of muscle fibers (Figs. 3.1–3.4) [1–4].

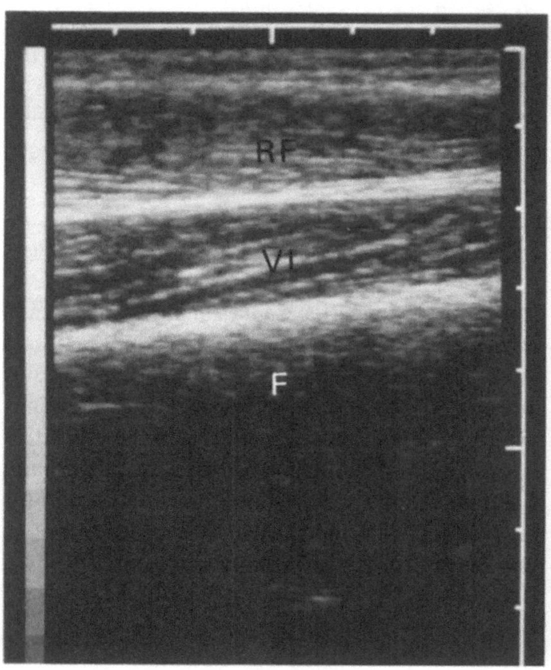

FIGURE 3.1. Longitudinal scan of the anterior region of the midthird of the thigh. The normal muscular echotexture consists of a hypoechoic background with fine oblique striae, which are prominent in the vastus intermedius. *F* = femur; *RF* = rectus femoris; *VI* = vastus intermedius.

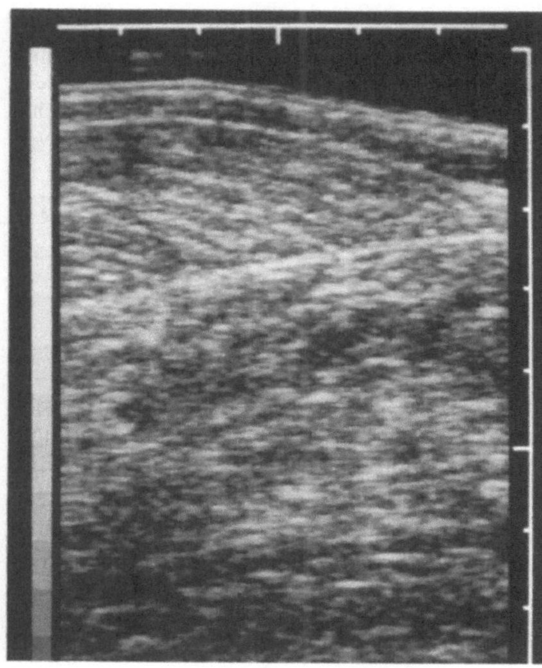

FIGURE 3.2. Longitudinal sonogram of the calf at the termination of the medial gastrocnemius shows fine oblique, internal, echogenic striae.

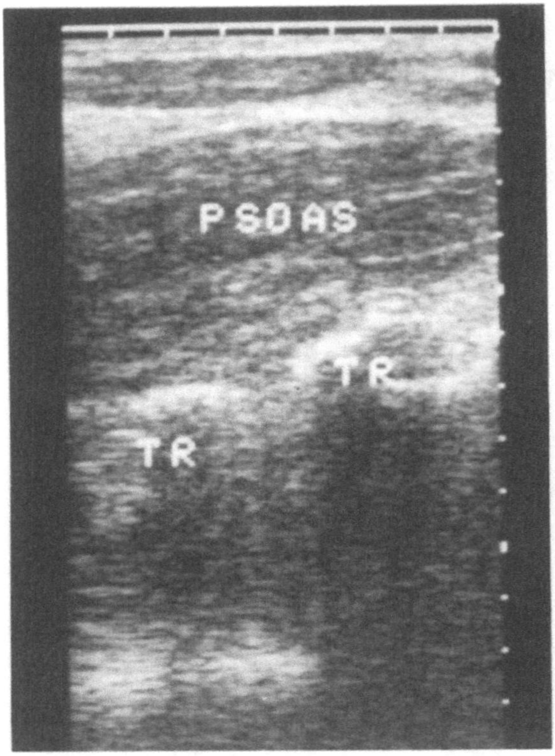

FIGURE 3.3. Longitudinal sonogram of psoas muscle in a young athlete shows the typical ultrasound appearance of skeletal musculature with longitudinally oriented, echogenic striae. *TR* = transverse processes of lumbar vertebrae.

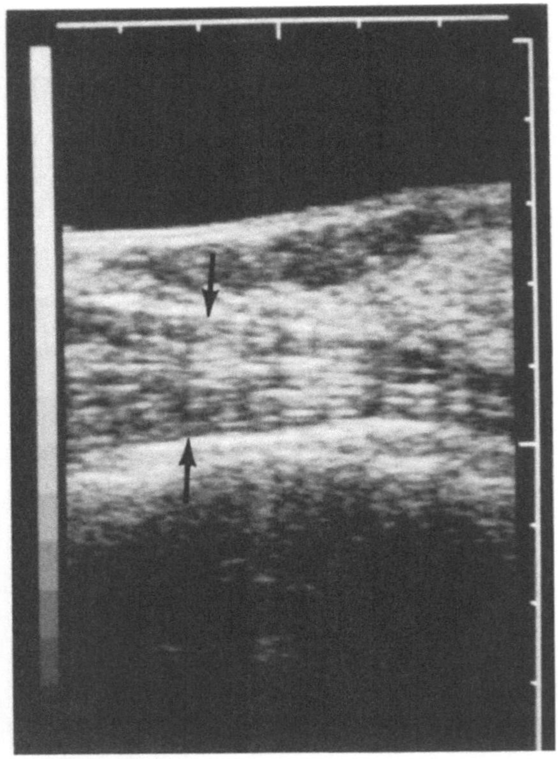

FIGURE 3.4. Oblique sonogram along the tail of the breast shows the pectoralis major muscle (*arrows*) with the striated pattern characteristic of skeletal musculature.

Transverse Sonograms

On transverse sections, the small connective septa appear as finely dotted echoes scattered throughout a generally hypoechoic background (Figs. 3.5–3.7). Occasionally, large intramuscular echogenic septa can be visualized, yielding a reticular pattern (Fig. 3.8) [1,2]. In this case, the identification of the connective septa on transverse scans may clarify misleading patterns seen on longitudinal sonograms (see Chapter 5). Perimuscular aponeuroses appear as brightly echogenic lines (Fig. 3.9). Small muscles of the distal extremities may not be well delineated on transverse scans, and therefore require specific muscular contraction for precise identification [2]. Fat separating groups of muscles is usually moderately echogenic.

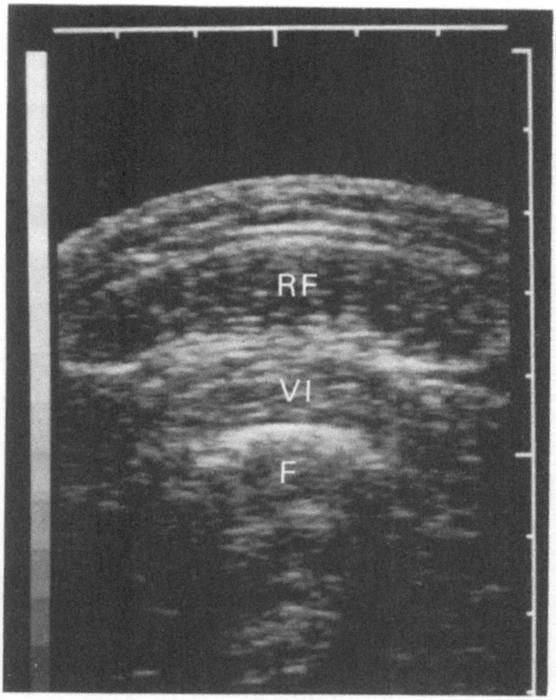

FIGURE 3.5. Transverse scan of the anterior region of the midthird of the thigh. The rectus femoris shows an overall hypoechogenicity, with scattered low-level echoes giving a mottled pattern. *F* = femur; *RF* = rectus femoris; *VI* = vastus intermedius.

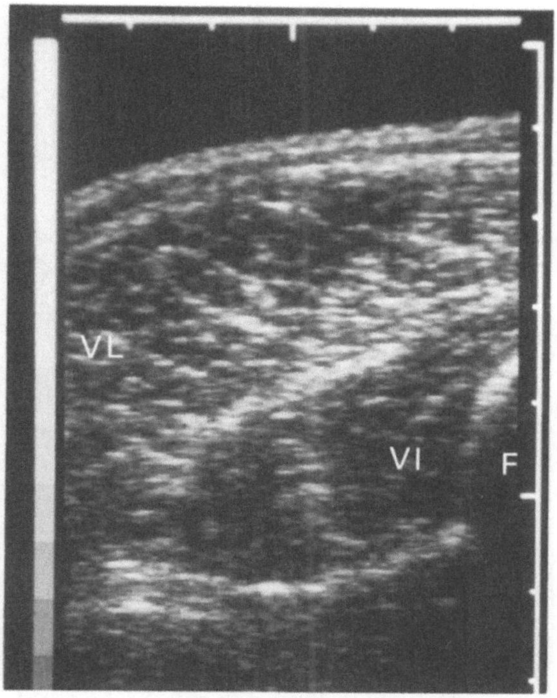

FIGURE 3.7. Transverse scan of the anterolateral region of the thigh shows the vastus lateralis and vastus intermedius. In addition to low-level dot echoes, small echogenic internal septations are visualized. *F* = femur; *VI* = vastus intermedius; *VL* = vastus lateralis.

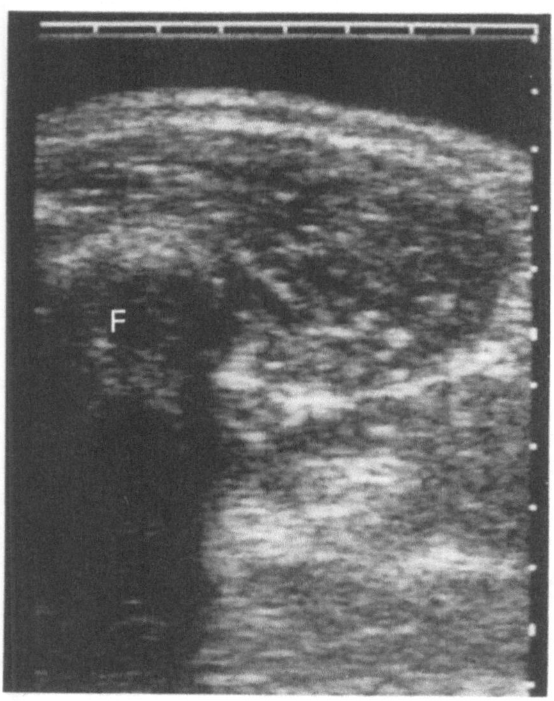

FIGURE 3.6. Transverse scan of the anterior region of the thigh focused on the vastus medialis. Note the echogenic dotted pattern over the background hypoechogenicity. *F* = femur.

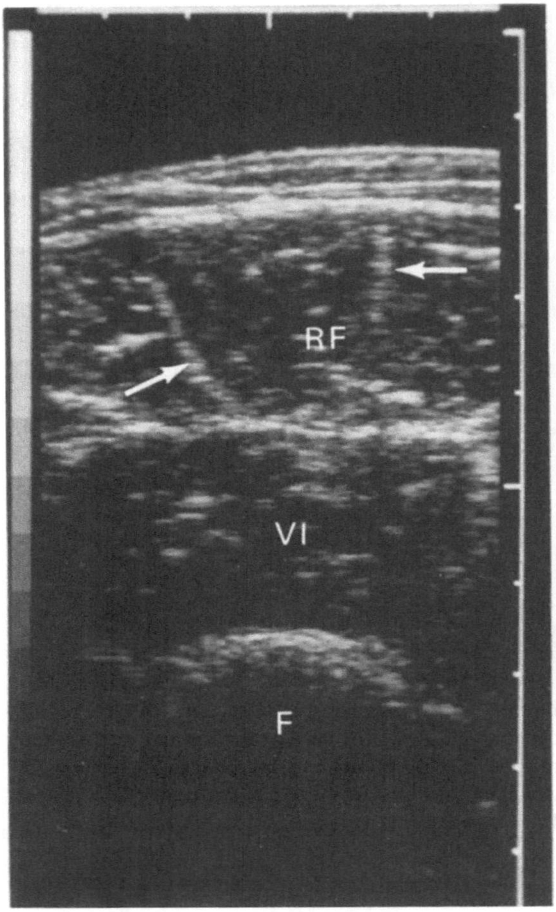

FIGURE 3.8. Transverse scan of the anterior region of the thigh shows prominent intramuscular septations in rectus femoris (*arrows*). *F* = femur; *RF* = rectus femoris; *VI* = vastus intermedius.

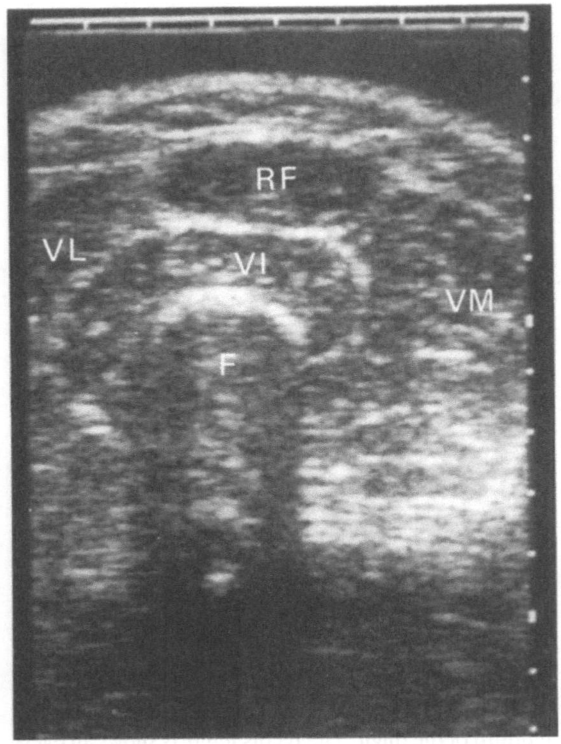

FIGURE 3.9. Transverse scan of the anterior region of the thigh. The echogenic perimuscular aponeuroses (epimysia) separating the various muscles of the quadriceps are clearly depicted. *F* = femur; *RF* = rectus femoris; *VI* = vastus intermedius; *VL* = vastus lateralis; *VM* = vastus medialis.

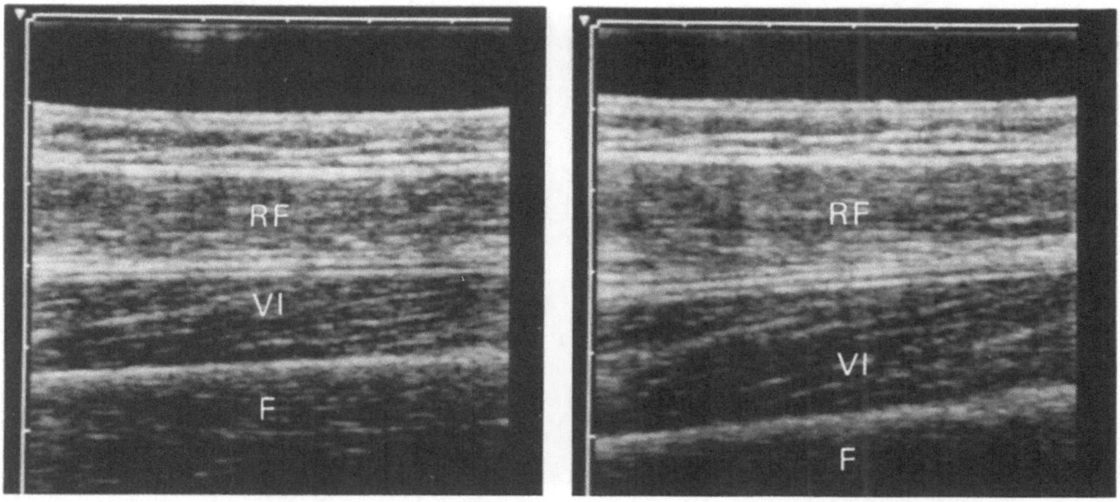

FIGURE 3.10. Real-time dynamic examination during muscular contraction. Longitudinal sonograms of the anterior region of the thigh. (a) Sonogram performed at rest. (b) Sonogram obtained during muscular contraction shows the increased thickness of the rectus femoris and vastus intermedius. Note the more oblique, echogenic striae in the vastus intermedius. *F* = femur; *RF* = rectus femoris; *VI* = vastus intermedius.

Dynamic Examination

During muscular contraction, the thickness of the muscular body increases. Alterations of the internal texture during contraction include a more pronounced obliquity of the echogenic striae and a more hypoechoic background (Figs. 3.10, 3.11).

Occasionally, small pulsating intramuscular arterial branches can be demonstrated (Fig. 3.12).

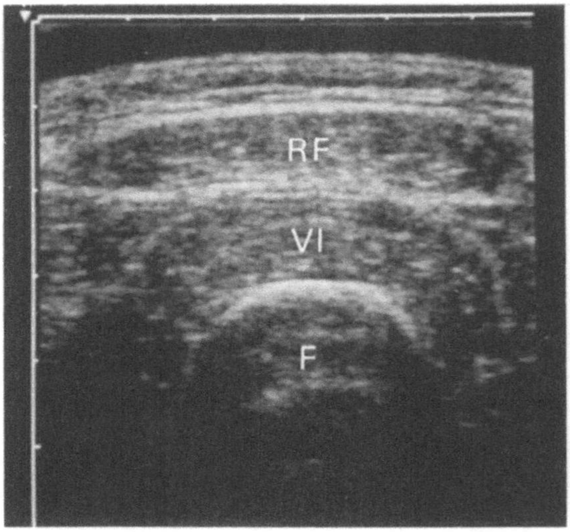

a

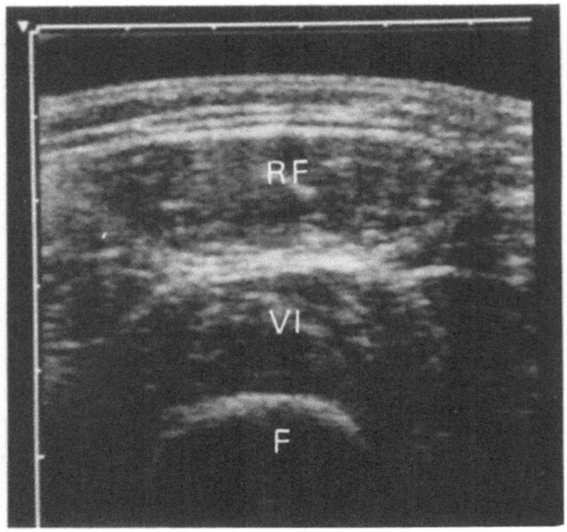

b

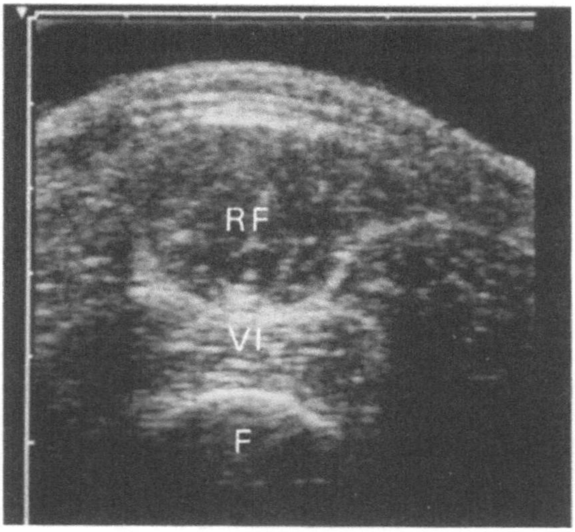

c

FIGURE 3.11. Real-time dynamic examination during muscular contraction. Transverse scans of the anterior region of the thigh. (a) Sonogram performed at rest. (b) Sonogram obtained during moderate contraction. Muscles are more hypoechoic. (c) Sonogram performed at full contraction. The rectus femoris has become rounded and markedly hypoechoic. *F* = femur; *RF* = rectus femoris; *VI* = vastus intermedius.

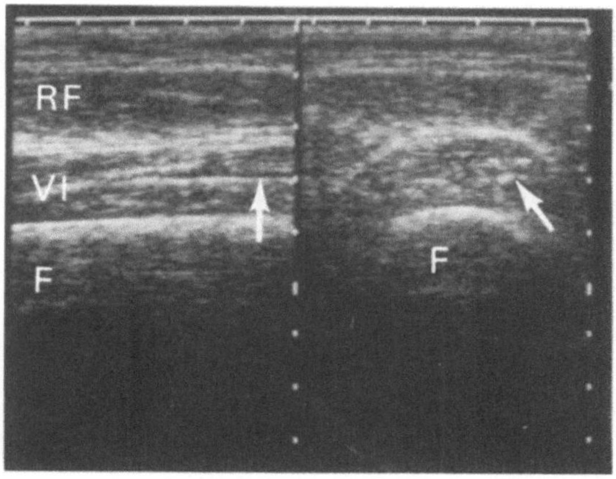

a b

FIGURE 3.12. Sonograms of the anterior region of the thigh show an intramuscular arterial branch (*arrow*). (a) Longitudinal scan. (b) Transverse scan. *F* = femur; *RF* = rectus femoris; *VI* = vastus intermedius.

References

1. Fornage BD, Touche DH, Segal P, Rifkin MD: Ultrasonography in the evaluation of muscular trauma. J Ultrasound Med 1983;2:549–554.
2. Fornage B: L'échographie des muscles et des tendons. In: Delorme G, Tessier JP (eds): Manuel d'Electroradiologie. Vol. 3, Échographie Normale. Paris, Masson, 1988.
3. Lenkey JL, Skolnick ML, Slasky BS, Campbell WL: Evaluation of the lower extremities. J Clin Ultrasound 1981;9:413–416.
4. Slasky BS, Lenkey JL, Skolnick ML, Campbell WL, Cover KL: Sonography of soft tissues of extremities and trunk. Semin Ultrasound 1982;3:288–330.

4
Normal Ultrasound Anatomy of Tendons

The sonographic examination of superficial tendons is best performed with the use of (*a*) a standoff pad, which should be thin because the transducer is of high frequency and short focused, (*b*) real-time linear-array electronic transducers, and (*c*) probe frequencies of 7.5 or 10 MHz [2–7]. Tendons are small organs and the examination technique should be meticulous to avoid artifacts (see Chapter 5). The sonographic appearance of normal tendons is roughly the same whatever the anatomic location.

Longitudinal Sonograms

Tendons are echogenic [4–9]. When high-frequency transducers are used, tendons show a markedly fibrillar texture on longitudinal sonograms (Figs. 4.1–4.5), a characteristic pattern that has also been demonstrated in other mammals (Fig. 4.6). Major tendons occasionally show echogenic margins, which have been assumed to represent the peritenon (Fig. 4.7) [5]. Normal tendons do not vary significantly in echogenicity but the surrounding tissues, particularly fat, may show significant topographic as well as individual variations [5]. Most often, tendons are slightly more echogenic than the adjacent fat, whether it be the fat pad of the knee posterior to the patellar tendon or the pre-Achilles fat pad (Fig. 4.8) [1,5]. Tendons appear markedly echogenic when they are surrounded by muscles, which are hypoechoic (Fig. 4.5).

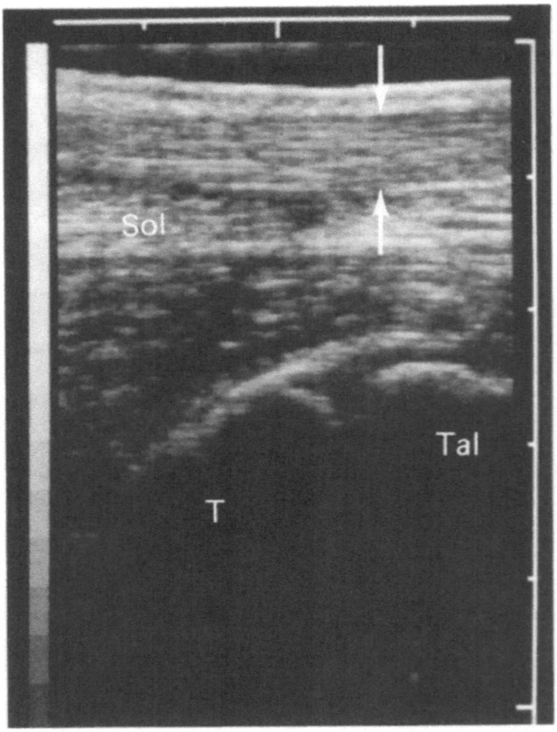

FIGURE 4.1. Longitudinal scan of the Achilles tendon at its origin (*arrows*) using a 7.5-MHz linear-array transducer. The tendon is moderately echogenic and shows a typical fibrillar texture. *Sol* = distal end of soleus; *T* = tibia; *Tal* = talus.

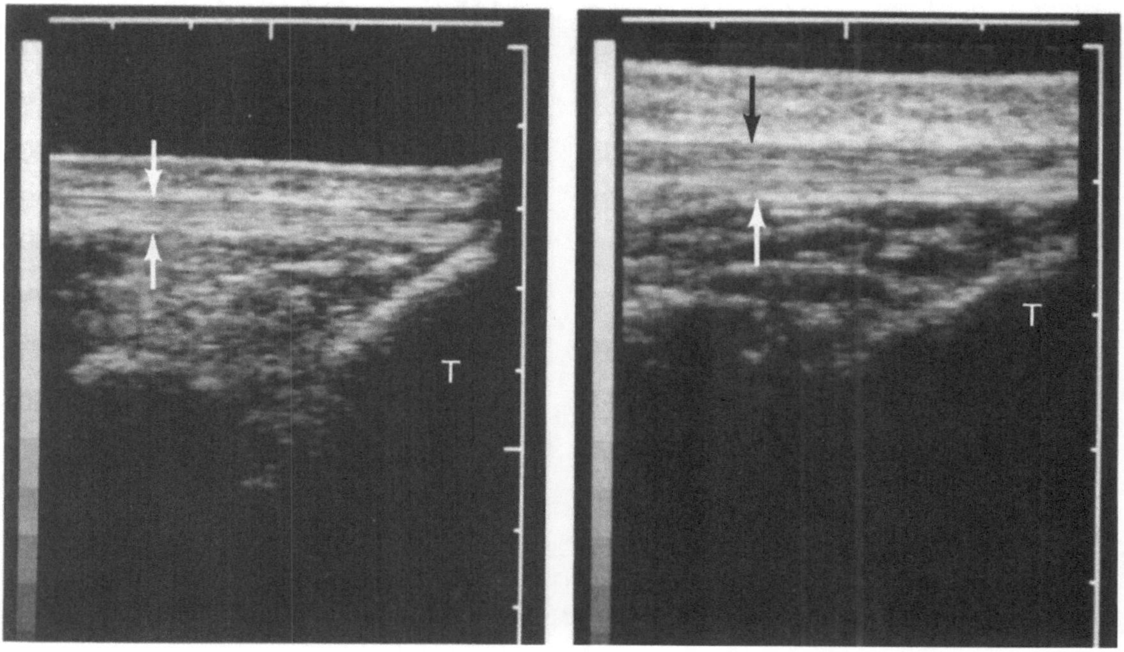

FIGURE 4.2. Longitudinal scans of the patellar tendon (*arrows*) show a moderately echogenic tendon with a fibrillar texture. (a) Scan obtained using a 5-MHz transducer. (*b*) Scan obtained using a 7.5-MHz transducer. Note the hypoechoic fat pad of the knee posterior to the tendon. *T* = tibia.

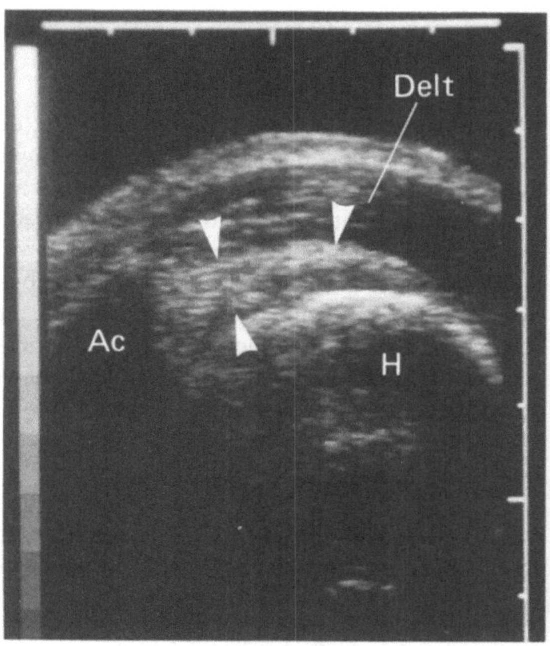

FIGURE 4.3. Coronal scan of the rotator cuff. The echogenic tendinous cuff (*arrowheads*) lies between the humeral head and the hypoechoic deltoid muscle. *Ac* = acromion; *Delt* = deltoid; *H* = humeral head.

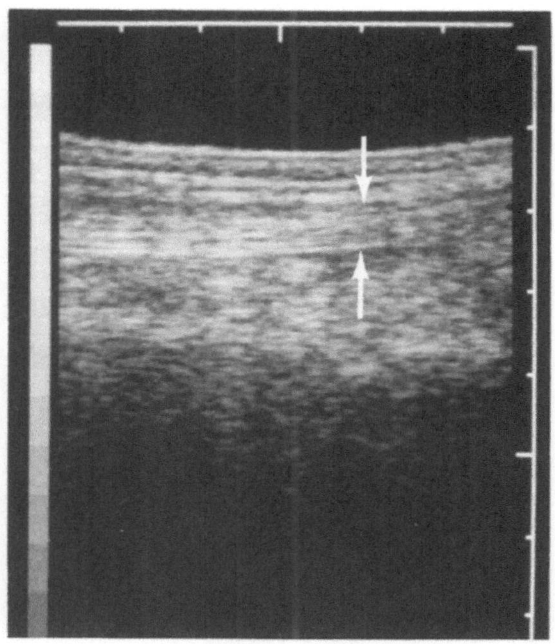

FIGURE 4.4. Longitudinal sonogram of the quadriceps tendon. The tendon (*arrows*) is echogenic and shows a fibrillar texture. Posteriorly, the fat is moderately echogenic.

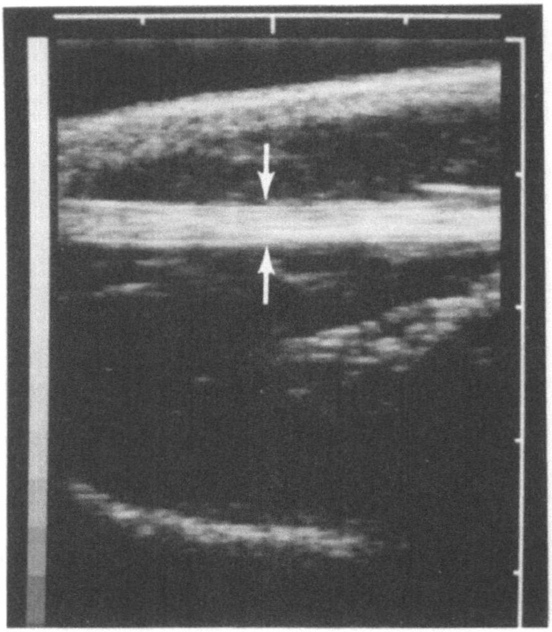

FIGURE 4.5. Longitudinal sonogram of the tendon of the flexor pollicis longus (*arrows*) in the thenar eminence. The tendon is markedly echogenic in relation to the surrounding muscles.

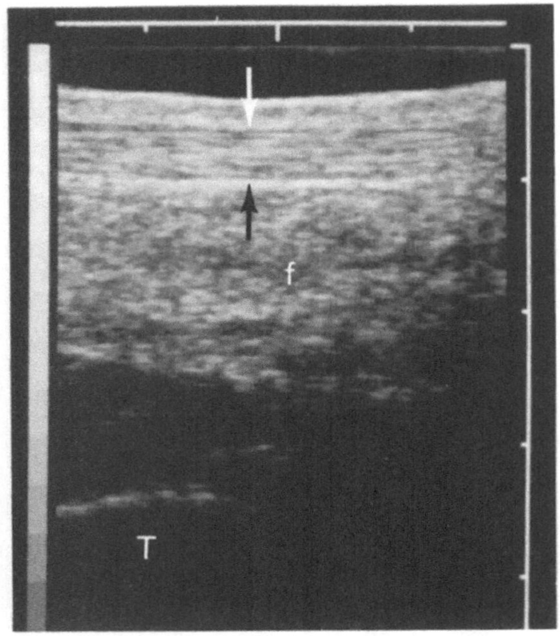

FIGURE 4.7. Longitudinal sonogram of the Achilles tendon. The tendon margins are more echogenic (*arrows*). The tendon and the pre-Achilles fat pad display the same echogenicity. *f* = pre-Achilles fat pad; *T* = lower extremity of tibia.

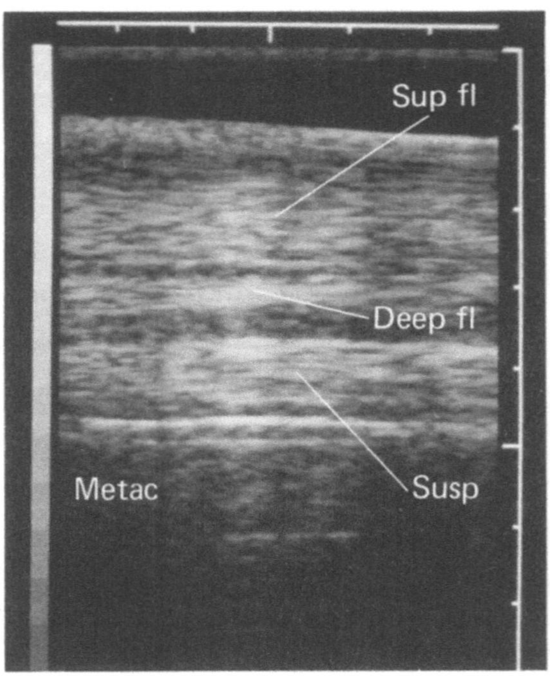

The attachments of tendons onto bones (e.g., of the patellar tendon onto the patella and tibial tuberosity; of the Achilles tendon onto the calcaneus; of the rotator cuff onto the humeral head; and of the common extensor and common flexor tendons of the forearm onto the radial or ulnar condyles, respectively) are readily demonstrable in relation to the bone surfaces, which are hyperechoic (Fig. 4.9). Synovial bursae can be demonstrated at selected sites (see also Chapter 6).

Transverse Sonograms

Transverse sections of tendons are also echogenic provided the scanning plane is strictly perpendicu-

FIGURE 4.6. Longitudinal sonogram of the posterior metacarpal region in a horse's forelimb shows the superficial digital flexor tendon (*Sup fl*), the deep flexor tendon (*Deep fl*), and the suspensory ligament (*Susp*) posterior to the third metacarpal bone (*Metac*).

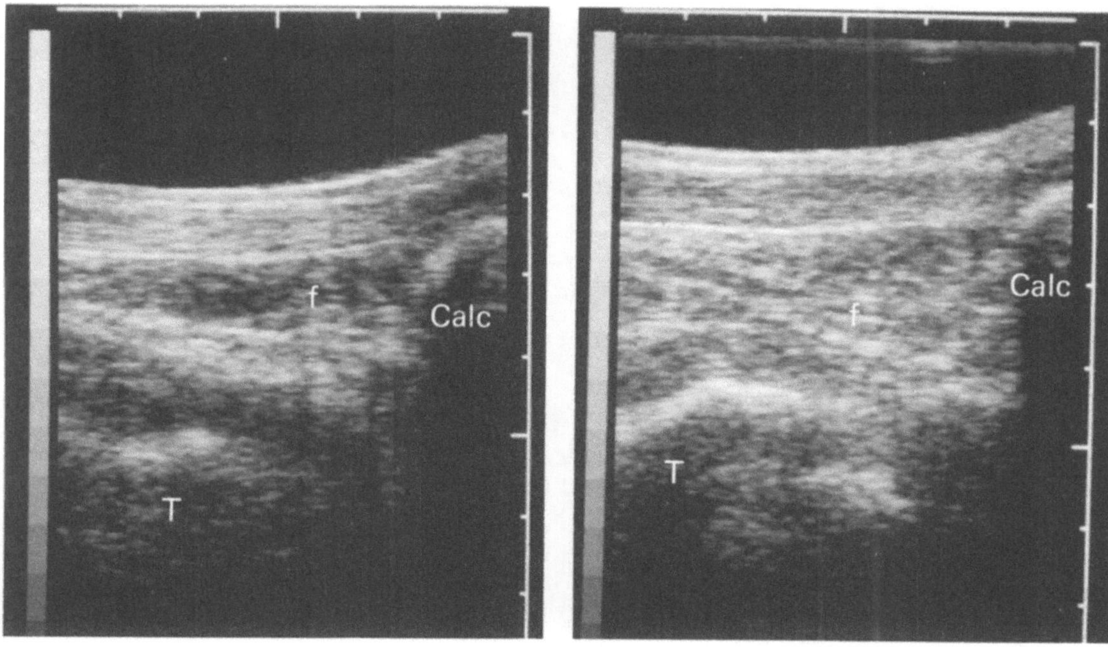

FIGURE 4.8. Longitudinal sonograms of the Achilles tendon showing variations in echogenicity of the pre-Achilles fat pad. (a) The fat is markedly hypoechoic. (b) In another patient, the tendon and the fat display similar echogenicity. *Calc* = calcaneus; *f* = pre-Achilles fat pad; *T* = lower extremity of tibia.

lar to the tendon axis (see Chapter 5) (Figs. 4.10–4.12). Large tendons show a fine punctate pattern corresponding to their fibrillar texture. Transverse scans are useful in measuring the maximum width and thickness of tendons. The normal patellar tendon is 4 to 6 mm thick and 15 to 25 mm wide. The Achilles tendon, which is less flattened than the patellar tendon, is 5 to 7 mm thick (mean $\pm$ SD = 5.3 $\pm$ 0.8 mm) and 12 to 15 mm wide (mean $\pm$ SD = 12.9 $\pm$ 1.8 mm). It must be noted that the greatest diameter of its elliptical transverse section is oriented obliquely forward and medially (Fig. 4.12) [5]. This orientation may result in an overestimation of its thickness when measurement is by a strictly sagittal scan (see also Chapter 5).

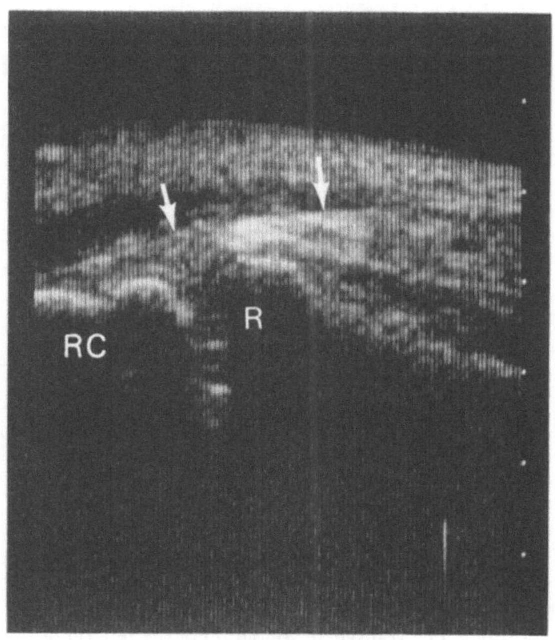

FIGURE 4.9. Coronal sonogram of the lateral aspect of the extended elbow shows the attachment of the echogenic common extensor tendon (*arrows*) onto the radial condyle. *R* = radial head; *RC* = radial condyle. (From Fornage BD, Rifkin MD: Ultrasound examination of tendons. Radiol Clin North Am 1988;26:87–107, with permission.)

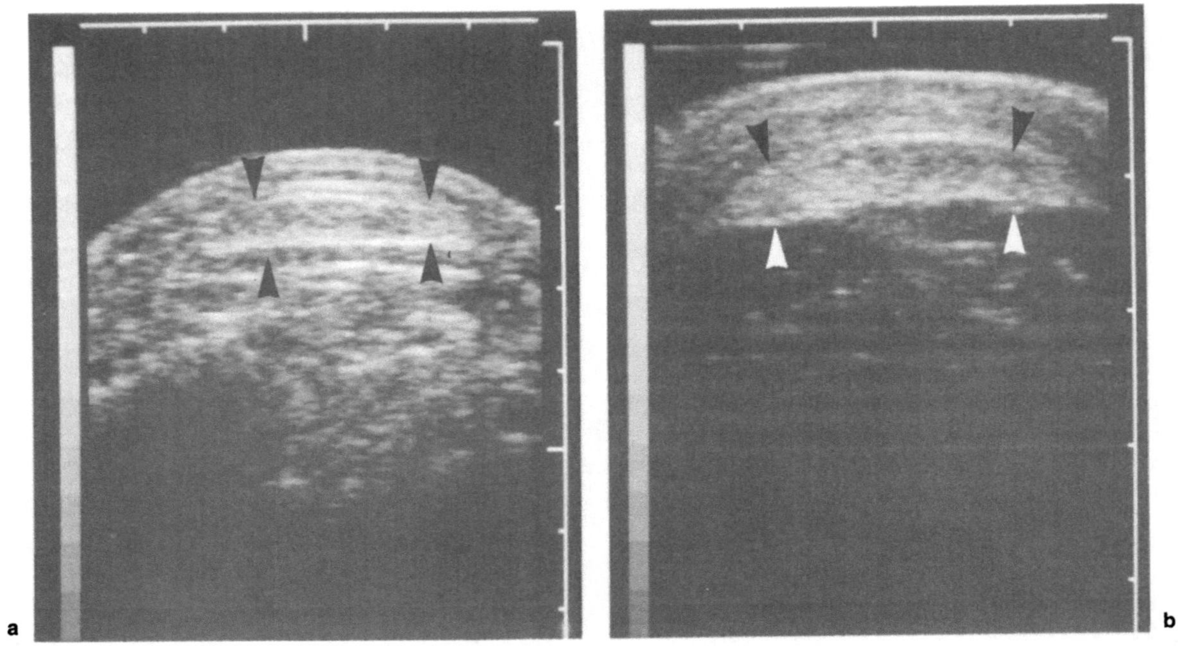

FIGURE 4.10. Transverse scans of the patellar tendon. (a) Sonogram obtained using a 5-MHz transducer. (b) Sonogram obtained using a 7.5-MHz transducer (another patient). In both cases, the tendon is flattened and echogenic (*arrowheads*).

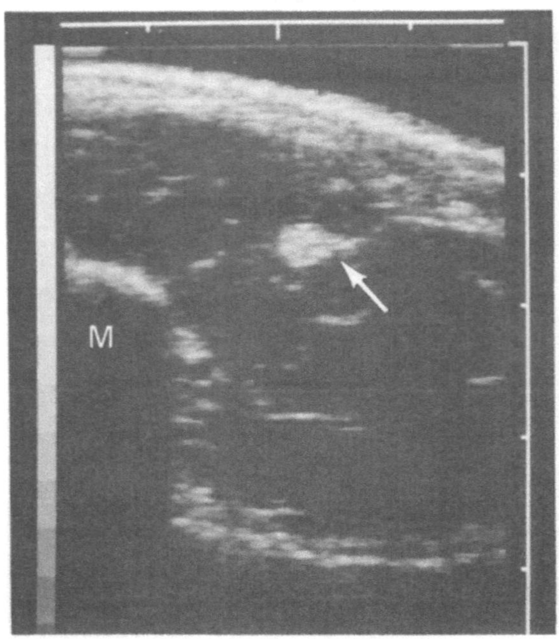

FIGURE 4.11. Transverse sonogram of the tendon of the flexor pollicis longus in the thenar eminence. The tendon (*arrow*) is markedly hyperechoic compared with the surrounding muscles. M = first metacarpal bone.

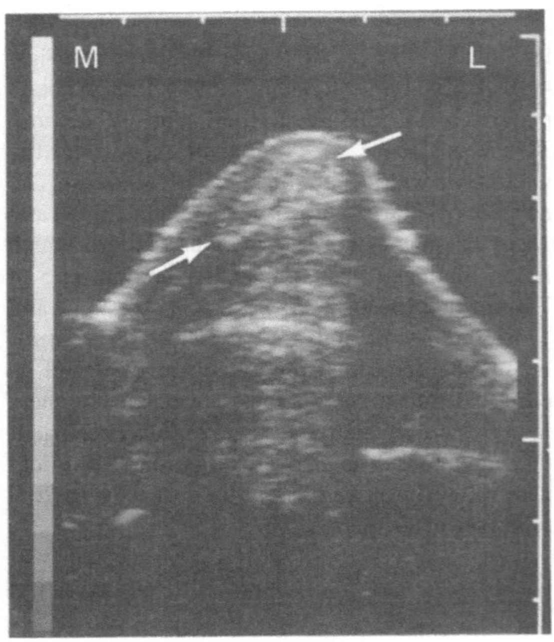

FIGURE 4.12. Transverse scan of the Achilles tendon. The tendon (*arrows*) is more echogenic than the pre-Achilles fat pad. The major axis of its cross section is oriented obliquely forward and medially. M = medial; L = lateral.

Dynamic Study

The sliding movements of tendons can be monitored by dynamic examination in longitudinal sections. However, the normal synovial sheaths cannot be visualized with currently available equipment.

References

1. Dillehay GL, Deschler T, Rogers LF, Neiman HL, Hendrix RW: The ultrasonographic characterization of tendons. Invest Radiol 1984;19:338–341.
2. Fornage BD, Rifkin MD, Touche DH, Segal PM: Sonography of the patellar tendon: Preliminary observations. AJR 1984;143:179–182.
3. Fornage BD, Schernberg FL, Rifkin MD: Ultrasound examination of the hand. Radiology 1985;155:785–788.
4. Fornage BD, Rifkin MD: Ultrasound examination of the hand. Radiology 1986;160:853–854.
5. Fornage BD: Achilles tendon: US examination. Radiology 1986;159:759–764.
6. Fornage B: L'échographie de la main: technique et anatomie ultrasonore normale. Journal d'Echographie et de Médecine Ultrasonore 1986;7:193–203.
7. Fornage BD, Rifkin MD: US examination of tendons. Radiol Clin North Am 1988;26:87–107.
8. Middleton WD, Edelstein G, Reinus WR, Melson GL, Murphy WA: Ultrasonography of the rotator cuff: Technique and normal anatomy. J Ultrasound Med 1984;3:549–551.
9. Middleton WD, Reinus WR, Totty WG, Melson GL, Murphy WA: US of the biceps tendon apparatus. Radiology 1985;157:211–215.

5
Artifacts and Pitfalls in Sonography of Muscles and Tendons

Some specific artifacts and pitfalls encountered in the ultrasound examination of muscles and tendons are presented in this chapter.

Muscles

Reverberation Artifacts

Reverberation artifacts originate in highly reflective interfaces, such as muscular aponeuroses, but also, occasionally, from the interface between the standoff pad and the skin (Figs. 5.1, 5.2). They reproduce more deeply the actual causative interface, and project at distances that are multiples of that between the probe and the reflector. They are readily identified by their displacement when the distance between the transducer and the reflector is modified (e.g., by increasing the pressure on the probe).

Pitfalls Related to Intramuscular Connective Septa

Large intramuscular septa are not rare and may result in misleading patterns. Sagittally oriented septa can be responsible for ill-defined echogenic areas when they are included in sagittal sonograms (Figs. 5.3, 5.4). Conversely, a frontally oriented septum may appear on a sagittal sonogram as a hyperechoic linear structure (Fig. 5.5). These echogenic patterns may mimic fibrous lesions.

Other Pitfalls

Some muscular regions, including the deep portion of soleus (Fig. 5.6) and the upper portion of adductors, are markedly hypoechoic. These normal patterns should not be misinterpreted as hypoechoic traumatic lesions.

Deep veins in the leg may resemble small, flattened, fluid-filled collections on longitudinal scans. Transverse sonograms as well as compression maneuvers are diagnostic (Fig. 5.7).

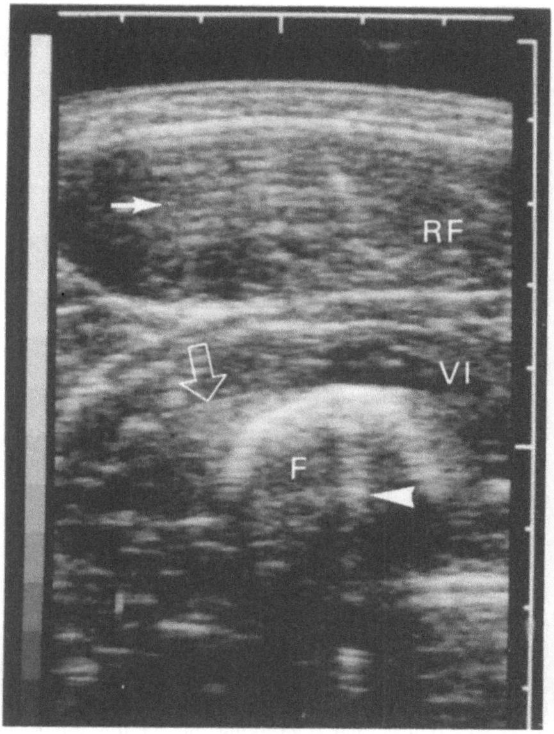

FIGURE 5.1. Reverberation artifacts associated with the interface between the standoff pad and the skin. Transverse scan of the anterior region of the thigh. Artifacts project in the rectus femoris (*arrow*). They include reverberations from the skin and from the anterior aponeurosis of the rectus femoris, and echoes resulting from back-and-forth propagation of the beam between those two interfaces. Note a bow-shaped, or "Chinese hat" artifact (*open arrow*) and a comet-tail artifact (*arrowhead*), both associated with the strong reflection on the anterior aspect of the femur. *F* = femur; *RF* = rectus femoris; *VI* = vastus intermedius.

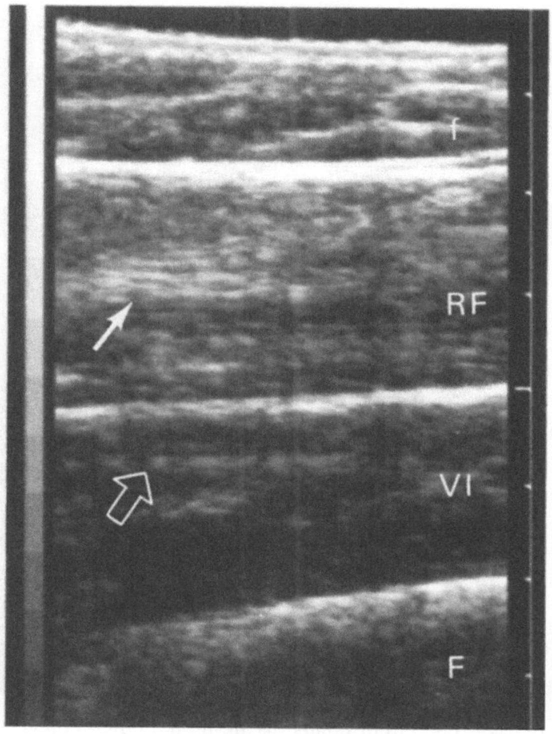

FIGURE 5.2. Reverberation artifacts. Longitudinal scan of the anterior region of the thigh. The *solid arrow* points to reverberation artifacts resulting from back-and-forth ultrasound propagation between the skin and the highly reflective anterior aponeurosis of the rectus femoris. The *open arrow* points to a reverberation artifact from the anterior aponeurosis of the rectus femoris. *F* = femur; *f* = subcutaneous fat; *RF* = rectus femoris; *VI* = vastus intermedius.

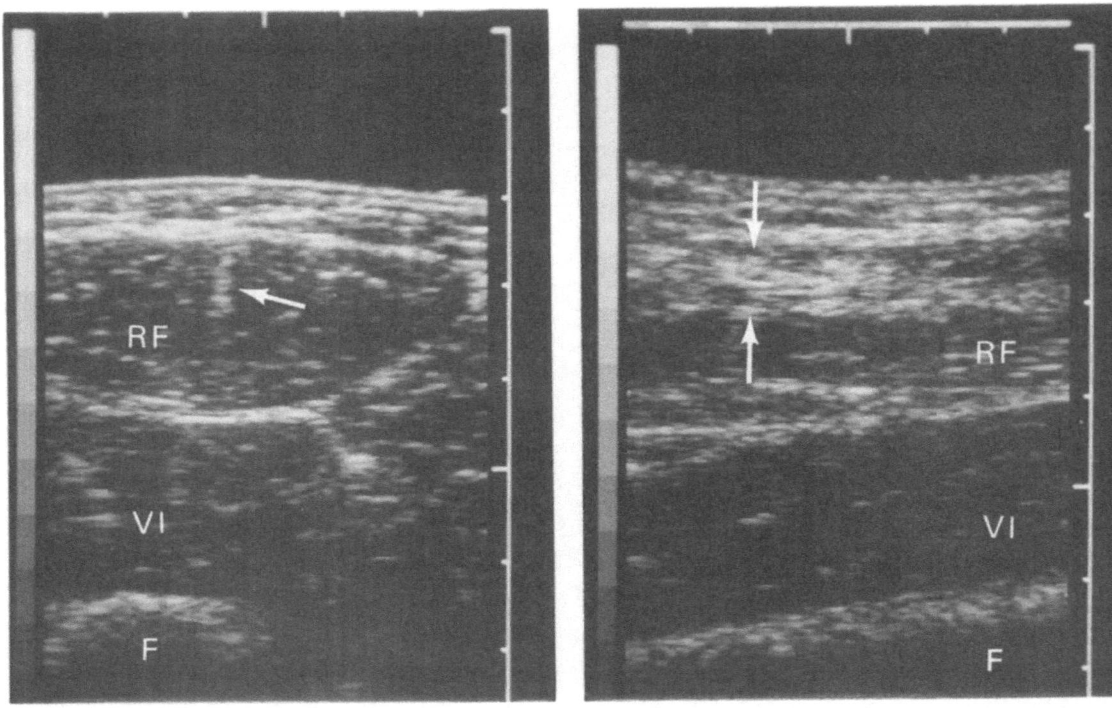

a b

FIGURE 5.3. Pitfall related to an intramuscular connective septation. Sonograms of the anterior region of the thigh. (a) Transverse sonogram shows a sagittally oriented septum (*arrow*) inside the rectus femoris muscle. (b) Longitudinal scan along the septum shows an ill-defined echogenic area (*arrows*), which mimics a fibrous scar. *F* = femur; *RF* = rectus femoris; *VI* = vastus intermedius.

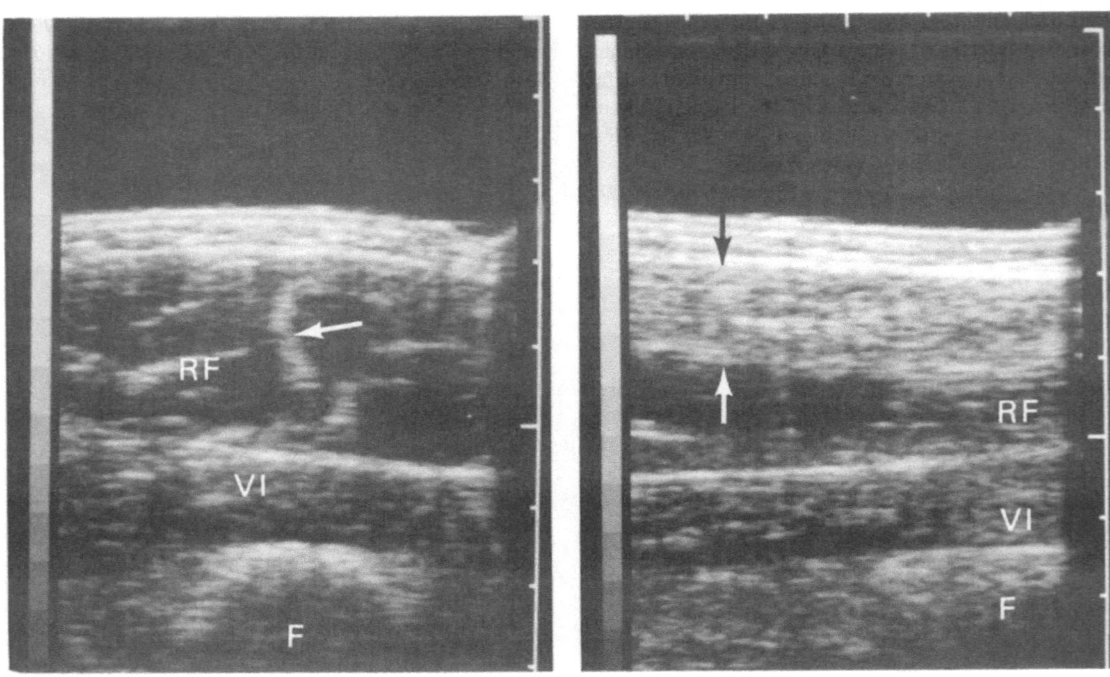

a b

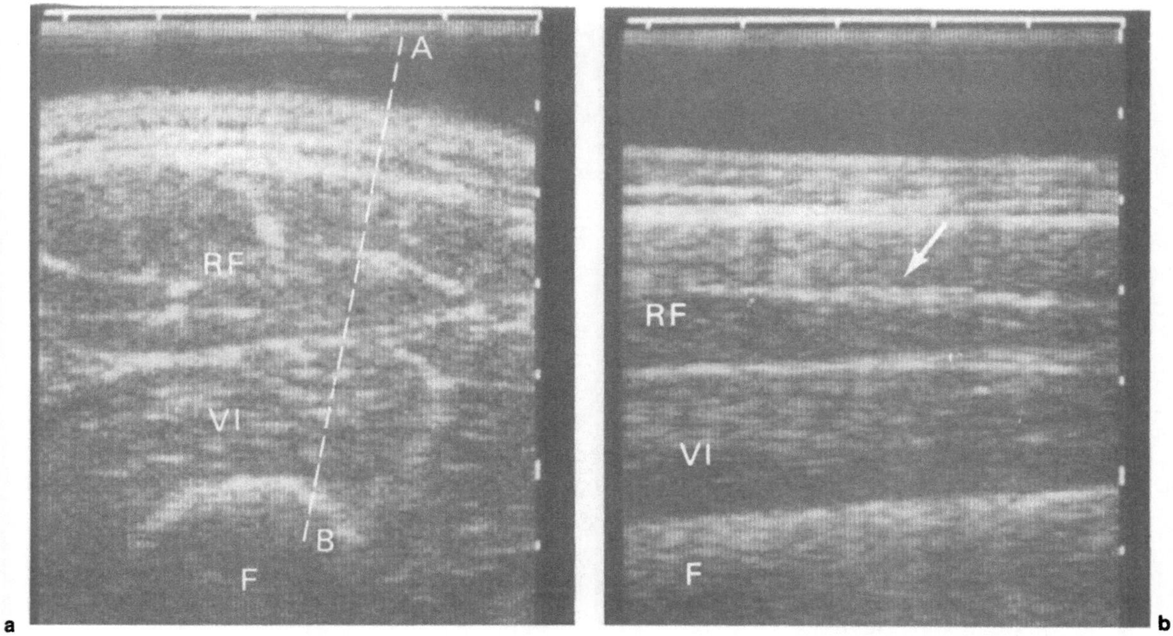

FIGURE 5.5. Pitfall related to an intramuscular septation. Sonograms of the anterior region of the thigh. (a) Transverse sonogram shows several connective septa in the rectus femoris, yielding a reticular pattern. (b) Longitudinal scan along AB. Section of a frontally oriented septum results in an echogenic line (*arrow*) parallel to the perimuscular aponeuroses. *F* = femur; *RF* = rectus femoris; *VI* = vastus intermedius.

FIGURE 5.6. Longitudinal sonogram of the calf at the termination of the medial gastrocnemius shows the marked hypoechogenicity of the anterior portion of the soleus (*arrow*). *Gast (med)* = medial gastrocnemius; *Sol* = soleus; *T* = tibia.

►

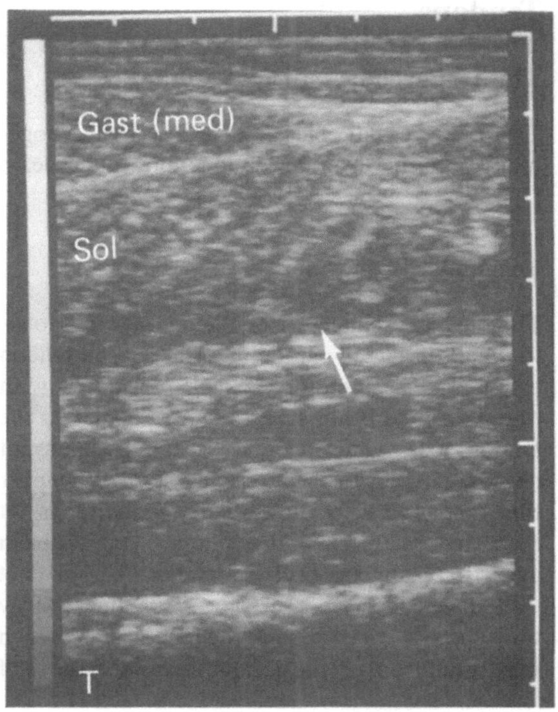

◄

FIGURE 5.4. Pitfall related to an intramuscular septation. Sonograms of the anterior region of the thigh. (a) Transverse sonogram shows several intramuscular septa, one of which is sagittally oriented (*arrow*). (b) Longitudinal sonogram along the septatum shows a large echogenic area (*arrows*). *F* = femur; *RF* = rectus femoris; *VI* = vastus intermedius.

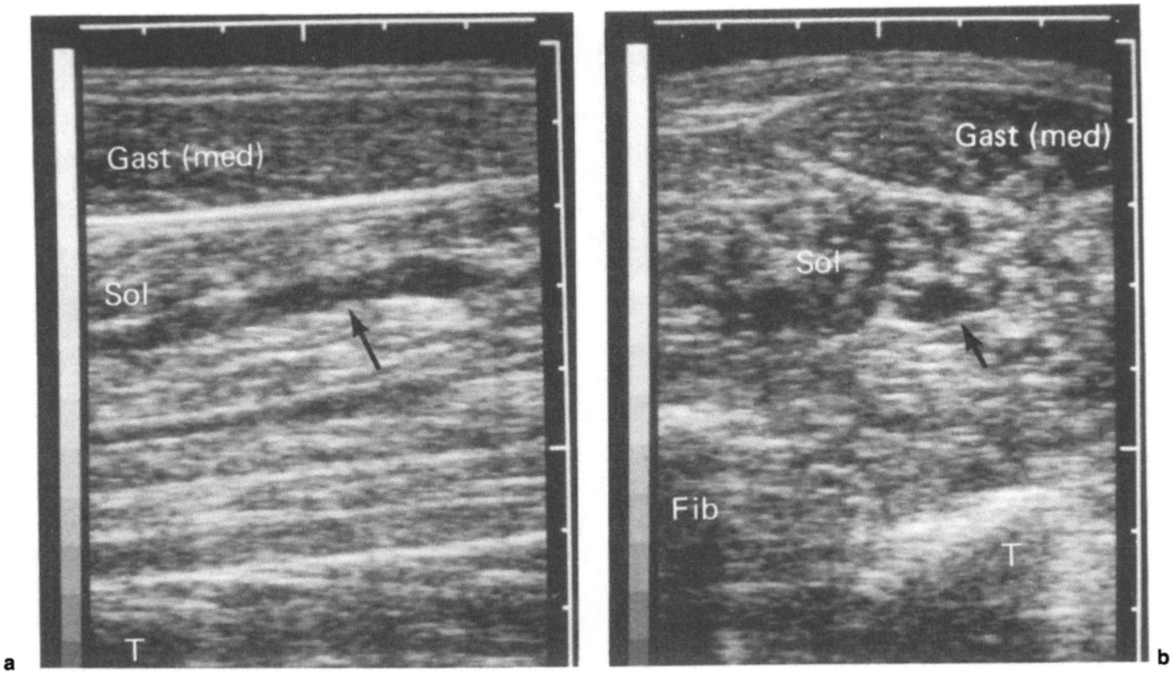

FIGURE 5.7. Pitfall of vascular origin. (a) Longitudinal sonogram of the left calf shows a thin, nonpulsatile, sonolucent structure with irregular margins (*arrow*) in the soleus. (b) Transverse sonogram confirms a deep vein (*arrow*). *Fib* = fibula; *Gast (med)* = medial gastrocnemius; *Sol* = soleus; *T* = tibia.

Tendons

The Hypoechoic Normal Tendon

This major pitfall consists of a false hypoechogenicity resulting from an oblique ultrasound beam [1–4].

Longitudinal Sonograms

The fibrillar architecture of tendons provides longitudinally oriented parallel interfaces. The tendon echotexture is therefore best demonstrated with the ultrasound beam strictly perpendicular to the tendon axis, whose course is usually, but not always, parallel to the overlying skin. Whenever this "golden rule" for ultrasound examination is not respected, scattering of the oblique beam results in an artifactually hypoechoic tendon. The artifact occurs whenever the surface of the linear-array transducer is not parallel to the tendon axis, which may result from (*a*) the oblique anatomic course of the tendon in relation to the skin or the tendon's physiologic curvature, or (*b*) the improper placement of the transducer when using a standoff pad (Figs. 5.8–5.10). Thus, flexor tendons of the fingers appear falsely hypoechoic on longitudinal scans because of their curved anatomic course in the concavity of phalanges [2–4]. However, a careful inspection of the scans shows that the tendons are echogenic in their short segments, where they lie parallel to the surface of the probe (Fig. 5.11). Aligning the flexor tendons through a slight flexion of the finger results in the display of their normal fibrillar echogenic texture (Fig. 5.12) [3].

The normal curvature of certain tendinous insertions also gives rise to artifactual hypoechogenicity. Again, the normal tendon echotexture is displayed when the surface of the linear-array transducer and the tendon segment are parallel to each other (Fig. 5.13). The patellar and Achilles tendons may be slightly curved at rest, but contraction of the quadriceps or the calf straightens the tendon and clears the artifact (Figs. 5.14, 5.15).

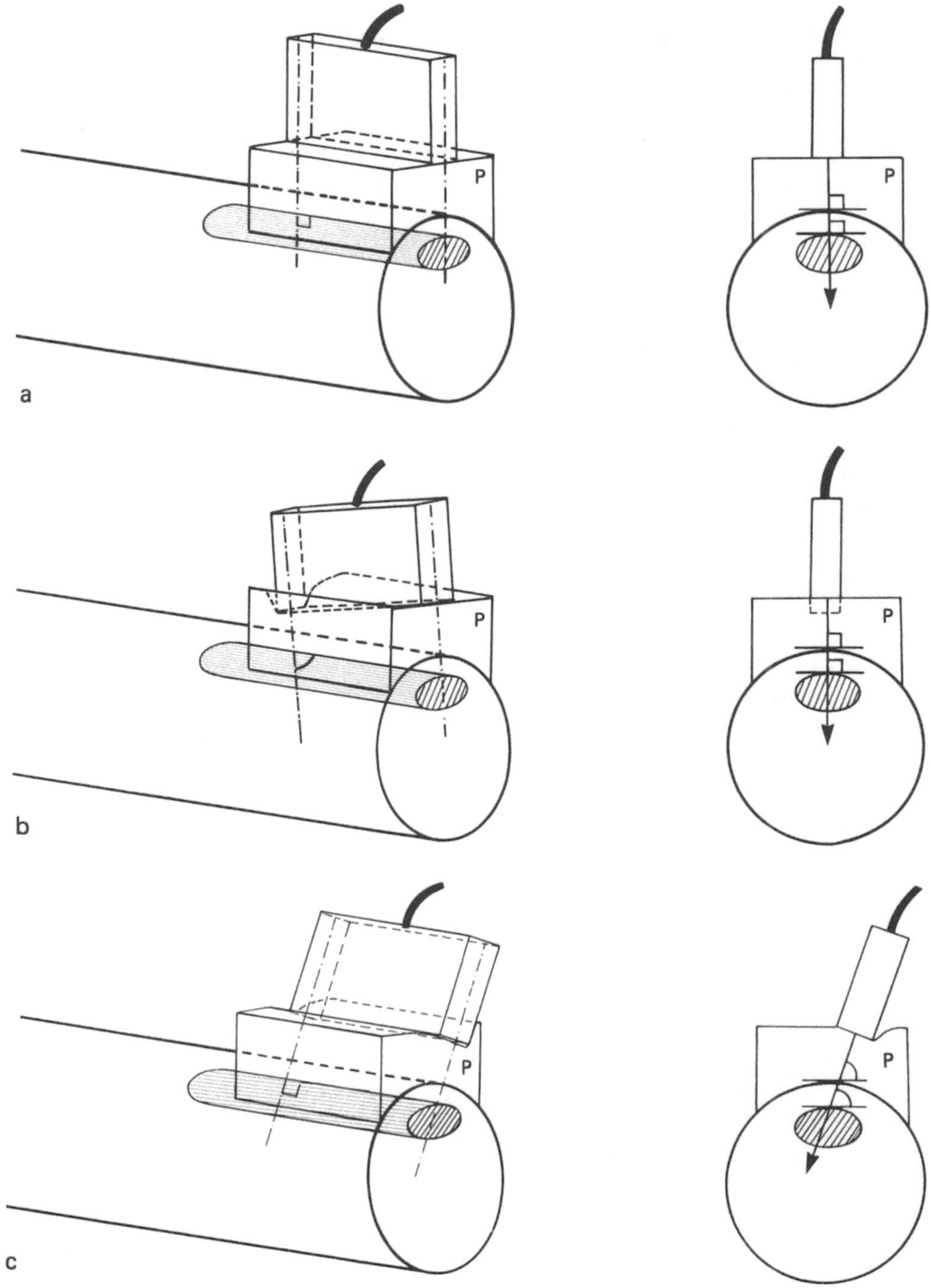

FIGURE 5.8. Artifacts related to the obliquity of the beam during real-time examination of tendons with the use of a linear-array transducer and a standoff pad. (a) Satisfactory technique. The beam is strictly perpendicular to the tendon. (b) Although the probe remains aligned with the overall axis of the extremity, an increased pressure at one end of the transducer results in the loss of parallelism between its surface and the tendon axis. The beam is oblique to the tendon. (c) The surface of the probe is parallel to the tendon, but the probe is at an angle laterally, so that the scanning plane is oblique to the skin and the tendon surface. *P* = standoff pad.

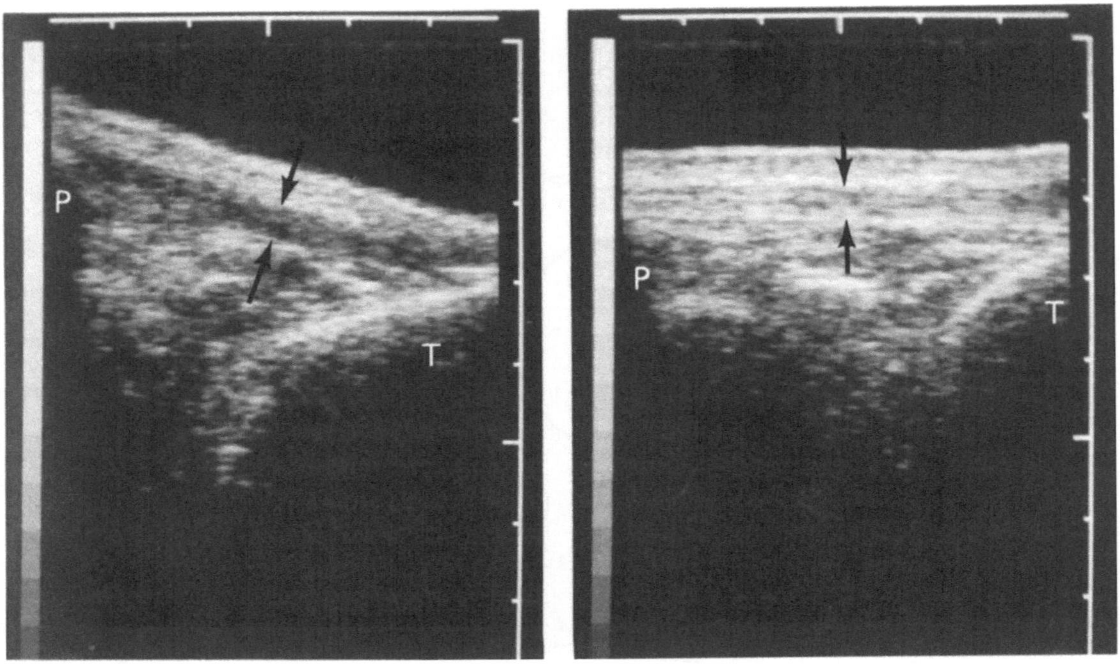

a b

FIGURE 5.9. Longitudinal sonograms of the patellar tendon show the false hypoechogenicity that results from the improper positioning of the linear-array transducer used in conjunction with a standoff pad. (a) The surface of the linear-array transducer is not parallel to the tendon. Hence, the beam is not perpendicular to the tendon, which appears markedly hypoechoic (*arrows*). (b) The surface of the probe is strictly parallel to the tendon, which now displays a normal echogenicity and fibrillar texture (*arrows*). *P* = patella; *T* = tibia.

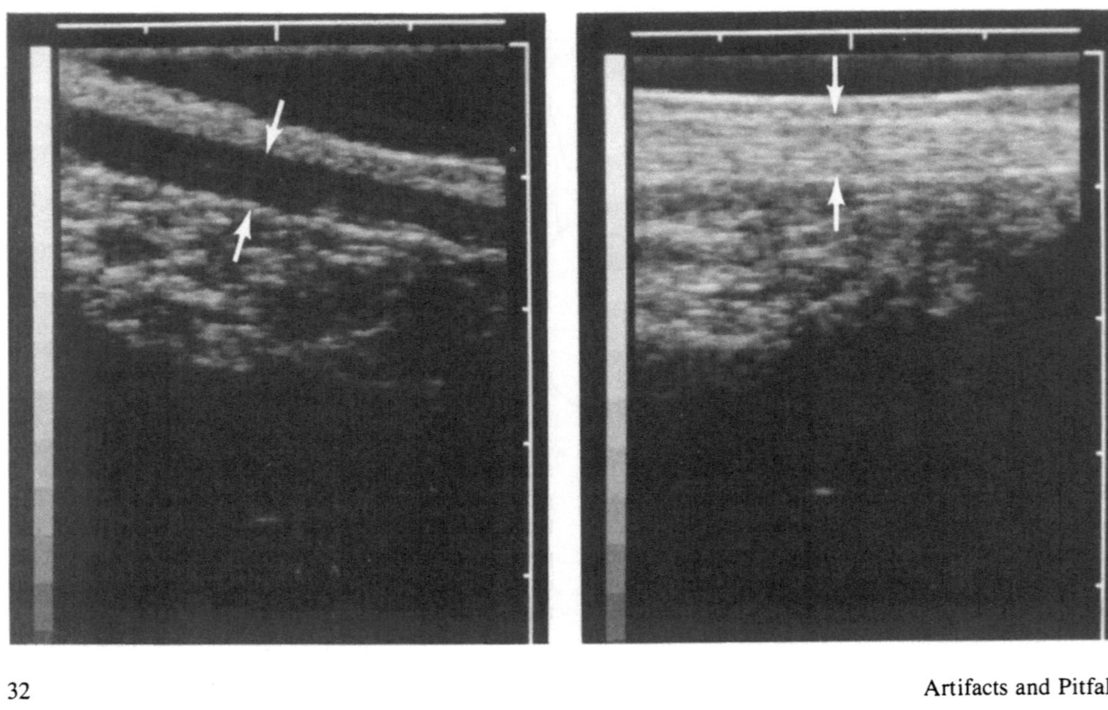

a b

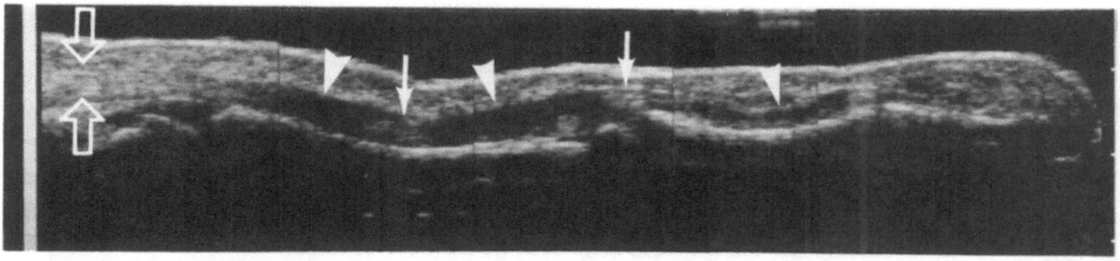

FIGURE 5.11. Longitudinal sonogram of the fourth finger shows the wavy course of the deep and superficial flexor tendons in the concavities of the first and second phalanges (montage of several partial views obtained with a 7.5-MHz probe). The tendon segments that lie oblique to the surface of the probe appear hypoechoic (*arrowheads*), contrasting with the short echogenic segments that are parallel to the transducer (*arrows*). Proximal to the metacarpophalangeal joint, the tendons lie parallel to the transducer and therefore display their normal echogenicity (*open arrows*).

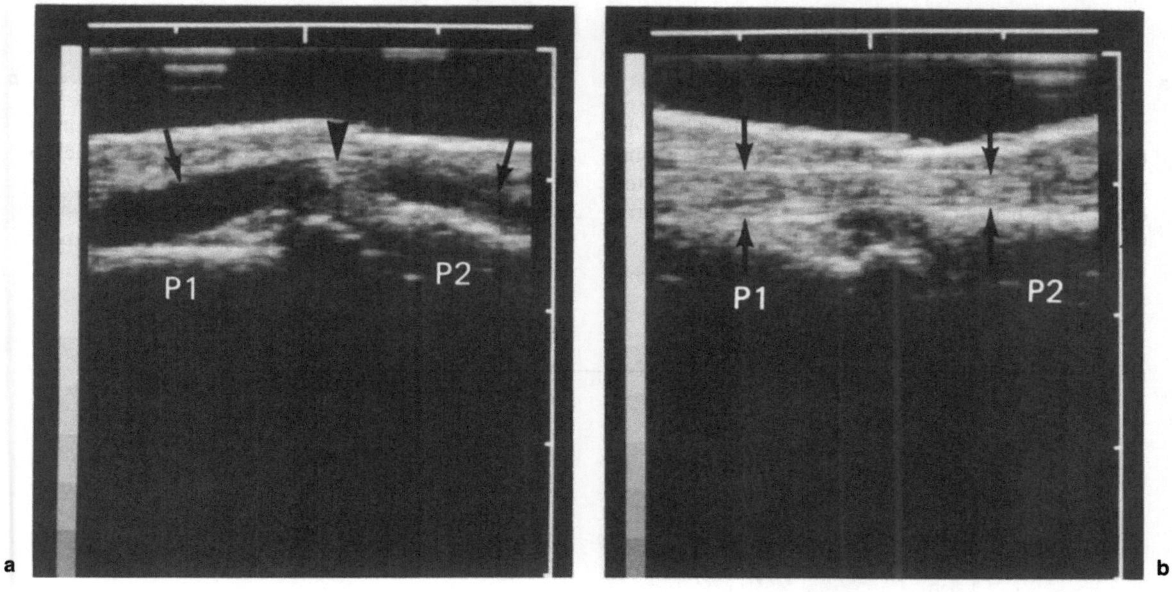

FIGURE 5.12. False hypoechogenicity of digital flexor tendons. Sagittal sonogram of the proximal interphalangeal joint. (a) With the finger extended, the flexor tendons are convex anteriorly. They show a normal echogenicity only in a short segment parallel to the surface of the linear-array transducer (*arrowhead*), while they are markedly hypoechoic in their curved segments (*arrows*). (b) A slight flexion of the finger aligns the tendons (*arrows*) to parallel with the surface of the transducer. A normal fibrillar echotexture is displayed. *P1* = proximal phalanx; *P2* = middle phalanx.

FIGURE 5.10. Longitudinal sonograms of the Achilles tendon show artifactual hypoechogenicity resulting from an improper examination technique. (a) The tendon is oblique to the beam and appears markedly hypoechoic (*arrows*). (b) The tendon is now strictly parallel to the linear-array transducer and shows a normal fibrillar echotexture (*arrows*).

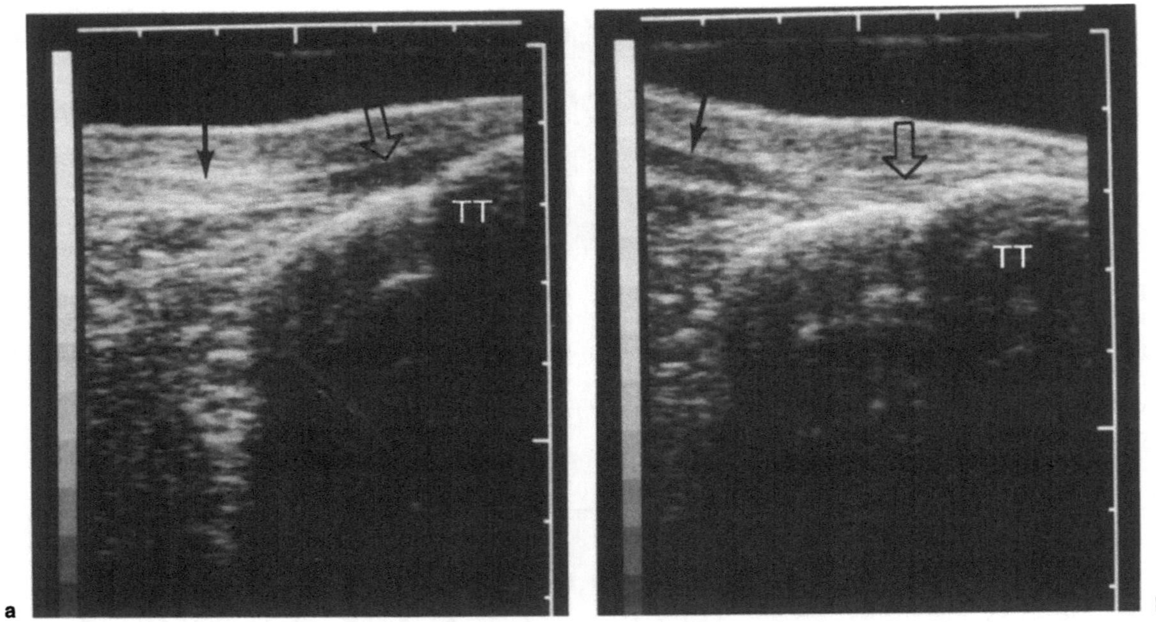

FIGURE 5.13. Artifactual tendon hypoechogenicity due to an improper examination technique: longitudinal sonograms of the lower attachment of the patellar tendon. (a) The patellar tendon is parallel to the linear-array transducer and is normally echogenic (*arrow*). Its distal end is slightly concave anteriorly and therefore appears markedly hypoechoic (*open arrow*). (b) The position of the transducer has been modified to place it parallel to the distal part of the tendon, which now displays a normal echogenicity (*open arrow*); the main tendon, which is now oblique to the beam, appears hypoechoic (*arrow*). *TT* = tibial tuberosity.

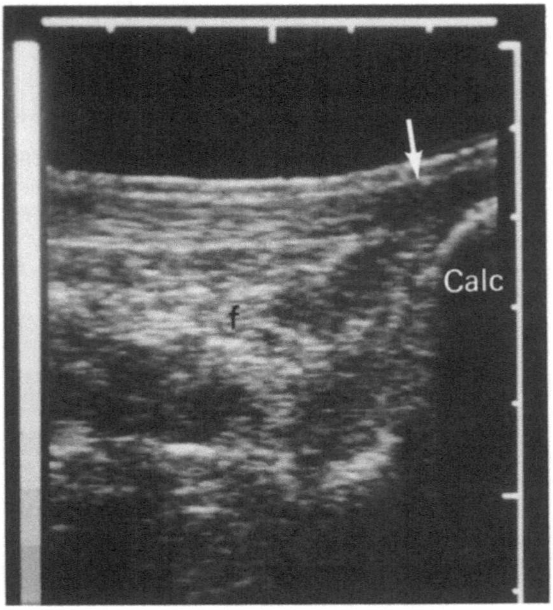

FIGURE 5.14. Artifactual hypoechogenicity of the distal end of the Achilles tendon related to its physiologic curvature. Longitudinal scan shows the falsely hypoechoic distal end of the tendon (*arrow*), contrasting with the normal fibrillar texture of the rest of the tendon. *Calc* = calcaneus; *f* = pre-Achilles fat pad.

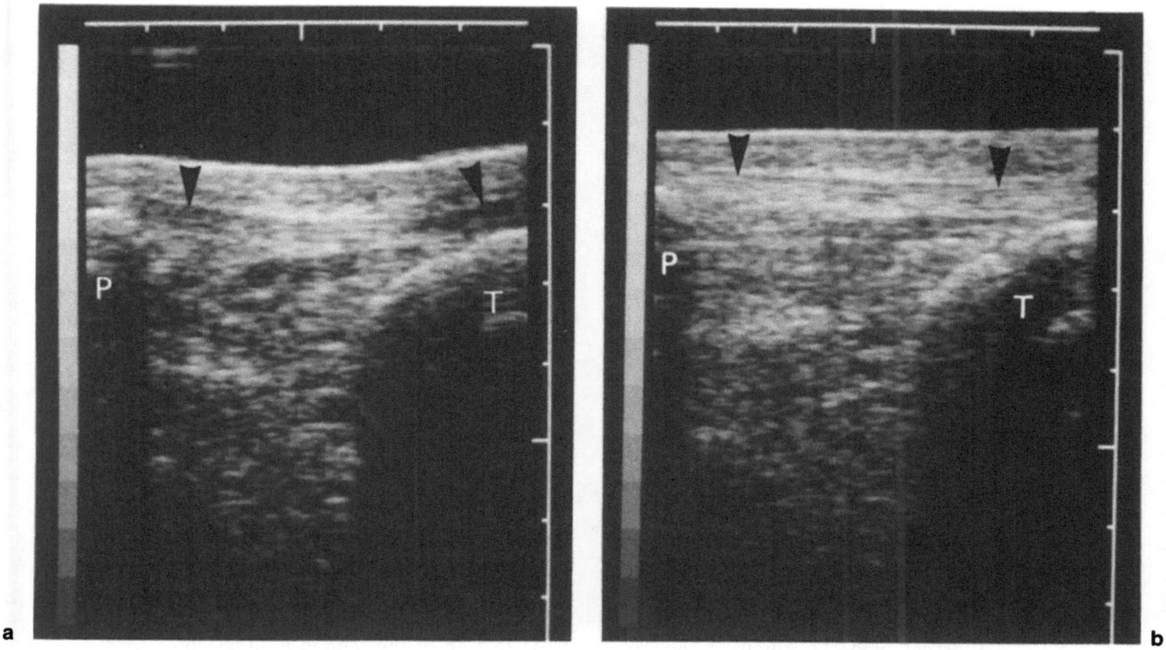

FIGURE 5.15. Artifactual hypoechogenicity of the patellar tendon related to its physiologic curvature. Longitudinal sonograms. (a) At rest, the obliquity of the proximal and distal thirds of the tendon in relation to the beam results in a false hypoechogenicity (*arrowheads*). Note that the midthird is normally echogenic. (b) Contraction of the quadriceps aligns the tendon so that it now shows normal echogenicity along its entire course (*arrowheads*). P = patella; T = tibia.

Transverse Sonograms

Any obliquity, even of a few degrees, of the transverse scanning plane in relation to the tendon axis results in a falsely hypoechoic cross section, again because of beam scattering. When the scanning plane is strictly perpendicular to the tendon axis, the transverse section shows a normal echogenicity (Fig. 5.16). The artifact is frequently encountered in tendons with a wavy course (Fig. 5.17) [2].

Taking this artifact into account, we have shown that all tendons have a similar sonographic appearance, with a fibrillar texture and a moderate to marked echogenicity. Since tendinitis is characterized by a thickened and hypoechoic tendon, awareness of false hypoechogenicity is crucial in the sonographic evaluation of tendon diseases, particularly in sports medicine. Furthermore, false hypoechogenicity may obscure minute lesions, such as partial ruptures.

That it is technically difficult to maintain the ultrasound beam strictly perpendicular to the tendon justifies the exclusive use of real-time sonographic units which allow easy adjustment of the probe position. When a standoff pad is used, the visualization of the skin surface as a distinct echogenic line indicates that the scan plane is orthogonal to the skin.

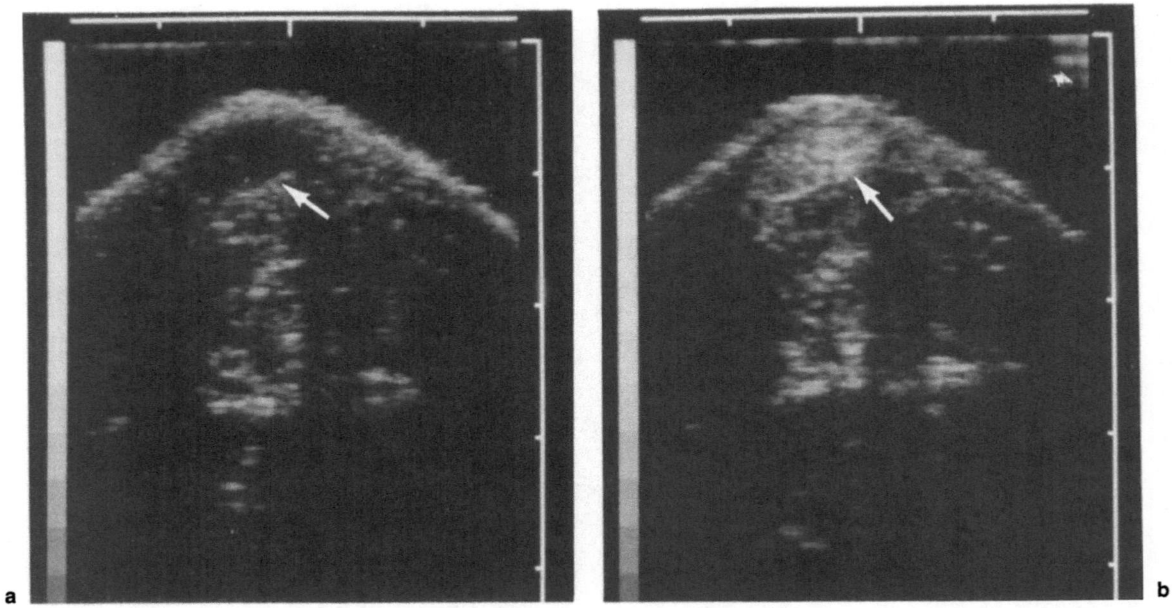

FIGURE 5.16. Artifactual hypoechogenicity of the transverse section of the Achilles tendon resulting from the tilt of the transverse scan plane. (a) The transverse scan plane is slightly oblique to the tendon axis. The oval cross section of the tendon is hypoechoic (*arrow*). (b) When the transverse scan plane is strictly perpendicular to the tendon axis, the cross section of the tendon is clearly echogenic (*arrow*).

Pitfall in Measuring the Thickness of Certain Tendons

The quadriceps, patellar, and Achilles tendons are flattened anteroposteriorly. While the quadriceps and patellar tendons are oriented in a frontal plane, the greatest diameter of the oval transverse section of the Achilles tendon is oriented more or less obliquely both anteriorly and medially. This orientation of the Achilles tendon results in the overestimation of its thickness on strictly sagittal scans (Fig. 5.18). That thickness is most accurately measured on transverse scans. The transverse scans also help determine the correct placement of the longitudinal scan plane for obtaining a true midsagittal sonogram of the tendon [1]. This peculiar orientation of the Achilles tendon also affects the measurement of its thickness on lateral radiographic views (Fig. 5.19).

Other Pitfalls

Intratendinous sesamoid bones are typically hyperechoic and cast an acoustic shadow (see Atlas section: Knee, other scans). Their characteristic locations are well known.

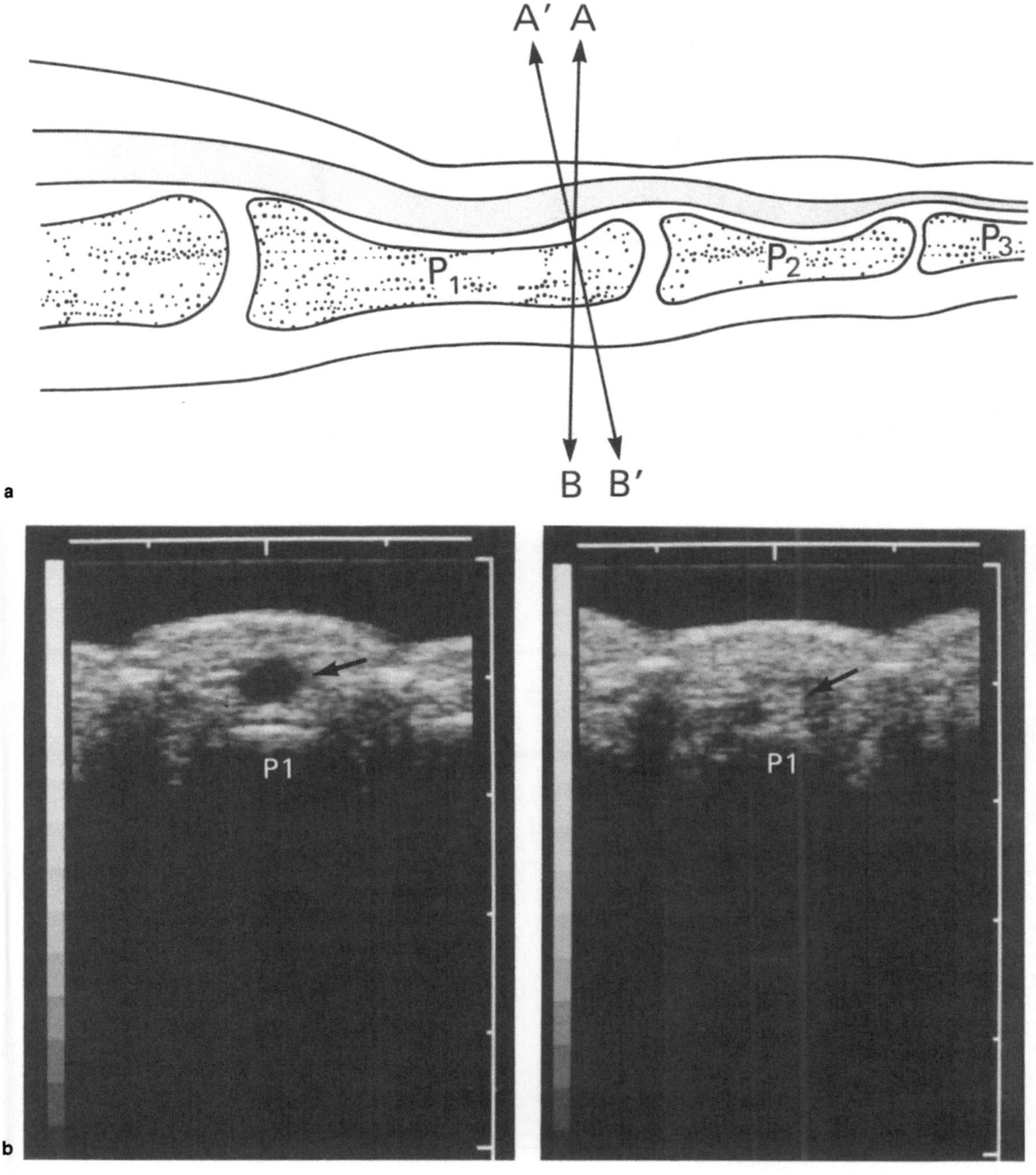

FIGURE 5.17. Transverse sonograms of the proximal phalanx of the third finger. (a) Diagram showing the orientation of the scan planes in b and c. (b) Transverse scan along *AB*. The scan is perpendicular to the overall axis of the finger but not to the flexor tendons, the cross section of which is markedly hypoechoic (*arrow*). (c) Transverse scan along *A'B'*. The transverse section of the flexor tendons is highly echogenic (*arrow*). *P1* = proximal phalanx.

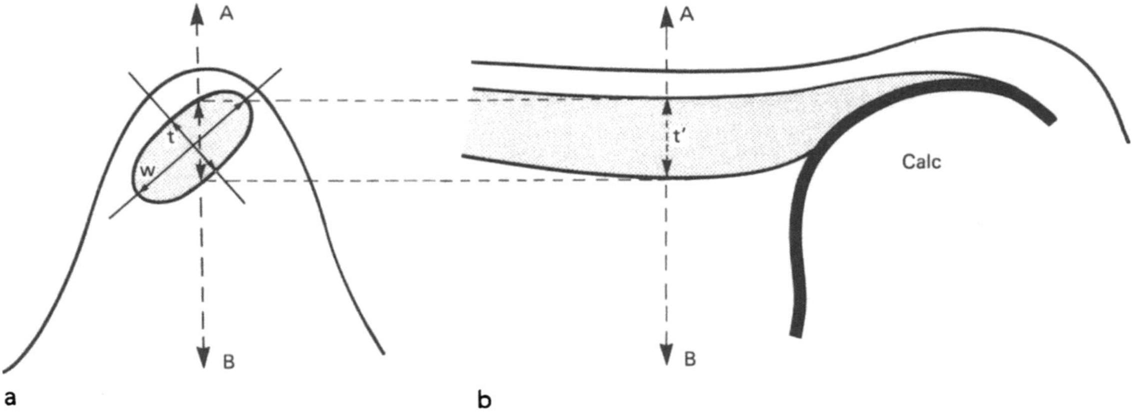

FIGURE 5.18. Diagram showing the possible overestimation of the thickness of the Achilles tendon on strictly sagittal sonograms due to the oblique orientation of the tendon plane. (a) Transverse scan along *AB* shows the oval cross section of the tendon with its oblique greatest diameter. (b) Sagittal section along *AB*. *Calc* = calcaneus; *t* = thickness; *t'* = overestimated thickness on strictly sagittal sections; *w* = width.

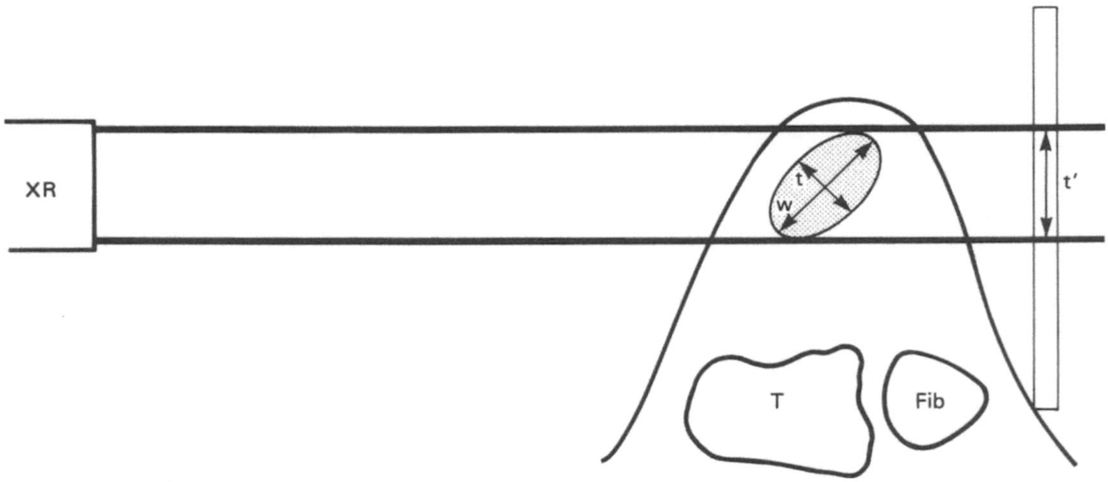

FIGURE 5.19. Diagram showing the possible overestimation of the thickness of the Achilles tendon on strictly lateral radiographic views due to the oblique tendon plane. *Fib* = fibula; *T* = tibia; *t* = thickness; *t'* = overestimated thickness on the radiographic projection; *w* = width.

Internal Fixation Devices

Internal fixation material may have been inserted temporarily or permanently. These devices include screws, plates, wires, rods, nails, and pins. Such metallic objects are readily located by sonography as hyperechoic structures characterized by a "comet-tail" artifact, which arises from reverberations inside the metal reflector (Fig. 5.20) [8–10]. However, only the proximal surface of the metallic object can be visualized. Sonography has proved highly effective in the detection and three-dimensional localization of various types of foreign bodies in soft tissues of the extremities [5–9].

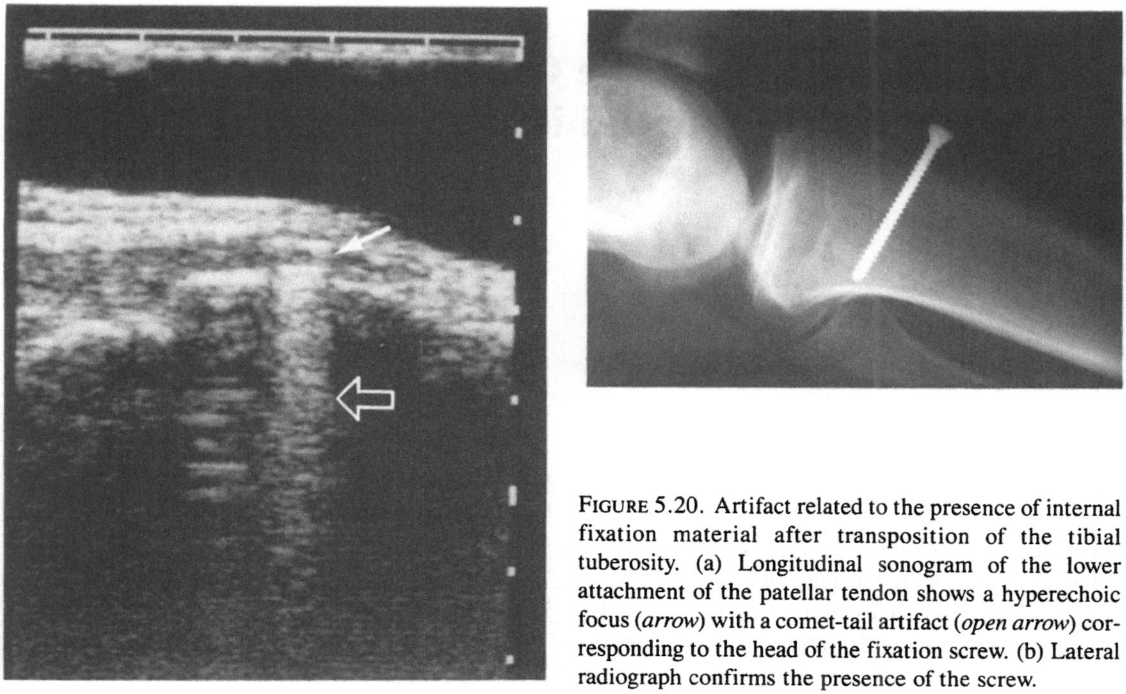

a b

FIGURE 5.20. Artifact related to the presence of internal fixation material after transposition of the tibial tuberosity. (a) Longitudinal sonogram of the lower attachment of the patellar tendon shows a hyperechoic focus (*arrow*) with a comet-tail artifact (*open arrow*) corresponding to the head of the fixation screw. (b) Lateral radiograph confirms the presence of the screw.

In Summary

In most cases, the combination of longitudinal and transverse scanning enables the recognition of an artifact or pitfall. The transition from one type of scan to the other also allows selection of the optimal scan plane for the region of interest.

References

1. Fornage BD: Achilles tendon: US examination. Radiology 1986;159:759–764.
2. Fornage BD: The hypoechoic normal tendon: A pitfall. J Ultrasound Med 1987;6:19–22.
3. Fornage BD, Rifkin MD: Ultrasound examination of the hand. Radiology 1986;160:853–854.
4. Fornage B: L'échographie de la main: technique et anatomie ultrasonore normale. Journal d'Echographie et de Medecine Ultrasonore 1986;7:193–203.
5. Fornage BD, Schernberg FL: Sonographic diagnosis of foreign bodies of the distal extremities. AJR 1986;147:567–569.
6. Fornage BD, Schernberg FL: Sonographic preoperative localization of a foreign body in the hand. J Ultrasound Med 1987;6:217–219.
7. Fornage BD, Nasca S, Durville A: Sonographic detection and three-dimensional localization of a metallic foreign body in soft tissues in a dog. Br Vet J 1987;143:278–282.
8. Fornage BD: Preoperative sonographic localization of a migrated transosseous stabilizing wire in the hand. J Ultrasound Med 1987;6:471–473.
9. Gooding GAW, Hardiman T, Sumers M, Stess R, Graf P, Grunfeld C: Sonography of the hand and foot in foreign body detection. J Ultrasound Med 1987;6:441–447.
10. Ziskin MC, Thickman DI, Goldenberg NJ, Lapayowker MS, Becker JM: The comet tail artifact. J Ultrasound Med 1982;1:1–7.

6
Sonographic Patterns of the Other Anatomic Components of the Extremities

Bones

Bones serve as anatomic landmarks in the ultrasound examination of muscles and particularly of tendons. Like gas, bones represent an obstacle to sound propagation. Yet the intense reflection of the beam on a bone surface silhouettes its near-contours, which appear as brightly echogenic lines (Fig. 6.1); occasionally, pathologic alteration of bones can be visualized. Sonography is now routinely used in the evaluation of the hip in the neonate [11,12,15]. Sonography can visualize sesamoid bones. It can also demonstrate internal surgical fixation devices (see also Chapter 5).

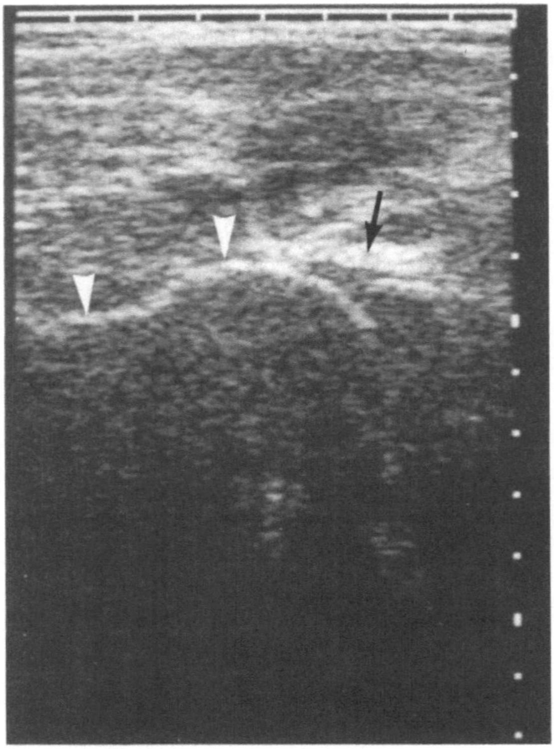

Joints

Without the use of high-frequency transducers, the normal articular capsule usually cannot be distinguished from adjacent soft tissues; exceptions include the hip, where it may be visualized as an echogenic line (Fig. 6.1) [13]. Retropatellar synovial plicae have been demonstrated as echogenic areas by a 7.5-MHz probe [3]. The articular cartilage is in some locations—for example, around the femoral condyles (Fig. 6.2) [1]—well depicted as a smooth, sonolucent layer of even thickness. Menisci of the knee are seen as echogenic structures (Fig. 6.2) [4]. Periarticular ligaments are barely distinguished, even in large joints.

Synovial bursae are, for the most part, only virtual cavities. With the use of high-frequency transducers, the deep infrapatellar bursa can be visualized posterior to the lower third of the patellar tendon (Fig. 6.3), and the deep (subtendinous) calcaneal bursa is occasionally seen anterior to the Achilles tendon (Fig. 6.4).

FIGURE 6.1. Transverse, slightly oblique sonogram of the anterior region of the right hip delineates the anterior aspect of the femoral head and neck (*arrowheads*). Note the echogenic articular capsule (*arrow*).

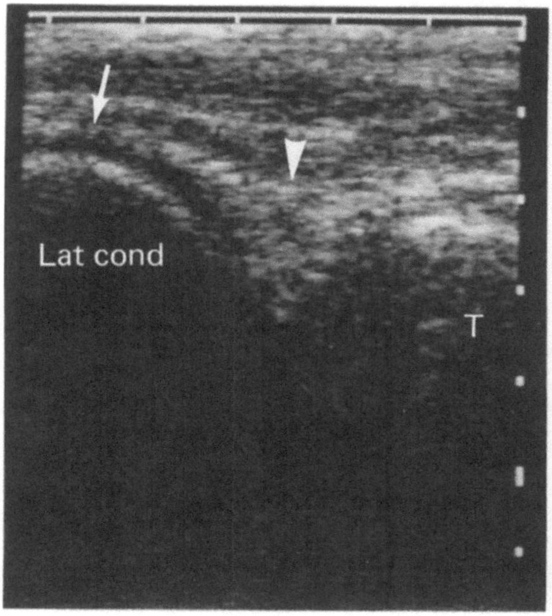

FIGURE 6.2. Longitudinal sonogram of the posterolateral aspect of the knee shows the lateral femoral condyle and the posterior aspect of the tibial plateau. The articular cartilage of the condyle appears as a sonolucent rim that is a few millimeters thick (*arrow*). Note the echogenic posterior horn of the lateral meniscus (*arrowhead*). *Lat cond* = lateral femoral condyle; *T* = tibia.

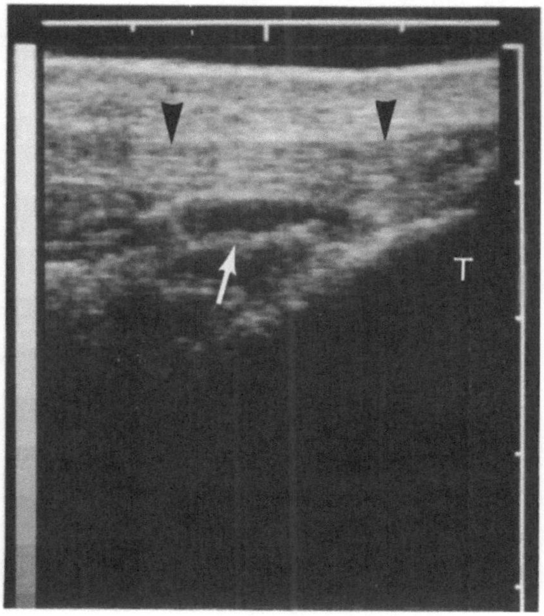

FIGURE 6.3. Deep infrapatellar bursa. Longitudinal sonogram of the patellar tendon shows the synovial bursa (*arrow*) posterior to the lower portion of the tendon (*arrowheads*). *T* = tibia. (From Fornage BD, Rifkin MD: Ultrasound examination of tendons. Radiol Clin North Am 1988;26:87–107, with permission.)

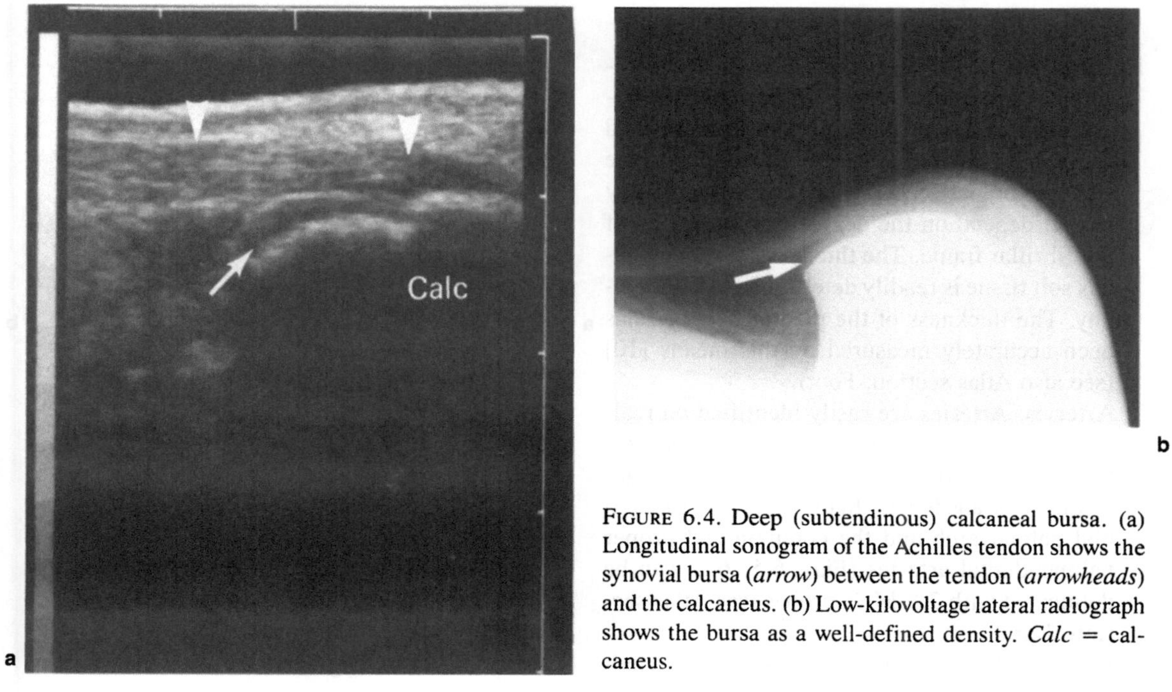

FIGURE 6.4. Deep (subtendinous) calcaneal bursa. (a) Longitudinal sonogram of the Achilles tendon shows the synovial bursa (*arrow*) between the tendon (*arrowheads*) and the calcaneus. (b) Low-kilovoltage lateral radiograph shows the bursa as a well-defined density. *Calc* = calcaneus.

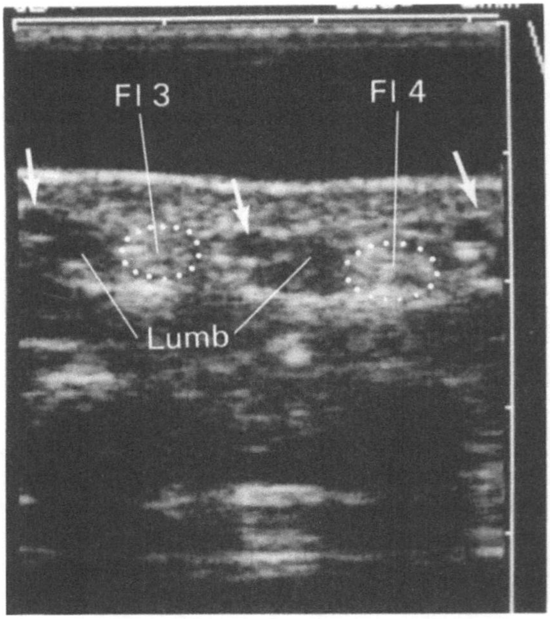

FIGURE 6.5. Transverse scan of the palm of the hand shows the transverse sections of the second, third, and fourth common palmar digital arteries (*arrows*). *Fl 3* = flexor tendons of the third finger; *Fl 4* = flexor tendons of the fourth finger; *Lumb* = lumbrical muscles.

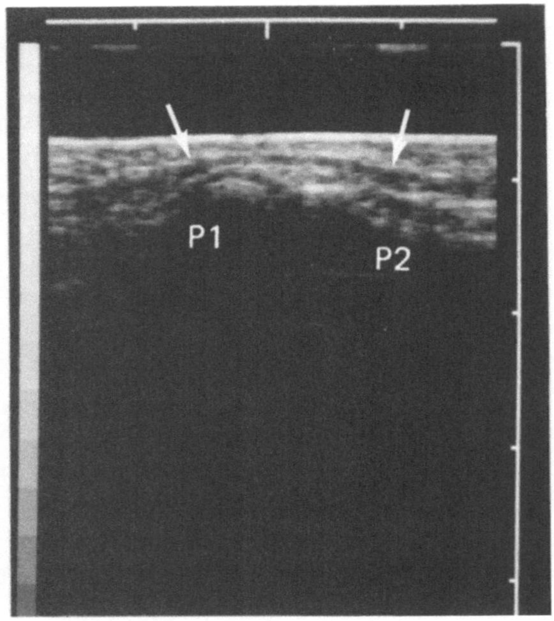

FIGURE 6.6. Coronal view of the proximal interphalangeal joint of the third finger shows a proper palmar digital artery (*arrows*). *P1* = proximal phalanx; *P2* = middle phalanx.

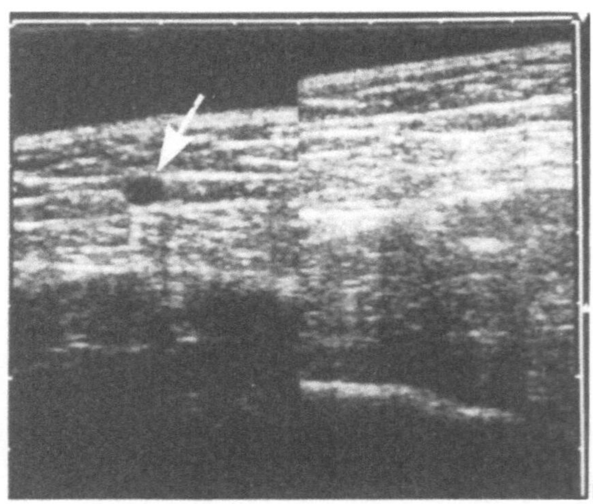

FIGURE 6.7. Transverse scans of the greater saphenous vein near its termination. (a) The vein (*arrow*) is clearly visualized in subcutaneous soft tissues. (b) A slight pressure on the transducer results in the vein's collapse.

Surrounding Soft Tissues

1. Subcutaneous tissue: Loose subcutaneous fibroadipose tissue may undergo variations in echogenicity similar to those previously described for fat between individuals and according to the site of examination [2,5]. The echogenicity may in part depend on the degree of reticulation of the fibrillar frame. The thickness of subcutaneous soft tissue is readily determined by sonography. The thickness of the plantar heel pad has been accurately measured by this means [10] (see also Atlas section: Foot).
2. Arteries: Arteries are easily identified on real-time examination. In the distal extremities, their visualization requires the use of high-frequency transducers. In the hand, the superficial palmar arch and the common and proper palmar digital arteries (Figs. 6.5, 6.6) can be delineated with 7.5-MHz probes (see Atlas section: Hand).
3. Superficial veins: Superficial veins are visualized only when a standoff pad is used. Their caliber varies according to gravity or compression maneuvers. They may collapse under a slight pressure on the probe (Fig. 6.7).
4. Nerves: It has long been assumed that normal nerves could not be demonstrated sonographi-

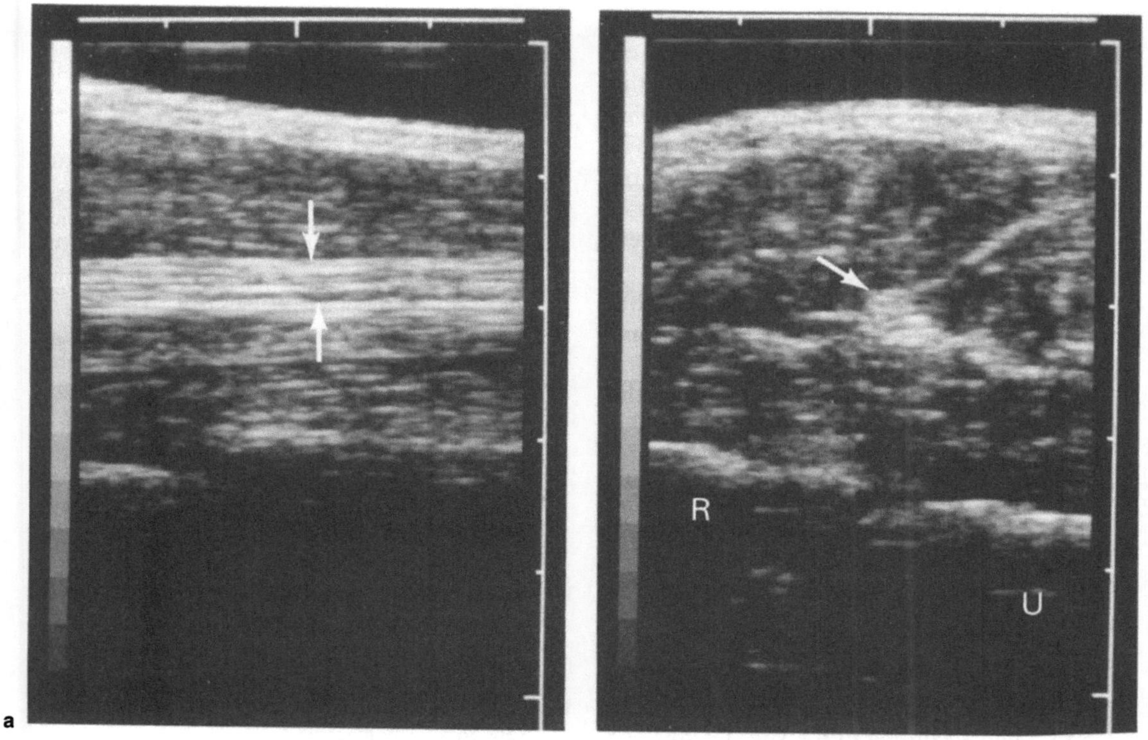

FIGURE 6.8. Median nerve. (a) Longitudinal sonogram of the anterior region of the midthird of the forearm. The median nerve is markedly echogenic and shows a characteristic fibrillar texture (*arrows*). It lies between the flexor digitorum superficialis and the flexor digitorum profundus muscles. It can be confirmed by its immobility during the flexion/extension movements of the fingers. (b) Transverse scan shows the echogenic cross section of the nerve (*arrow*) surrounded by the hypoechoic muscles. *R* = radius; *U* = ulna. (Part b from Fornage BD: Peripheral nerves of the extremities: Imaging with US. Radiology 1988;167:179–182, with permission.)

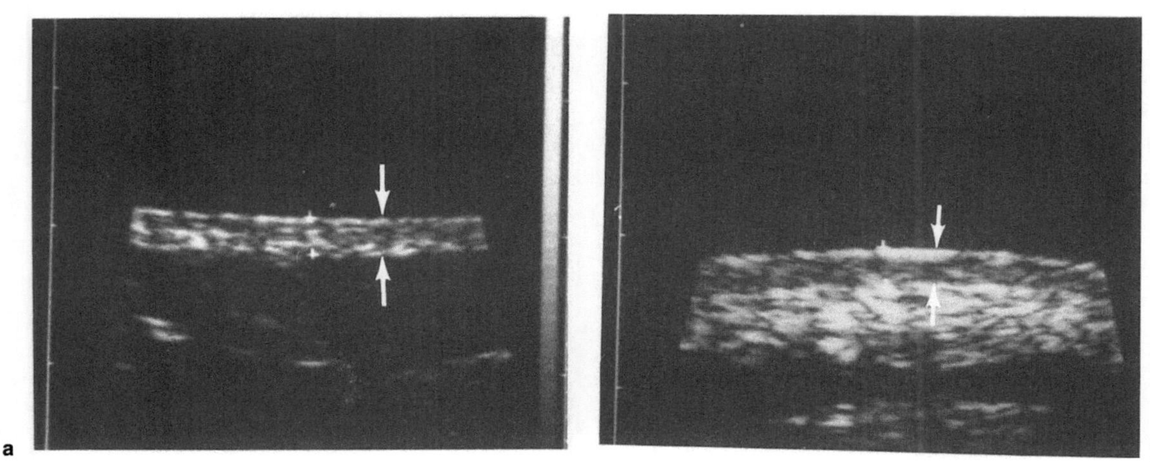

FIGURE 6.9. Sonogram of the skin obtained with a 10-MHz sector probe. (a) Sonogram performed on the skin of the buttock. Dermis (*arrows*) has a midlevel echogenicity. The subcutaneous fat is markedly hypoechoic. (b) Sonogram of the skin at the back of the neck. Dermis (*arrows*) is less echogenic than the hypodermis.

cally. Recently, the recurrent laryngeal nerve has been reported as a hypoechoic structure on imaging with a 10-MHz transducer [14]. In the extremities, we have been able to sonographically identify the ulnar nerve at the elbow and in the forearm, the median nerve in the forearm (Fig. 6.8) and carpal tunnel, the sciatic nerve in the posterior thigh, and the lateral popliteal nerve at the fibular neck (see Atlas section: Forearm, Wrist, Thigh) [7,9]. Nerves are highly echogenic and have a fibrillar texture resembling that of the tendons. Their demonstration is confirmed by their immobility during the flexion/extension movements of the surrounding muscles.

5. Skin: The skin can be displayed with high-frequency transducers and a water-path system. When a 5-MHz probe is used, the skin is typically seen as two echogenic parallel lines. However, with the use of a 7.5- or 10-MHz transducer, one can evaluate its texture and measure its thickness [6]. The thickness of the "total skin" (epidermis and dermis) varies from 1.5 to 5 mm according to the site. Dermis and hypodermis may vary in relative echogenicity (Fig. 6.9).
6. Nail: The nail and the subungual space can be evaluated with high-frequency transducers (see Atlas section: Hand) [7,8].

References

1. Aisen AM, McCune WJ, MacGuire A, et al: Sonographic evaluation of the cartilage of the knee. Radiology 1984;153:781–784.
2. Behan M, Kazam E: The echographic characteristics of fatty tissues and tumors. Radiology 1978; 129:143–151.
3. Derks WHJ, De Hooge P, Van Linge B: Ultrasonographic detection of the patellar plica in the knee. J Clin Ultrasound 1986;14:355–360.
4. Dragonat P, Claussen C: Sonographische Meniskusdarstellungen. ROFO 1980;133:185–187.
5. Fornage B: Aspects échographiques des lipomes superficiels. J Radiol 1984;65:853–857.
6. Fornage BD, Deshayes JL: Ultrasound of normal skin. J Clin Ultrasound 1986;14:619–622.
7. Fornage BD: L'échographie de la main: technique et anatomie ultrasonore normale. Journal d'Echographie et de Médecine Ultrasonore 1986;7:193–203.
8. Fornage BD: Glomus tumors in the fingers: Diagnosis with US. Radiology 1988;167:183–185.
9. Fornage BD: Peripheral nerves of the extremities: Imaging with US. Radiology 1988;167:179–182.
10. Gooding GAW, Stess RM, Graf PM, Grunfeld C: Heel pad thickness: Determination by high-resolution ultrasonography. J Ultrasound Med 1985;4: 173–174.
11. Graf R: A Guide to Sonography of the Infant Hip. New York, Thieme, 1987.
12. Morin C, Harcke HT, McEwen GD: The infant hip: Real-time US assessment of acetabular development. Radiology 1985;157:673–677.
13. Seltzer SE, Finberg HJ, Weissman BN: Arthrosonography: Technique, sonographic anatomy, and pathology. Invest Radiol 1980;15:19–28.
14. Solbiati L, De Pra L, Ierace T, Belloti E, Derchi LE: High-resolution sonography of the recurrent laryngeal nerve: Anatomic and pathologic considerations. AJR 1985;145:989–993.
15. Zieger M, Schulz RD: Method and results of ultrasound in hip studies. Ann Radiol (Paris) 1986;29: 383–386.

7
Other Imaging Techniques for Muscles and Tendons

Sartoris recently described musculoskeletal imaging as an "evolving subspecialty" [52], and hence the place of each of the imaging modalities is under assessment. Of the newer modalities, magnetic resonance (MR) imaging is doubtless the most popular in this arena, borne out by the profusion of published studies. In contrast, the comparatively limited number of reports on high-resolution sonography of soft tissues would suggest that this recent approach has been neglected in many institutions. To allow a perspective on the preceding chapters, this chapter briefly reviews the utilization of other modalities, both old and new, and also highlights recent experimental developments in ultrasound imaging of muscles and tendons.

Plain Radiography

Plain radiographs provide the most detailed images of bones and are therefore still required whenever the soft tissue changes may be associated with a bone lesion.

Low-Kilovoltage Radiography

Radiographic contrast depends on effective atomic number, electron density, and specific gravity of the absorbing tissue, and on the energy spectrum of incident x-rays. Only the last parameter is operator dependent. Below 30 kV, the attenuation of x-rays is predominantly by photoelectric interaction, which varies as the cube of the atomic number of the absorbing tissues. This accounts for the optimal differential attenuation (or contrast) obtained with low-kilovoltage technique in soft tissues, which principally consist of water and fat [38].

The low-kilovoltage technique is easily performed and does not require sophisticated equipment, although the use of a mammographic unit is recommended (Fig. 7.1) [60]. However, the contrast resolution is still limited. Low-kilovoltage radiographs are helpful in the diagnosis of calcifications and foreign bodies of minute size or low density, particularly in very superficial soft tissues and the distal extremities [14,15,17,21].

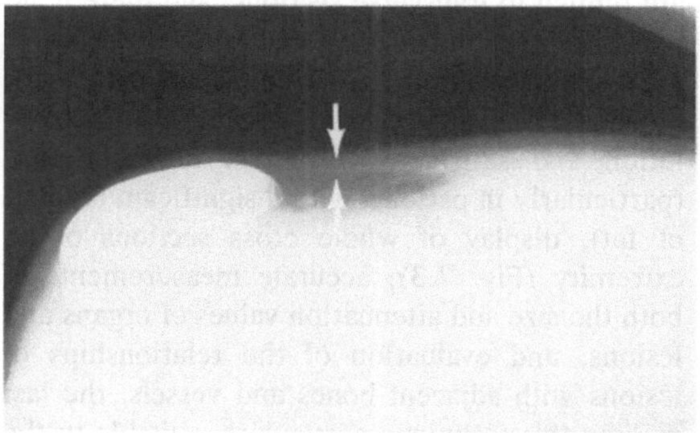

FIGURE 7.1. Low-kilovoltage (30-kV) radiograph of the knee made using a mammographic unit. Lateral view of the flexed knee shows the patellar tendon (*arrows*), which is well delineated by the subcutaneous fat anteriorly and by the fat pad of the knee posteriorly.

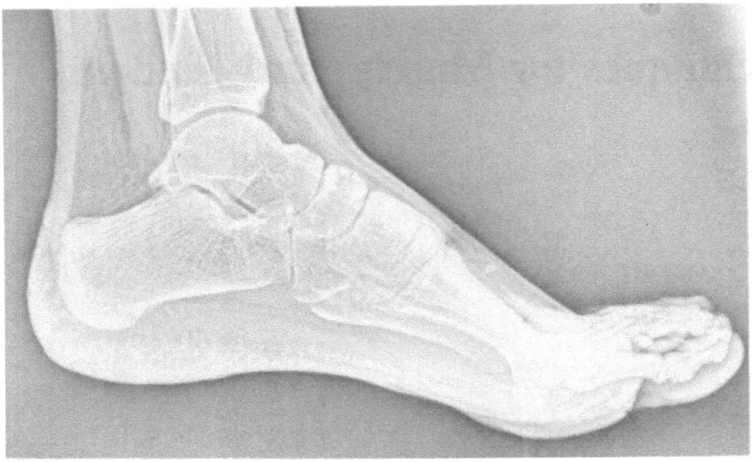

FIGURE 7.2. Lateral xeroradiography of the foot. Bony structures and soft tissues are seen on the same exposure. Note the clear depiction of the Achilles tendon.

Xeroradiography

Xeroradiography, like plain radiography, only offers a planar projection of soft tissues. However, it provides both wide contrast latitude, allowing the demonstration of bone and soft tissue densities on a single exposure, and sharp detail contrast due to the "edge effect" (Fig. 7.2) [61].

Computed Tomography

Computed tomography (CT) units are costly and, at most institutions, heavily utilized. Routine scans are limited to transverse sections, and there is no real-time capability. Also, radiation exposure is greater than with other routine radiographic techniques. The advantages of CT are high spatial resolution and satisfactory contrast differentiation (particularly in patients with a significant amount of fat), display of whole cross sections of an extremity (Fig. 7.3), accurate measurements of both the size and attenuation values of organs and lesions, and evaluation of the relationships of lesions with adjacent bones and vessels, the last making this technique particularly suitable to the staging of malignant tumors [10,19,25,27,34,37, 48,66]. Few reports have been made on CT evaluation of the distal extremities [47,54,55]. CT has also proved useful in the evaluation and follow-up of neuromuscular diseases [7,24,48,58].

Magnetic Resonance Imaging

Early clinical studies have shown MR to have excellent contrast resolution in soft tissues [5,6,39]. Muscles, tendons, fat, and vessels are readily differentiated because of their different signal intensities [4,13,35,40,63]. A significant advantage of MR over CT is the routine production of transverse, sagittal, coronal, and oblique sections (Fig. 7.4), which makes this modality outstanding in the three-dimensional evaluation of tumors [1,43,59]. Surface coils currently provide impressive spatial resolution, enabling detailed analysis of soft tissues of the extremities [2,3,8,11,18,23,29–33,36,44–46,49,64,65].

Current limitations of MR imaging include the lack of real-time imaging, its costliness, and the restricted siting of equipment. Severe patient claustrophobia and having a pacemaker or certain ferromagnetic foreign bodies are well-known contraindications to MR examination.

Angiography

Angiography, whether conventional or digital, is still indicated in the preoperative local staging of malignant tumors of the extremities (Fig. 7.5) [42,62].

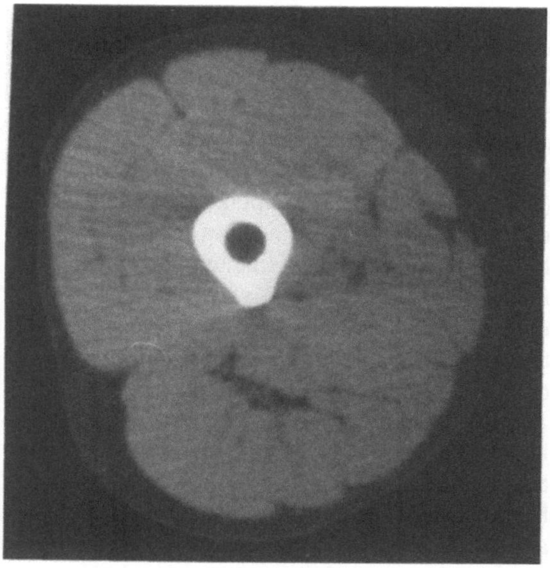

FIGURE 7.3. CT scan of the thigh. The cross section of the whole extremity allows evaluation of the relationship of the various muscles.

Tenography

The direct intratendinous injection of a few milliliters of water-soluble contrast medium has been used to visualize tendons and ligaments [22]. An easier, indirect tenographic technique in which contrast material is injected into the tendon sheath has recently been used in the assessment of tenosynovitis in the wrist and ankle [12,20,57]. This later technique, or "synoviography," provides the most detailed images of synovial sheaths.

Radionuclide Imaging

The use of radiopharmaceuticals allows both morphologic and functional imaging studies. Bone scanning is based on the uptake of a radiotracer (technetium-99m diphosphonate or pyrophosphate) by bone and any tissue with ossification; the uptake occurs because of the radiotracer's affinity

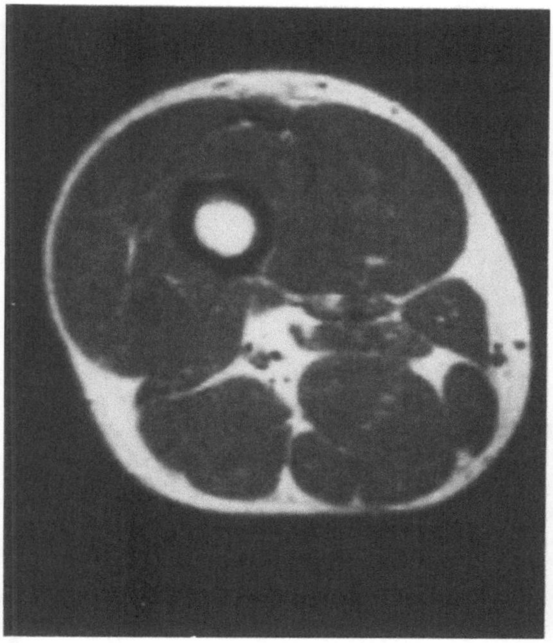

a

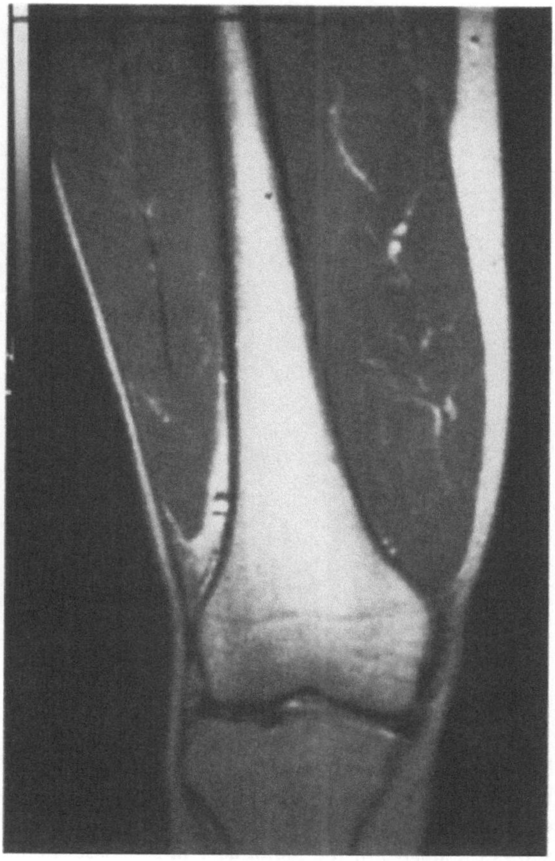

b

FIGURE 7.4. MR images of the lower extremity. (a) Transverse MR image of the lower third of the thigh clearly demonstrates the different muscles (1.5-T equipment; TR 400 ms, TE 50 ms). (b) Coronal view of the thigh and the knee (TR 1500 ms, TE 50 ms).

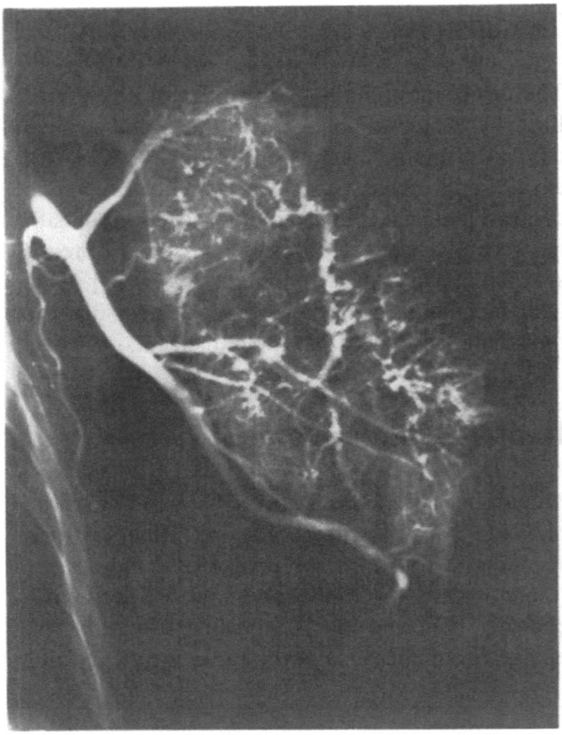

FIGURE 7.5. Arteriography of angiosarcoma of the shoulder. Preoperative arteriogram obtained at 4 s shows the tumor neovascularization.

for calcium salts. In the three-phase technique, early scans are isotope angiograms, blood-pool images demonstrate the radiotracer in soft tissues of the extremities ("parenchymographic" phase), and delayed images obtained about 3 hours after injection represent the standard bone scan (Fig. 7.6) [51]. The focal uptake of "bone-seeking" radiopharmaceuticals in muscles due to an elevated local concentration of calcium salts has been reported in cases of myositis ossificans or rhabdomyolysis [28,41,53,56]. On the other hand, soft tissue abscesses may demonstrate specific uptake of gallium 67 [9].

Since radionuclide scanning provides no etiologic information, is costly because of the radiotracer, and requires a gamma camera, it has not been routinely used in the evaluation of muscles and tendons. However, it must be noted that bone scanning has proved extremely valuable in sports medicine by demonstrating early stress fractures [51].

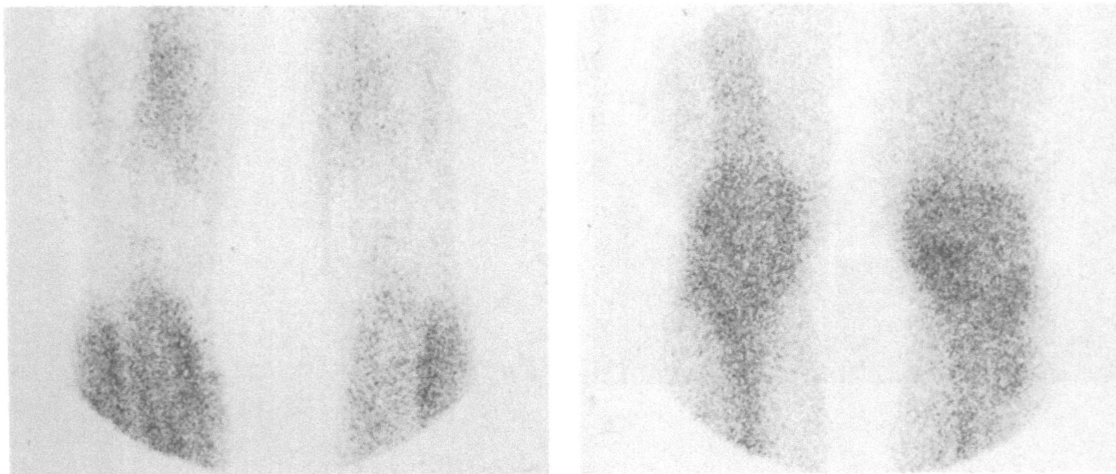

FIGURE 7.6. Radionuclide imaging using technetium-99m-labeled diphosphonate. Anterior views of both knees. (a) Blood-pool, or tissue-phase, image shows the uptake of the radiotracer in soft tissues and lack of bone definition. (b) Delayed image shows the bone activity as usually demonstrated on conventional bone scans.

Thermography

Two different techniques of thermography—liquid-crystal plate thermography and infrared telethermography (Fig. 7.7)—have been used. The thermal distribution at the surface of the skin has been evaluated by several investigators in patients with trauma-related conditions on the assumption that the superficial heat-mapping would correlate with the deep vascular distribution of underlying muscles. Thermography has proved to be inaccurate in this application and has been abandoned.

Prospects in Ultrasonography

Experimental developments in ultrasound imaging of the extremities concern ultrasound transmission imaging and computerized sonography.

Ultrasound Transmission Imaging

Conventional sonography is based on the analysis of reflected beams. In ultrasound transmission imaging, sonograms are obtained from transmitted beams: the beams scattered in passing through the object are collimated and focused on an array of receiving transducers, creating a video signal [26]. The images are planar projections, comparable to plain radiographs. Bones are silhouetted, while muscles, tendons, and neurovascular bundles are clearly demonstrated with a spatial resolution of approximately 1 mm in the hand. Real-time imaging is available.

Computerized Axial Sonography

In computerized axial sonography, four (or six) linear-array transducers are placed in a quadrangular (or hexagonal) pattern around the object, which is immersed in water. Incremental rotation provides a 360° scan. The computer processing of the echoes uses a matrix reconstruction similar to that utilized in CT. This modality, like CT, provides entire cross sections of the extremity [50]. It also delineates the bones without acoustic shadowing (Fig. 7.8).

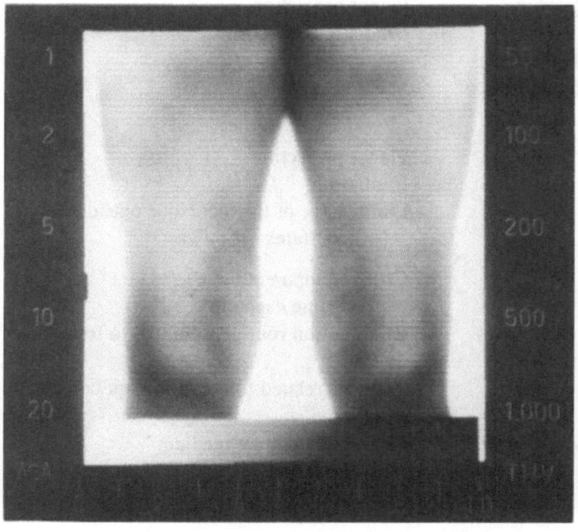

FIGURE 7.7. Gray-scale infrared telethermography of the posterior aspect of the thighs. Hot areas are displayed in the black-on-white mode.

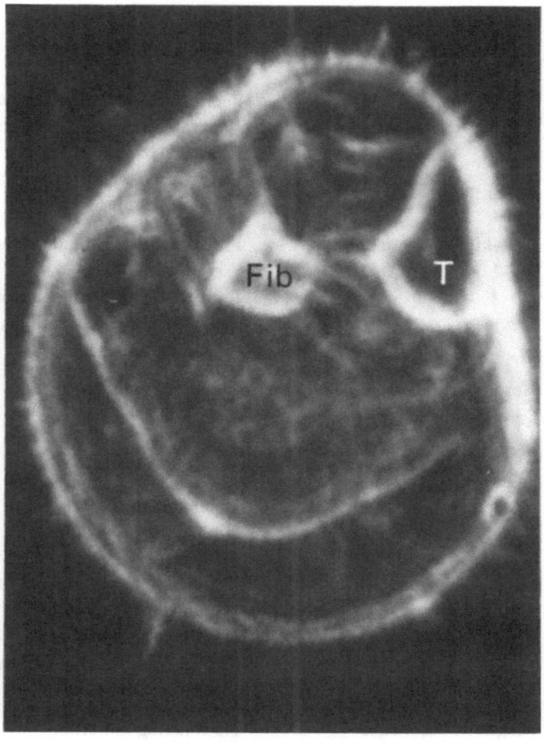

FIGURE 7.8. Computerized axial sonography. Transverse scan of the leg. Note the absence of the acoustic shadowing that is associated with bone structures on conventional sonography. *Fib* = fibula; *T* = tibia. (Courtesy of Dr. Hans-Dieter Rott.)

In Summary

Sonography is at present the only real-time cross-sectional imaging technique and it is associated with the best cost/risk/information ratio (Table 7.1). Sonography should be the first examination performed in the evaluation of trauma, inflammation, or even tumor of muscles and tendons of the extremities, as well as in many other diseases that involve adjacent soft tissues. Indeed, sonography is growing more and more popular among orthopedic surgeons and specialists in sports medicine, who have found the cost-effective diagnostic tool and the objective follow-up test that were needed for the evaluation of muscles and tendons [16].

TABLE 7.1. Advantages and Limitations of Imaging Modalities Available in Muscle and Tendon Evaluation

Imaging modality	Advantages	Limitations
Sonography	No ionizing radiation Easily performed Low cost Equipment widespread High spatial resolution Real-time imaging Full multiplanar capability Three-dimensional evaluation Accurate measurement of lesions Best cost/risk/information ratio Biopsy needle guidance	Partial views of the extremities Requires a well-trained operator
Plain radiography	Easily performed Low cost Technique and equipment widespread Accurate evaluation of bones Characterization of fat Detection of calcifications and radiopaque foreign bodies	Limited contrast resolution for soft tissues Planar projection Irradiation
Low-kilovoltage radiography	Performed with standard radiographic equipment Improved contrast resolution (characterization of fat, calcifications, and radiopaque foreign bodies)	Planar projection Irradiation
Xeroradiography	Enhanced contrast	Planar projection Irradiation Limited use of the technique outside the United States
Computed tomography	Good spatial resolution High contrast resolution Provides whole sections of the extremities Detailed analysis of bones and vessels Distance measurements Density measurements Can be performed in patients with plaster cast	Costly equipment No real-time capability Examination routinely limited to transverse scans Artifacts related to metal foreign bodies Irradiation Parenteral contrast medium
Magnetic resonance imaging	No ionizing radiation Optimal contrast resolution Multiplanar capability Approach to chemical tissue analysis Can be performed in patients with plaster cast	Costly equipment and examination Real-time study not yet available Contraindicated in patients with pacemakers and certain ferromagnetic foreign bodies

(Continued)

TABLE 7.1. *Continued*

Imaging modality	Advantages	Limitations
Angiography	Analysis of vascular distribution (preoperative evaluation of malignant tumors)	Costliness Invasiveness Irradiation Parenteral contrast medium
Tenography		
Direct tenography		Planar projection Invasiveness Irradiation Local injection of contrast medium
Synoviography	Detailed imaging of synovial sheaths	Planar projection Invasiveness Irradiation Local injection of contrast medium
Radionuclide imaging	Association of morphologic and functional studies	Costliness Planar projection Injection of a radiotracer No real-time capability
Thermography	No ionizing radiation	Indirect technique Lack of diagnostic accuracy

References

1. Aisen AM, Martel W, Braunstein EM, McMillin KI, Phillips WA, Kling TF: MRI and CT evaluation of primary bone and soft-tissue tumors. AJR 1986;146:749–756.
2. Bellon EM, Sacco DC, Steiger DA, Coleman PE: Magnetic resonance imaging in "housemaid's knee" (prepatellar bursitis). Magn Reson Imaging 1987;5:175–177.
3. Beltran J, Noto AM, Mosure JC, Weiss KL, Zuelzer W, Christoforidis AJ: The knee: Surface-coil MR imaging at 1.5 T. Radiology 1986;159:747–751.
4. Beltran J, Noto AM, Herman LJ, Lubbers LM: Tendons: High-field-strength, surface coil MR imaging. Radiology 1987;162:735–740.
5. Berquist TH: Magnetic resonance imaging: Preliminary experience in orthopedic radiology. Magn Reson Imaging 1984;2:41–52.
6. Brady TJ, Gebhardt MC, Pykett IL, et al: NMR imaging of forearms in healthy volunteers and patients with giant-cell tumor of bone. Radiology 1982;144:549–552.
7. Bulcke JAL, Herpels V: Diagnostic value of CT scanning in neuromuscular diseases. Radiologe 1983;23:523–528.
8. Burk DL, Kanal E, Brunberg JA, Johnstone GF, Swensen HE, Wolf GL: 1.5-T surface-coil MRI of the knee. AJR 1986;147:293–300.
9. Datz FL, Lewis SE, Conrad MR, Maravilla A, Parkey RW: Pyomyositis diagnosed by radionuclide imaging and ultrasonography. South Med J 1980; 73: 649–651.
10. Egund N, Ekelund L, Sako M, Persson B: CT of soft-tissue tumors. AJR 1981;137:725–729.
11. Ehman RL, Berquist TH: Magnetic resonance imaging of musculoskeletal trauma. Radiol Clin North Am 1986;24:291–319.
12. Engel J, Luboshitz S, Israeli A, Ganel A: Tenography in De Quervain's disease. Hand 1981;13: 142–146.
13. Fisher MR, Dooms GC, Hricak H, Reinhold C, Higgins CB: Magnetic resonance imaging of the normal and pathologic muscular system. Magn Reson Imaging 1986;4:491–496.
14. Fornage B, Touche D, Deshayes JL, Segal Ph: Diagnostic des calcifications du tendon rotulien. Comparaison écho-radiographique. J Radiol 1984;65: 355–359.
15. Fornage BD, Schernberg FL: Sonographic diagnosis of foreign bodies of the distal extremities. AJR 1986;147:567–569.
16. Fornage B: L'échographie et la médecine du sport. Journal d'Echographie et de Médecine Ultrasonore 1986;7:111–112.
17. Fornage BD, Schernberg FL: Sonographic preoperative localization of a foreign body in the hand. J Ultrasound Med 1987;6:217–219.
18. Gallimore GW, Harms SE: Knee injuries: High-resolution MR imaging. Radiology 1986;160:457–461.

19. Genant HK, Cann CE, Chafetz NI, Helms CA: Advances in computed tomography of the musculoskeletal system. Radiol Clin North Am 1981;19:645–674.
20. Gilula LA, Oloff L, Caputi R, Destouet JM, Jacobs A, Solomon MA: Ankle tenography: A key to unexplained symptomatology. Part II. Diagnosis of chronic tendon disabilities. Radiology 1984;151: 581–587.
21. Gooding GAW, Hardiman T, Sumers M, Stess R, Graf P, Grunfeld C: Sonography of the hand and foot in foreign body detection. J Ultrasound Med 1987;6:441–447.
22. Grevsten S, Eriksson K: Tendinography for diagnosing injuries to tendons and ligaments. Acta Radiol [Diagn] 1979;20:447–457.
23. Hartzman S, Reicher MA, Bassett LW, Duckwiler GR, Mandelbaum B, Gold RH: MR imaging of the knee. Part II. Chronic disorders. Radiology 1987;162:553–557.
24. Hawley RJ, Schellinger D, O'Doherty DS: Computed tomographic patterns of muscles in neuromuscular diseases. Arch Neurol 1984;41:383–387.
25. Heelan RT, Watson RC, Smith J: Computed tomography of lower extremity tumors. AJR 1979;132:933–937.
26. Hentz VR, Marich KW, Dev P: Preliminary study of the upper limb with the use of ultrasound transmission imaging. J Hand Surg [Am] 1984;9A:188–193.
27. Hermann G, Rose JS: Computed tomography in bone and soft tissue pathology of the extremities. J Comput Assist Tomogr 1979;3:58–66.
28. Kaplan GN: Ultrasonic appearance of rhabdomyolysis. AJR 1980;134:375–377.
29. Kieft GJ, Bloem JL, Obermann WR, Verbout AJ, Rozing PM, Doornbos J: Normal shoulder: MR imaging. Radiology 1986;159:741–745.
30. Kneeland JB, Carrera GF, Middleton WD, et al: Rotator cuff tears: Preliminary application of high-resolution MR imaging with counter rotating current loop-gap resonators. Radiology 1986;160: 695–699.
31. Kneeland JB, Kellman GM, Middleton WD, et al: Diagnosis of diseases of the supraclavicular region by use of MR imaging. AJR 1987;148:1149–1151.
32. Koenig H, Lucas D, Meissner R: The wrist: A preliminary report on high-resolution MR imaging. Radiology 1986;160:463–467.
33. Kulkarni MV, Patton JA, Price RR: Technical considerations for the use of surface coils in MRI. AJR 1986;147:373–378.
34. Levine E, Lee KR, Neff JR, Maklad NF, Robinson RG, Preston DF: Comparison of computed tomography and other imaging modalities in the evaluation of musculoskeletal tumors. Radiology 1979;131: 431–437.
35. Li DKB, Adams ME, McConkey JP: Magnetic resonance imaging of the ligaments and menisci of the knee. Radiol Clin North Am 1986;24:209–227.
36. Lufkin RB, Sharpless T, Flannigan B, Hanafee W: Dynamic-range compression in surface-coil MRI. AJR 1986;147:379–382.
37. McLeod RA, Stephens DH, Beabout JW, Sheedy PF II, Hattery RR: Computed tomography of the skeletal system. Semin Roentgenol 1978;13:235–247.
38. Melson GL, Staple TW, Evens RG: Soft tissue radiographic technique. Semin Roentgenol 1973;8:19–24.
39. Moon KL Jr, Genant HK, Helms CA, Chafetz NI, Crooks LE, Kaufman L: Musculoskeletal applications of nuclear magnetic resonance. Radiology 1983;147:161–171.
40. Murphy WA, Totty WG, Carroll JE: MRI of normal and pathologic skeletal muscle. AJR 1986;146:565–574.
41. Patel R, Mishkin FS: Technetium-99m pyrophosphate imaging in acute renal failure associated with nontraumatic rhabdomyolysis. AJR 1986;147:815–817.
42. Paushter DM, Borkowski GP, Buonocore E, Belhobek GH, Marks KE: Digital subtraction angiography for preoperative evaluation of extremity tumors. AJR 1983;141:129–133.
43. Petasnick JP, Turner DA, Charters JR, Gitelis S, Zacharias CE: Soft-tissue masses of the locomotor system: Comparison of MR imaging with CT. Radiology 1986;160:125–133.
44. Reicher MA, Rauschning W, Gold RH, Bassett LW, Lufkin RB, Glen W Jr: High-resolution magnetic resonance imaging of the knee joint: Normal anatomy. AJR 1985;145:895–902.
45. Reicher MA, Bassett LW, Gold RH: High-resolution magnetic resonance imaging of the knee joint: Pathologic correlations. AJR 1985;145:903–909.
46. Reicher MA, Hartzman S, Bassett LW, Mandelbaum B, Duckwiler G, Gold RH: MR imaging of the knee. Part I. Traumatic disorders. Radiology 1987;162: 547–551.
47. Reiser M, Rupp N, Lehner K, Paar O, Gradinger R, Karpf PM: Demonstration of Achilles tendon on CT. Normal and pathological changes. ROFO 1985;143:173–177.
48. Riddlesberger MM Jr: Computed tomography of the musculoskeletal system. Radiol Clin North Am 1981;19:463–477.
49. Roger B, Chaise F, Girard F, Laval-Jeantet M, Travers V, Delepine G: Aspect du genou normal en

I.R.M. avec antenne de surface. J Radiol 1986;67: 193–199.
50. Rott HD, Breimesser FH, Rödl W: Imaging technics in muscular dystrophies. J Genet Hum 1985;33:397–403.
51. Rupani HD, Holder LE, Espinola DA, Engin SI: Three-phase radionuclide bone imaging in sports medicine. Radiology 1985;156:187–196.
52. Sartoris DJ: Musculoskeletal imaging: An evolving subspecialty. AJR 1987;148:1186–1187.
53. Siegel BA, Engel WK, Derrer EC: Localization of technetium-99m diphosphonate in acutely injured muscle: Relationship to muscle calcium deposition. Neurology 1977;27:230–238.
54. Solomon MA, Gilula LA, Oloff LM, Oloff J, Compton T: CT scanning of the foot and ankle: 1. Normal anatomy. AJR 1986;146:1192–1203.
55. Solomon MA, Gilula LA, Oloff LM, Oloff J: CT scanning of the foot and ankle: 2. Clinical applications and review of the literature. AJR 1986;146: 1204–1214.
56. Suzuki Y, Hisada K, Takeda M: Demonstration of myositis ossificans by ^{99m}Tc pyrophosphate bone scanning. Radiology 1974;111:663–664.
57. Teng MM, Destouet JM, Gilula LA, Resnick D, Hembree JL, Oloff LM: Ankle tenography: A key to unexplained symptomatology. Part I. Normal tenographic anatomy. Radiology 1984;151:575–580.
58. Termote JL, Baert A, Crolla D, Palmers Y, Bulcke JA: Computed tomography of the normal and pathologic muscular system. Radiology 1980;137:439–444.
59. Totty WG, Murphy WA, Lee JKT: Soft-tissue tumors: MR imaging. Radiology 1986;160:135–141.
60. Vazelle F, Rochcongar P, Masse M, Ramée A: La pathologie du tendon d'Achille. Un emploi inattendu du mammographe. J Radiol 1981;62:299–307.
61. Vukanovic S, Sidani AH, Ducommun JC, Curati WL: Xerography and ultrasonography in soft tissues pathology. J Belge Radiol 1981;64:309–319.
62. Watt I: Radiology in the diagnosis and management of bone tumours. J Bone Joint Surg [Br] 1985;67B:520–529.
63. Weinreb JC, Cohen JM, Maravilla KR: Iliopsoas muscles: MR study of normal anatomy and disease. Radiology 1985;156:435–440.
64. Weiss KL, Beltran J, Shamam OM, Stilla RF, Levey M: High-field MR surface-coil imaging of the hand and wrist. Part I. Normal anatomy. Radiology 1986;160:143–146.
65. Weiss KL, Beltran J, Lubbers LM: High-field MR surface-coil imaging of the hand and wrist. Part II. Pathologic correlations and clinical relevance. Radiology 1986;160:147–152.
66. Yeh HC, Rabinowitz JG: Ultrasonography of the extremities and pelvic girdle and correlation with computed tomography. Radiology 1982;143:519–525.

Part II
Atlas of Ultrasound Scans of the Extremities

Abbreviations

Abbreviation	Anatomic structure
Abd dig min	Abductor digiti minimi
Abd hal	Abductor hallucis
Abd pol long	Abductor pollicis longus
Ac	Acromion
Ach tend	Achilles tendon
Add	Adductors
Add brev	Adductor brevis
Add hal obl	Adductor hallucis (oblique head)
Add long	Adductor longus
Add mag	Adductor magnus
Anc	Anconeus
Ant tib art	Anterior tibial artery
Artic cart	Articular cartilage
Bic brac	Biceps brachii
Bic fem	Biceps femoris
Bic fem (lg)	Biceps femoris (long head)
Bic fem (s)	Biceps femoris (short head)
Brac	Brachialis
Brac art	Brachial artery
Brachiorad	Brachioradialis
Calc	Calcaneus
Cap	Capitate
Com palm dig art	Common palmar digital artery
Cr lig	Cruciate ligament
Cub	Cuboid
Delt	Deltoid
Dors ped art	Dorsalis pedis artery
Ext carp rad (long & brev)	Extensor carpi radialis (longus and brevis)
Ext carp rad brev	Extensor carpi radialis brevis
Ext carp rad long	Extensor carpi radialis longus
Ext carp uln	Extensor carpi ulnaris
Ext dig	Extensor digitorum
Ext dig brev	Extensor digitorum brevis
Ext dig long	Extensor digitorum longus
Ext dig min	Extensor digiti minimi
Ext hal long	Extensor hallucis longus
Ext ind	Extensor indicis
Ext pol brev	Extensor pollicis brevis
Ext pol long	Extensor pollicis longus
F	Femur

Abbreviation	Anatomic structure
Fem art	Femoral artery
Fib	Fibula
Flex carp rad	Flexor carpi radialis
Flex carp uln	Flexor carpi ulnaris
Flex dig brev	Flexor digitorum brevis
Flex dig long	Flexor digitorum longus
Flex dig min brev	Flexor digiti minimi brevis
Flex dig prof	Flexor digitorum profundus
Flex dig sup	Flexor digitorum superficialis
Flex hal brev	Flexor hallucis brevis
Flex hal long	Flexor hallucis longus
Flex pol long	Flexor pollicis longus
Gast (lat)	Gastrocnemius (lateral head)
Gast (med)	Gastrocnemius (medial head)
Glut max	Gluteus maximus
Grac	Gracilis
H	Humerus
Ham	Hamate
Interos	Interosseous
Lat cond	Lateral condyle
Lat men	Lateral meniscus
Lumb	Lumbrical
Lun	Lunate
Med	Median nerve
Med cond	Medial condyle
Med men	Medial meniscus
Metac	Metacarpal
Metat	Metatarsal
Navic	Navicular
P	Patella
P1	Proximal phalanx
P2	Middle phalanx
P3	Distal phalanx
Palm long	Palmaris longus
Pat tend	Patellar tendon
Per	Peronei (brevis and longus)
Per art	Peroneal artery
Per brev	Peroneus brevis
Per long	Peroneus longus
Pis	Pisiform
Pop	Popliteus
Pop art	Popliteal artery
Post tib art	Posterior tibial artery
Pron quad	Pronator quadratus

Abbreviation	Anatomic structure
Pron ter	Pronator teres
Prop palm dig art	Proper palmar digital artery
Quad tend	Quadriceps tendon
R	Radius
Rad art	Radial artery
RF	Rectus femoris
Sart	Sartorius
Scaph	Scaphoid
Sciat	Sciatic nerve
Semimemb	Semimembranosus
Semitend	Semitendinosus
Sol	Soleus
Sup palm arch	Superficial palmar arterial arch
Supin	Supinator
T	Tibia
T fasc lat	Tensor fasciae latae
Tal	Talus
Tib ant	Tibialis anterior
Tib post	Tibialis posterior
Trap	Trapezium
Tric (lat)	Triceps brachii (lateral head)
Tric (long)	Triceps brachii (long head)
Tric (med)	Triceps brachii (med head)
Triq	Triquetrum
TT	Tuberosity of tibia
U	Ulna
Uln art	Ulnar artery
VI	Vastus intermedius
VL	Vastus lateralis
VM	Vastus medialis

A. Upper Extremity

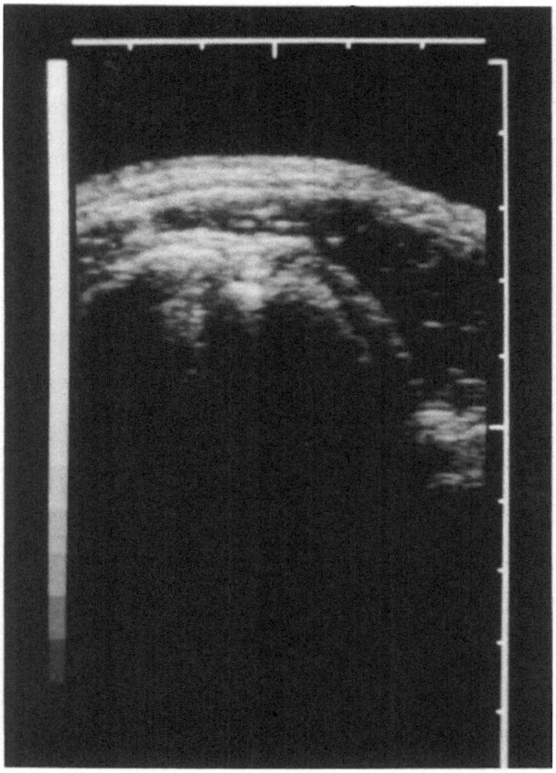

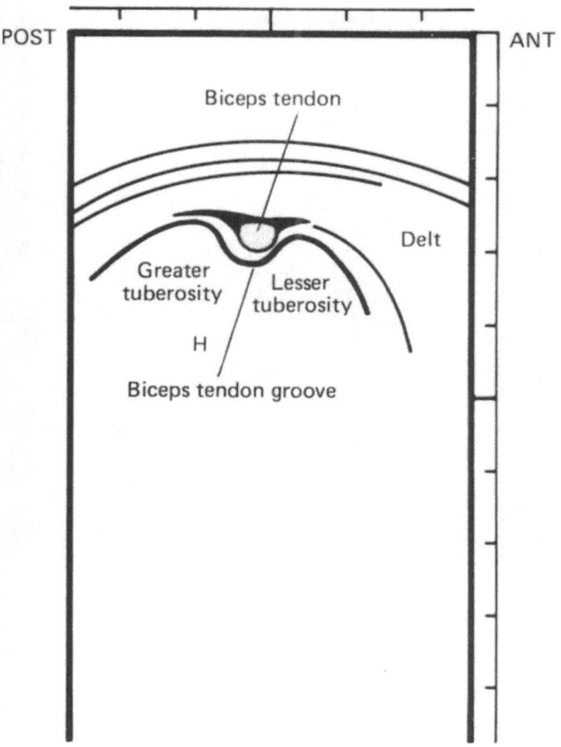
POST
ANT
Biceps tendon
Delt
Greater tuberosity
Lesser tuberosity
H
Biceps tendon groove

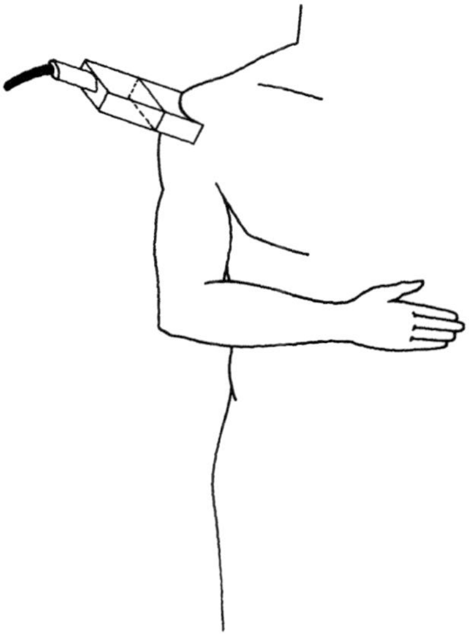

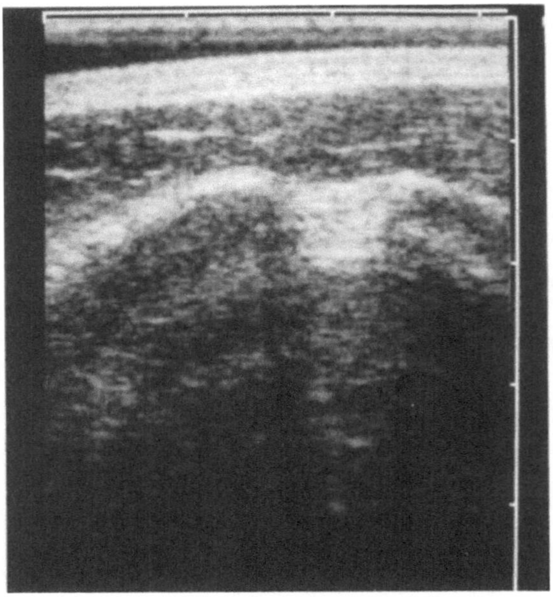

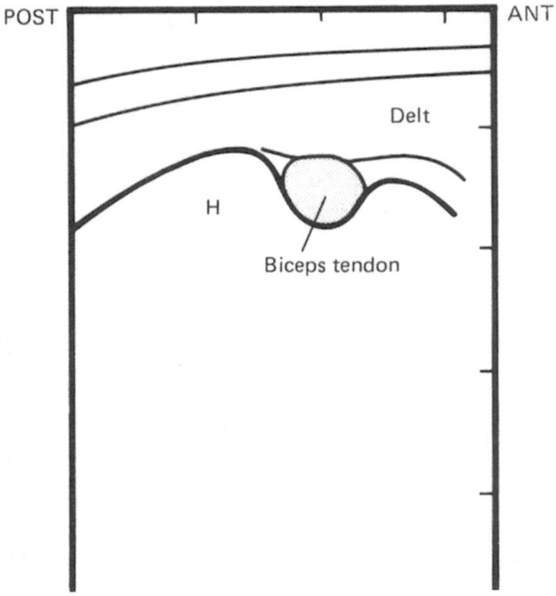
POST
ANT
Delt
H
Biceps tendon

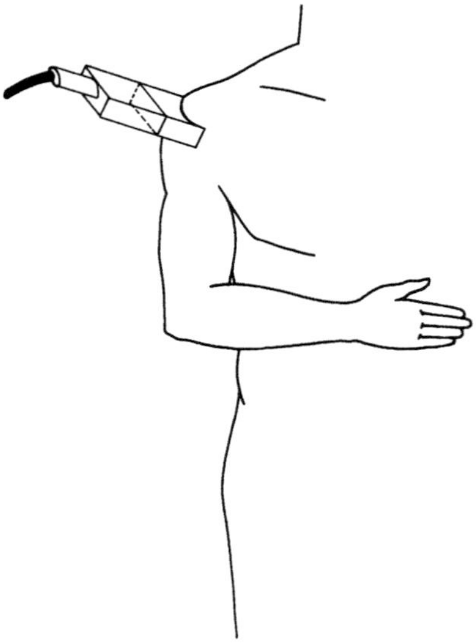

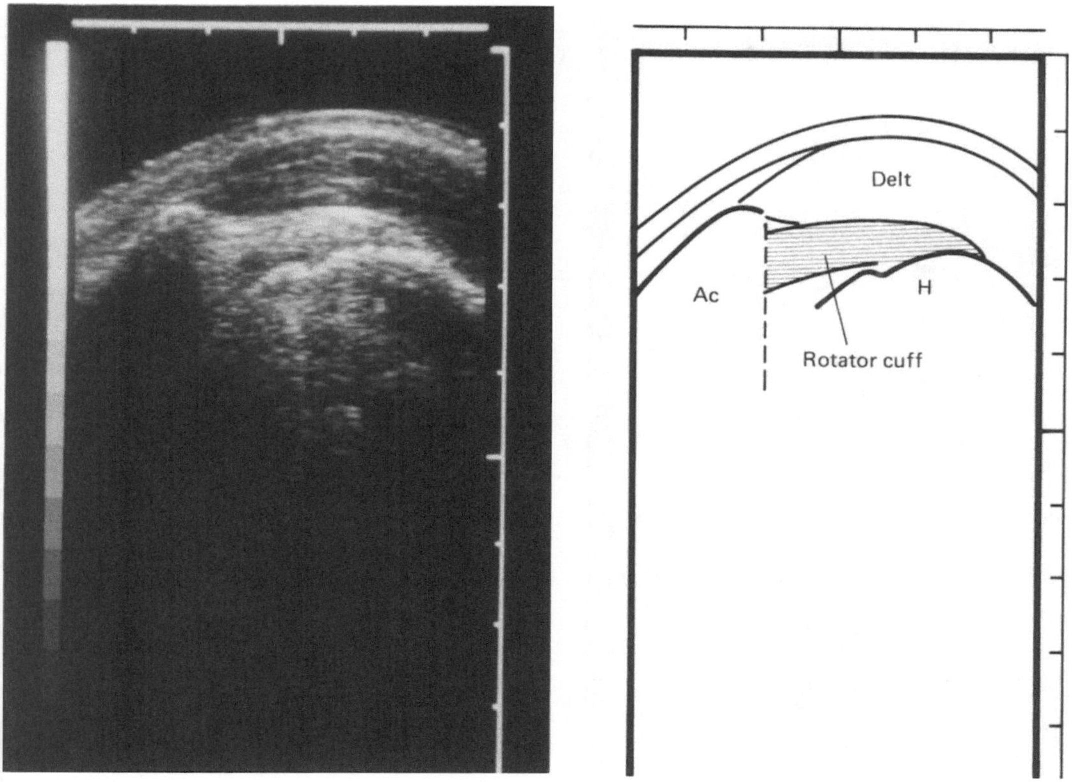
Delt
Ac
H
Rotator cuff

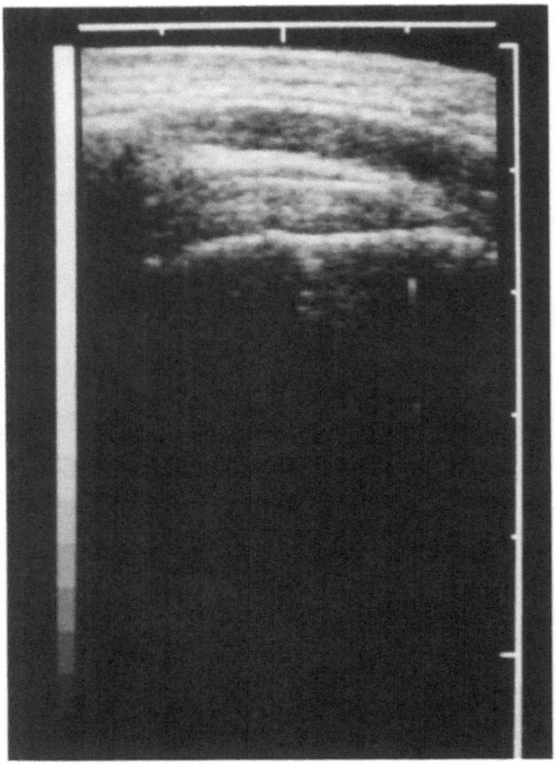

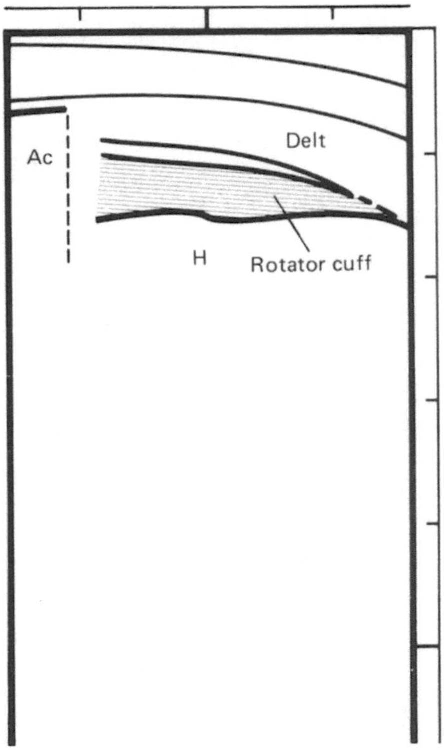
Delt
Ac
H
Rotator cuff

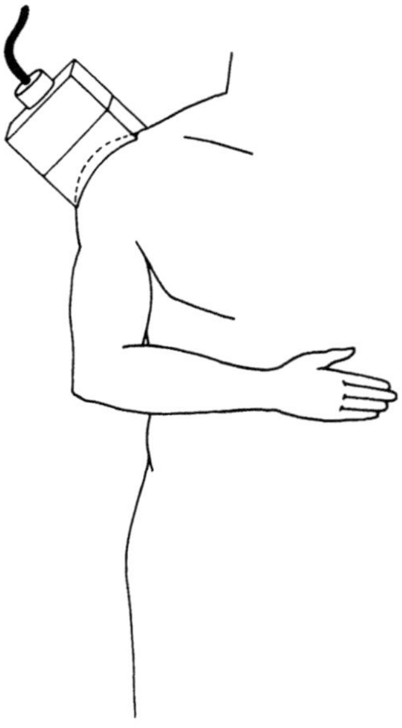

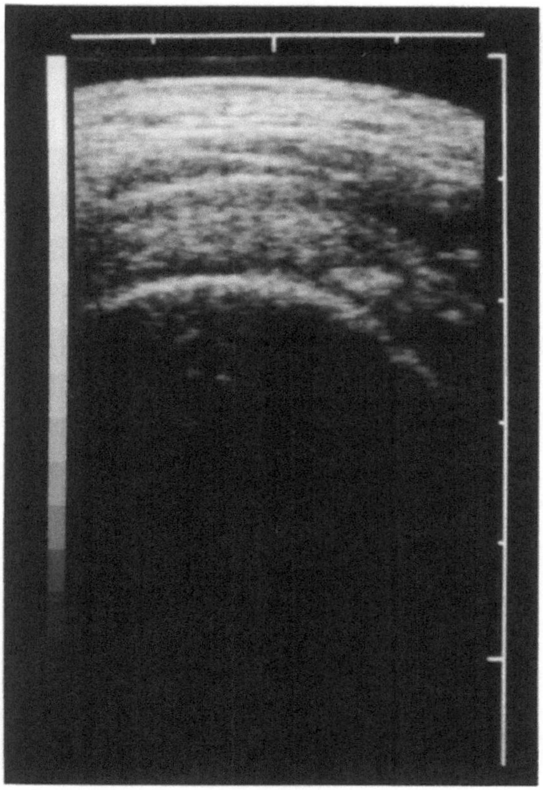

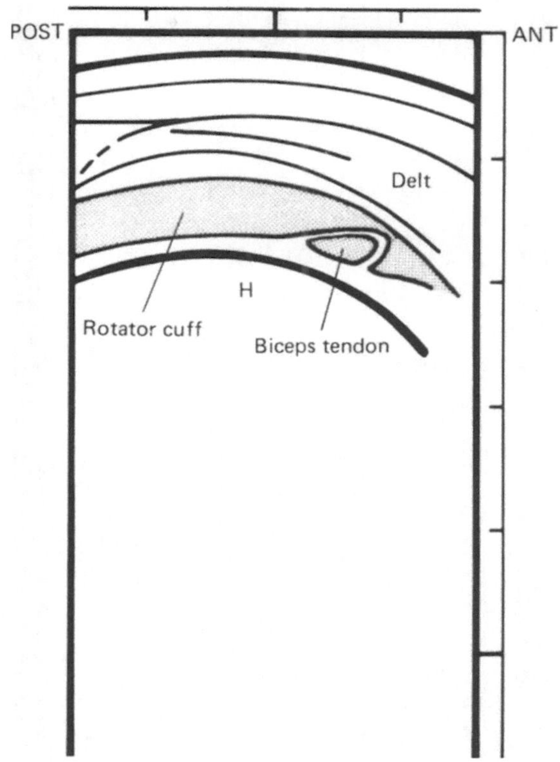
POST
ANT
Delt
H
Rotator cuff
Biceps tendon

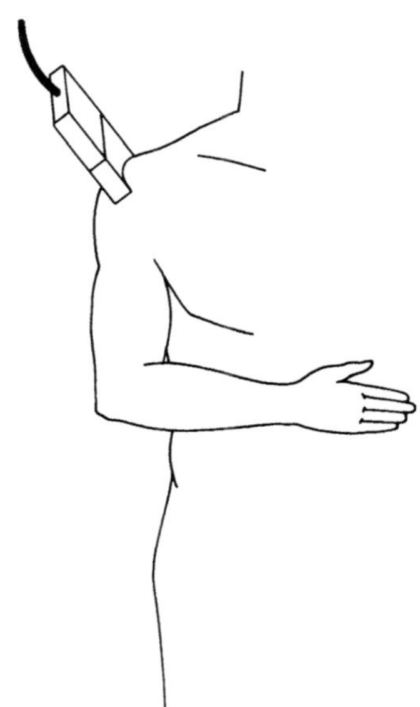

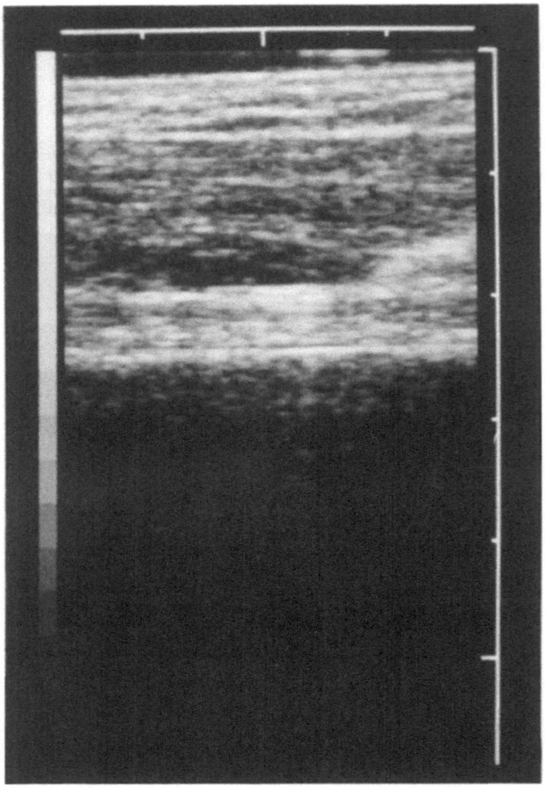

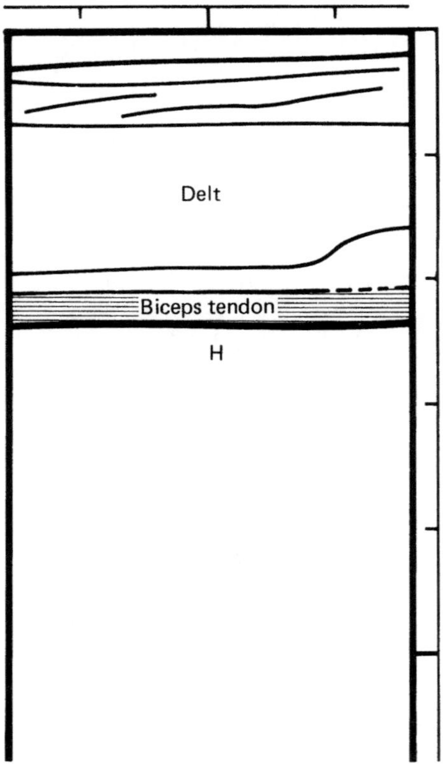

Delt
Biceps tendon
H

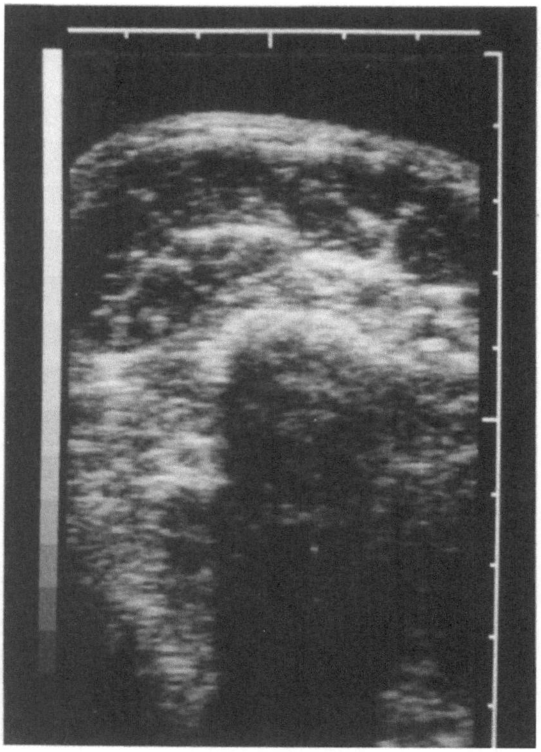

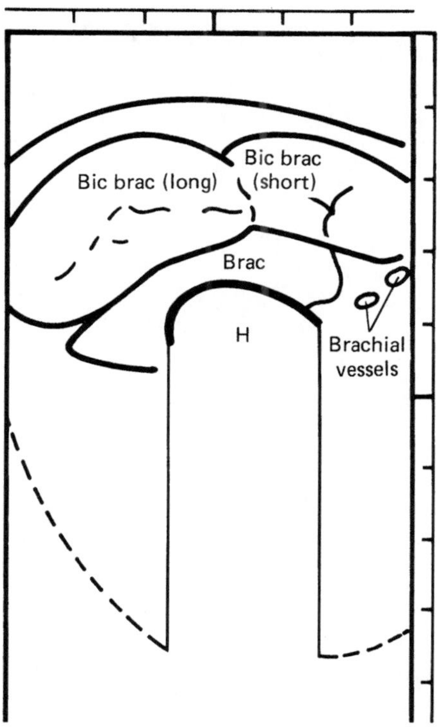
Bic brac (long)
Bic brac (short)
Brac
H
Brachial vessels

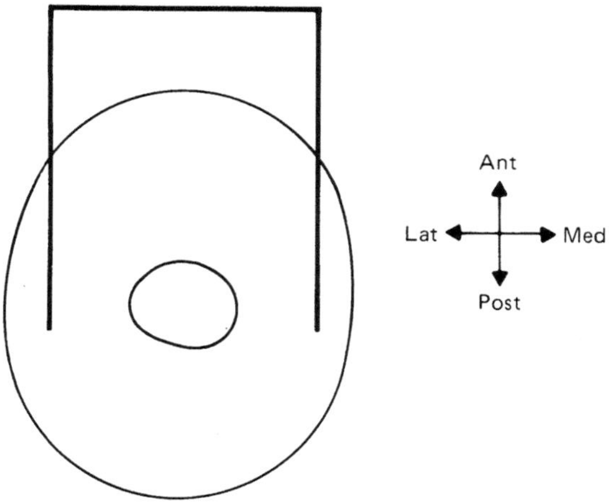
Ant
Lat
Med
Post

Arm, Upper Third

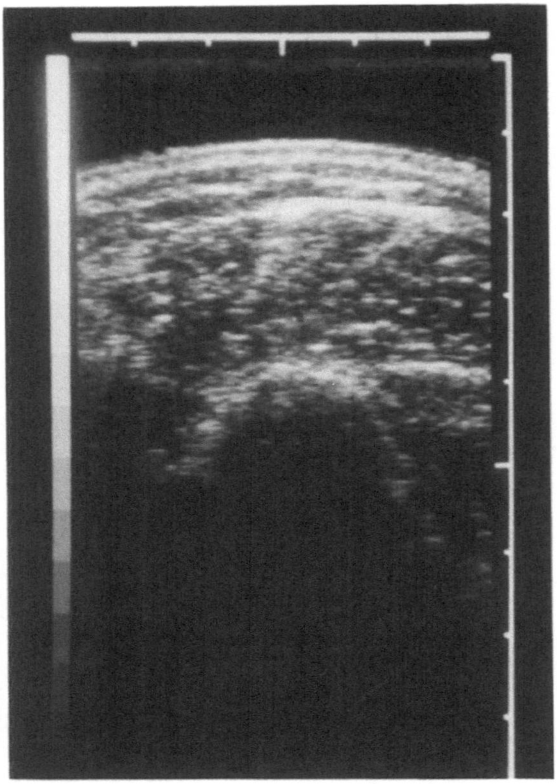

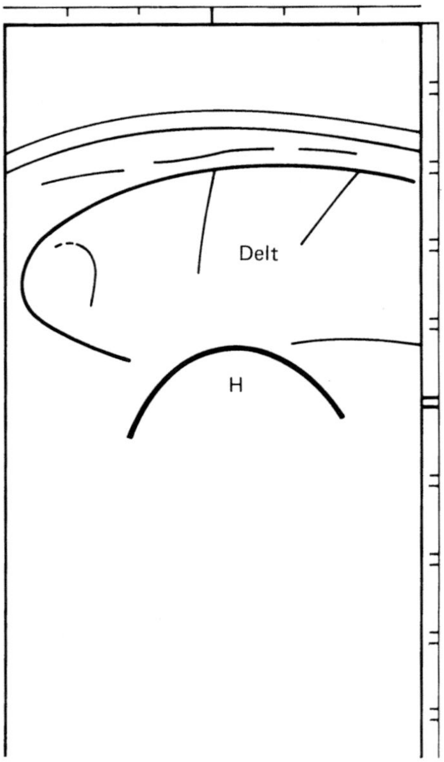

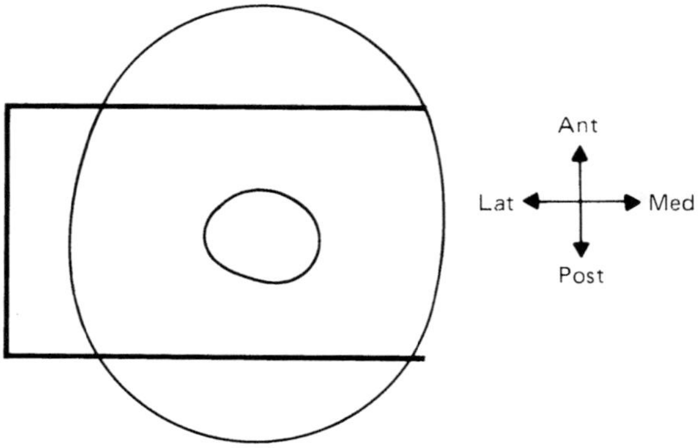

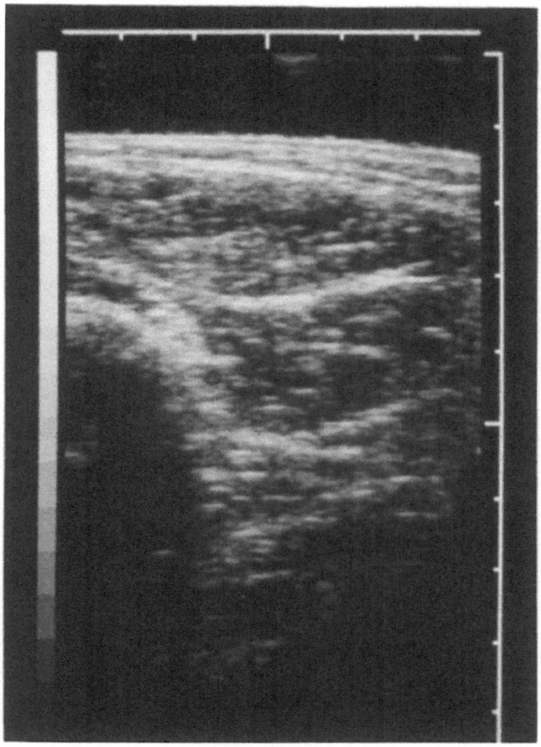

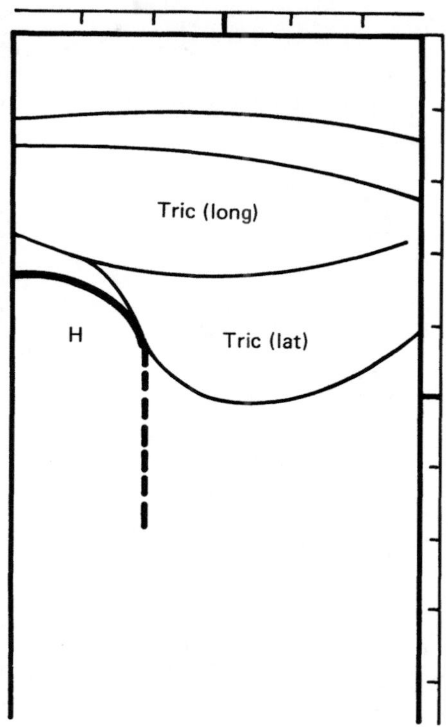
Tric (long)
H
Tric (lat)

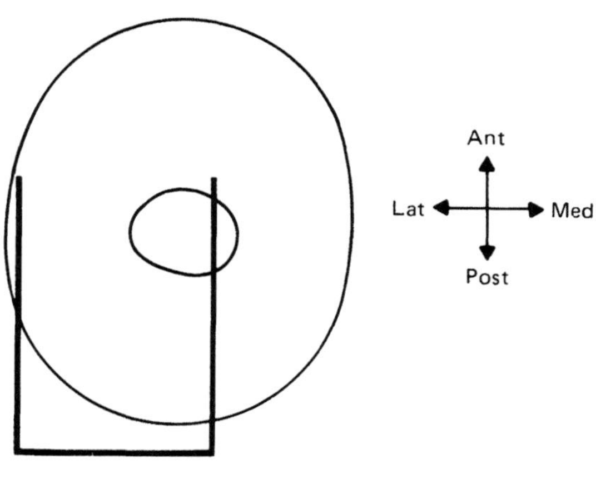
Ant
Lat
Med
Post

Arm, Upper Third

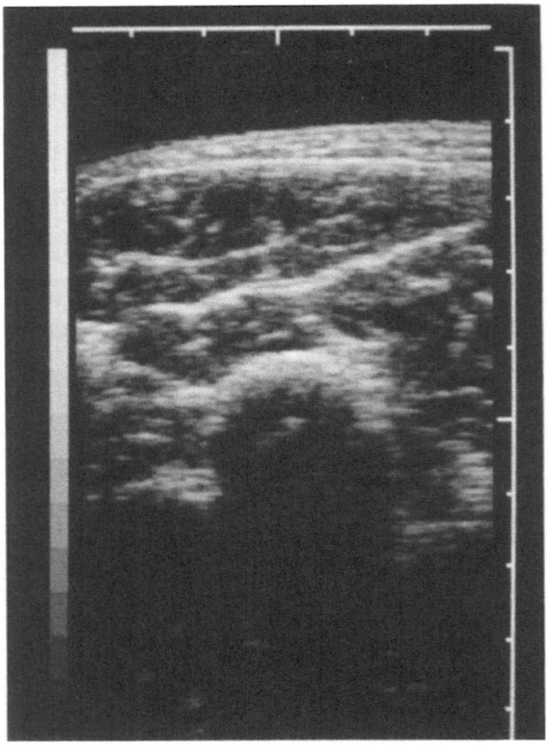

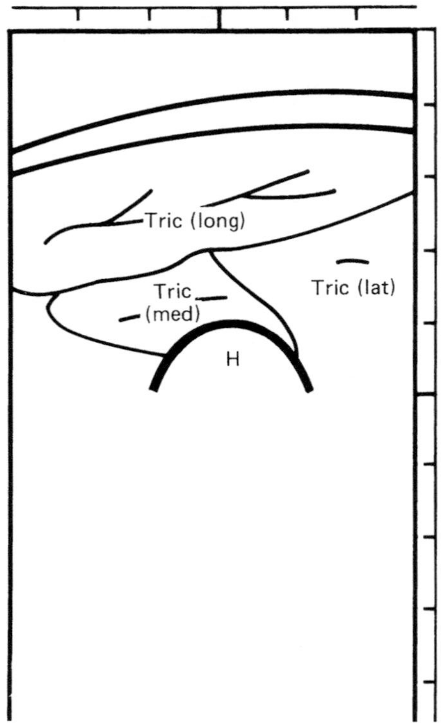

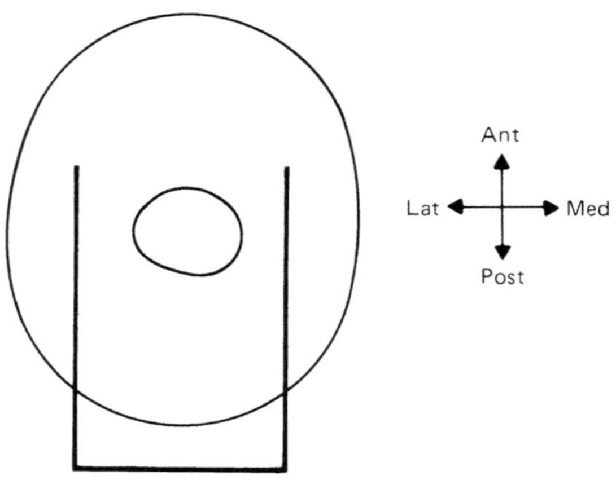

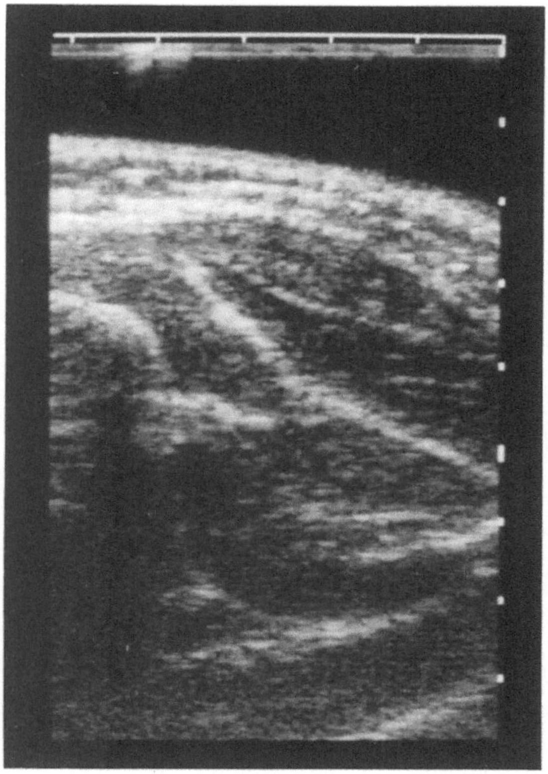

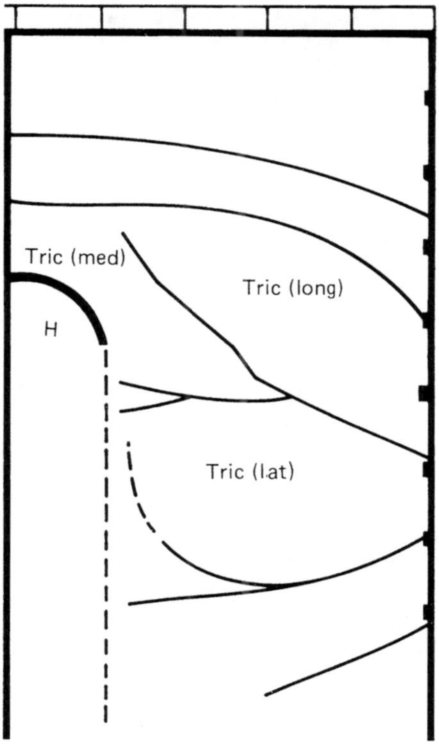
Tric (med)
Tric (long)
H
Tric (lat)

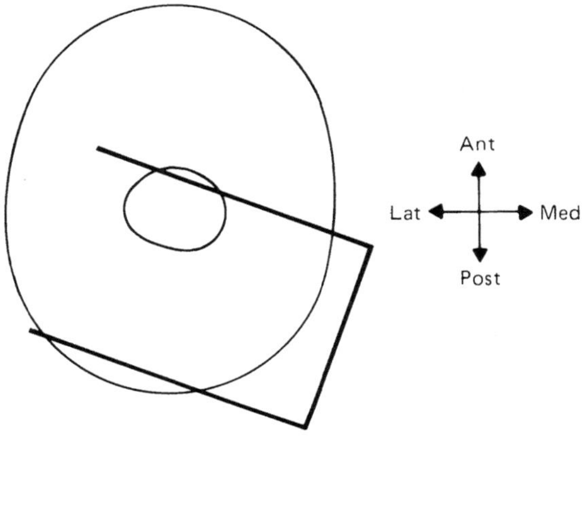
Ant
Lat
Med
Post

Arm, Upper Third

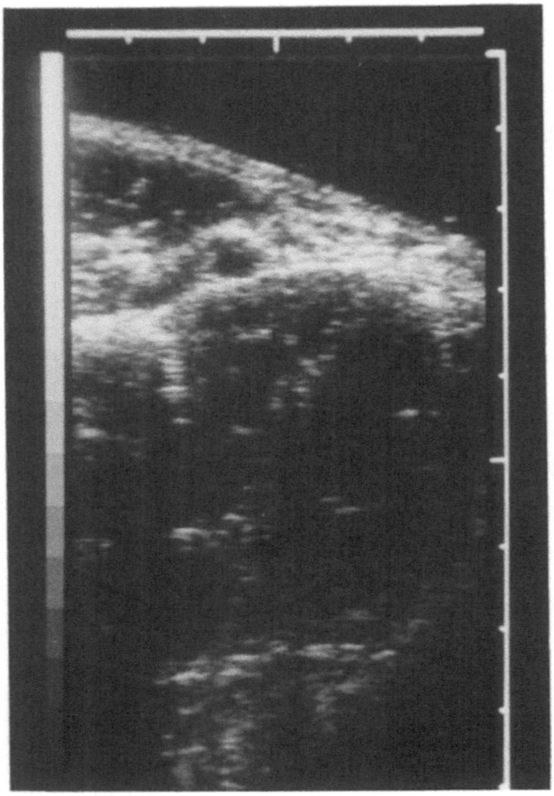

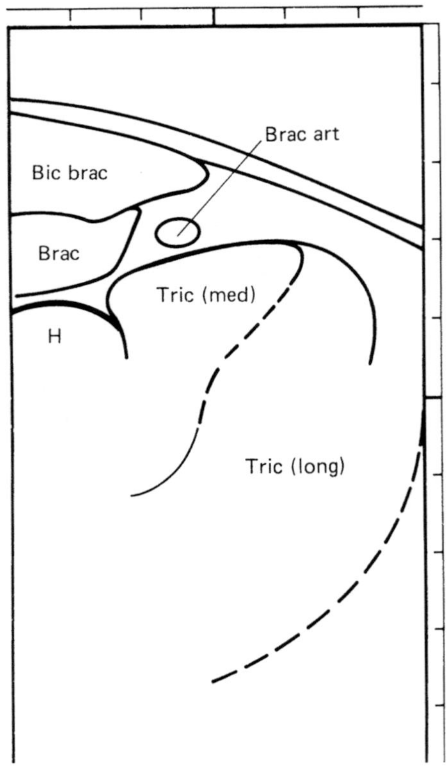

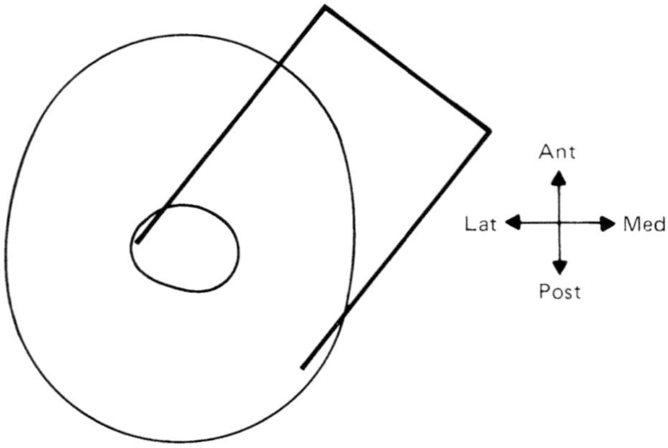

Arm, Midthird

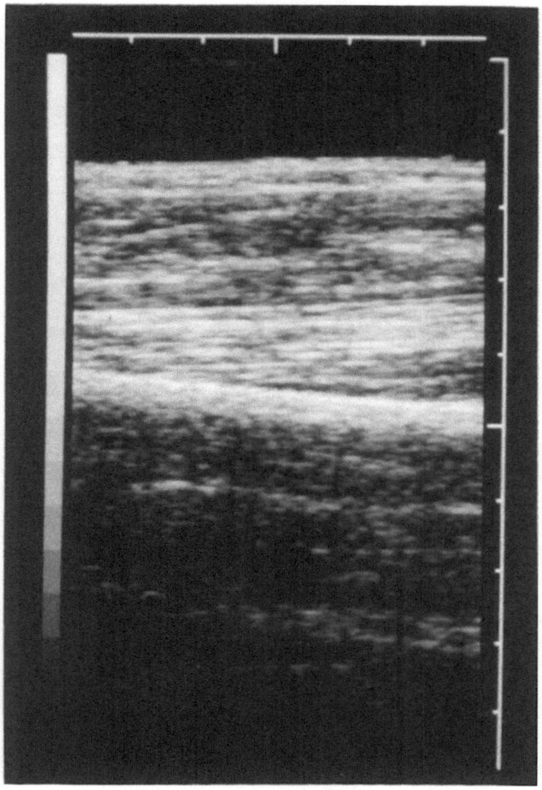

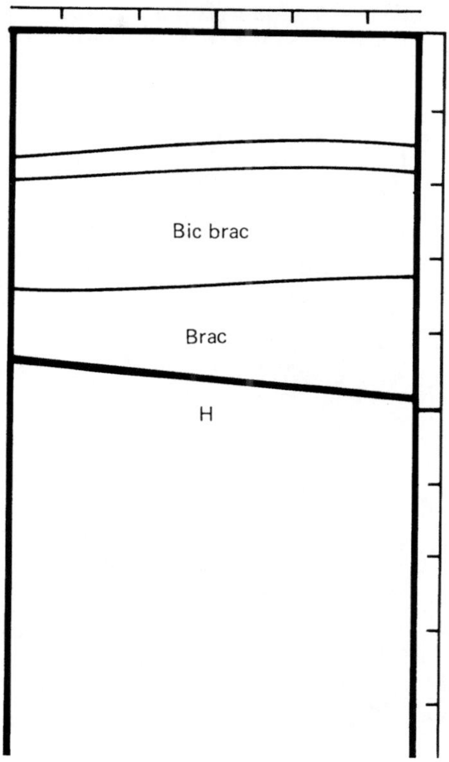

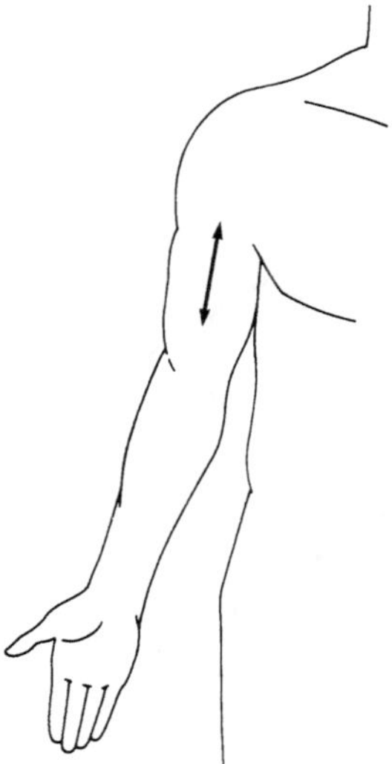

Arm, Midthird

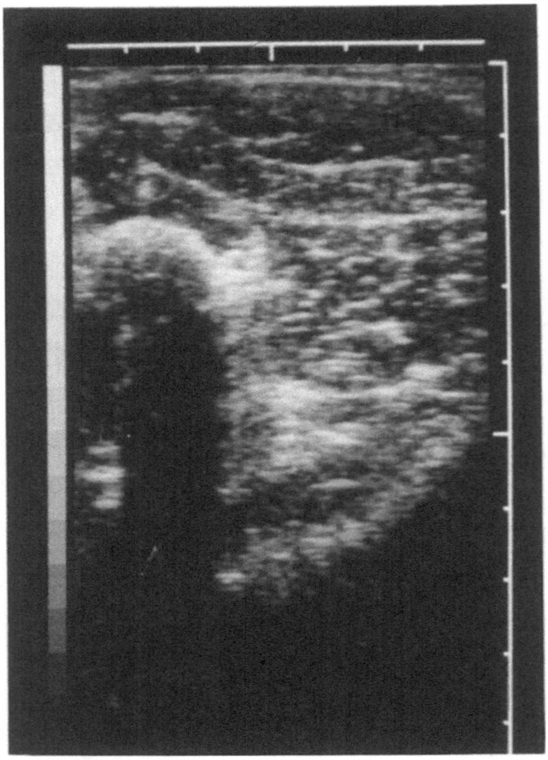

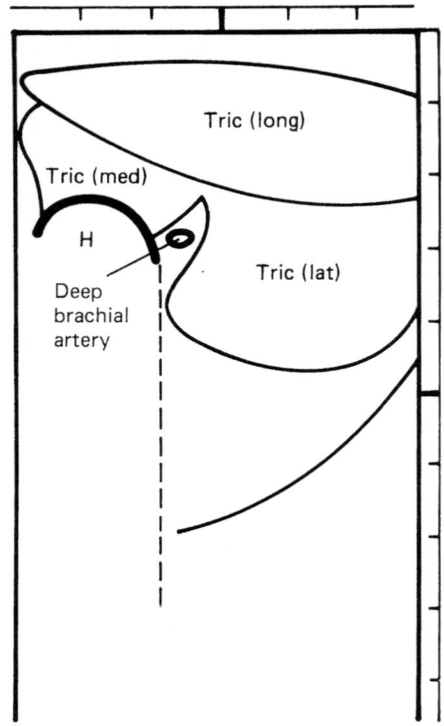

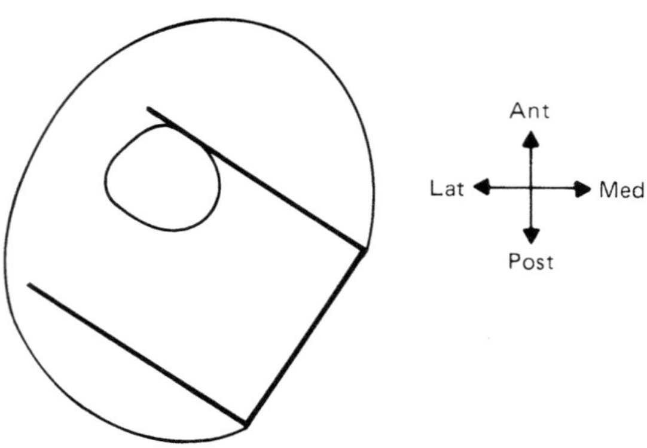

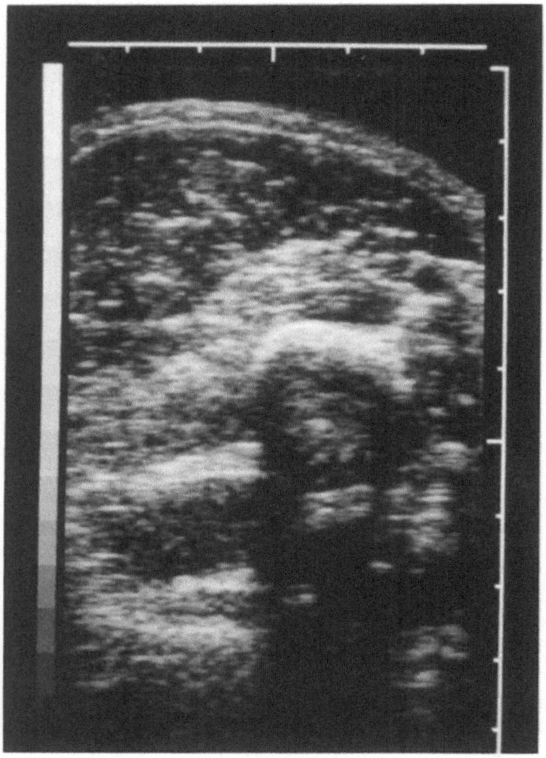

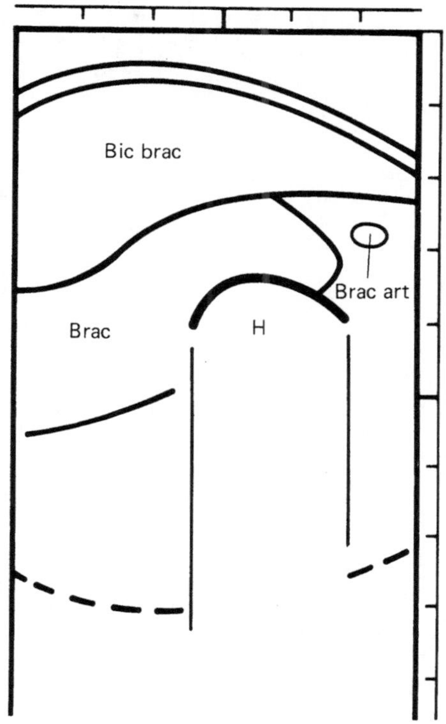
Bic brac
Brac art
Brac
H

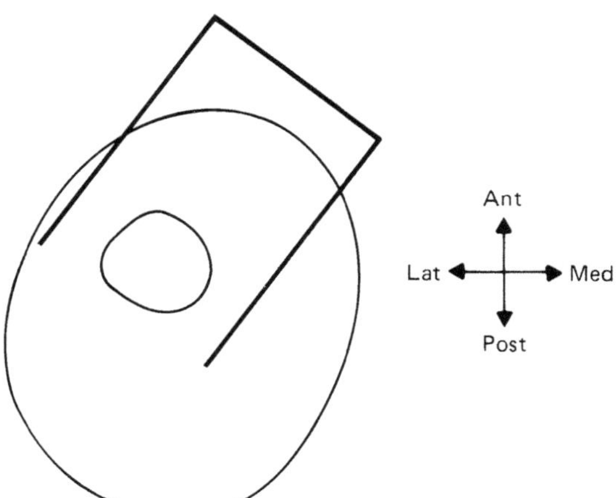
Ant
Lat
Med
Post

Arm, Lower Third

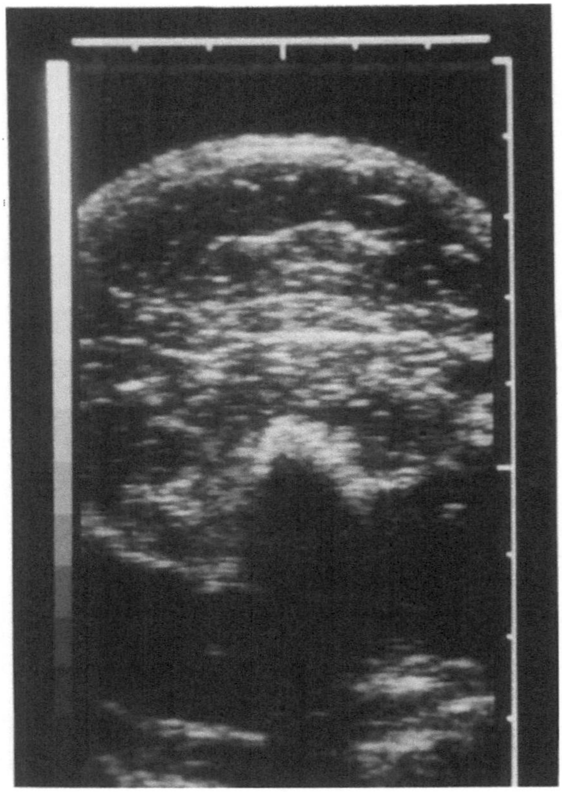

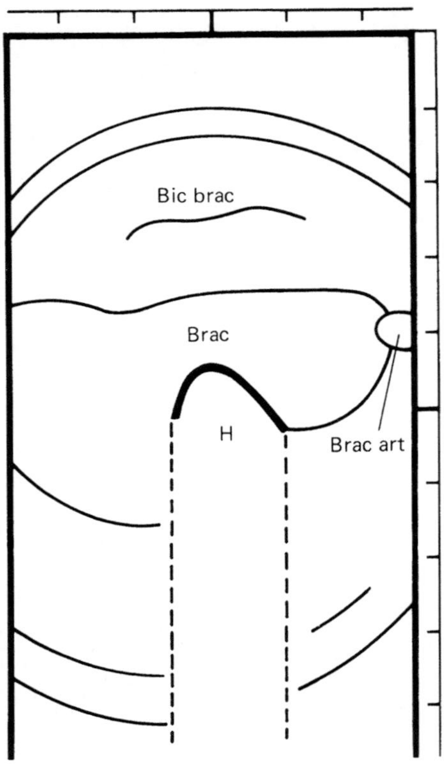

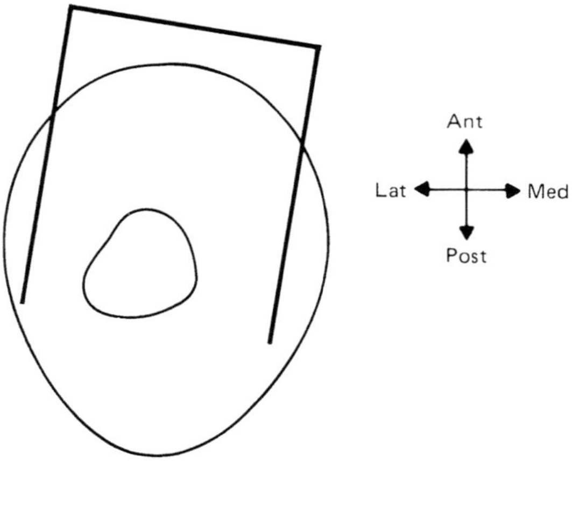

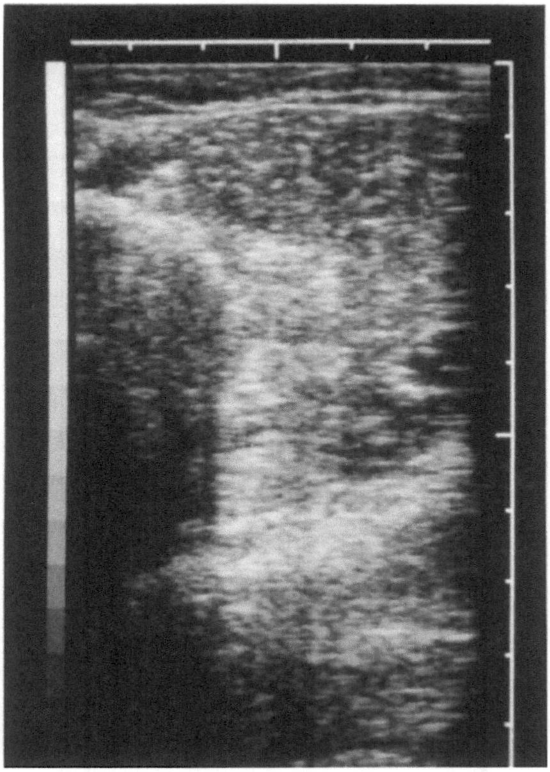

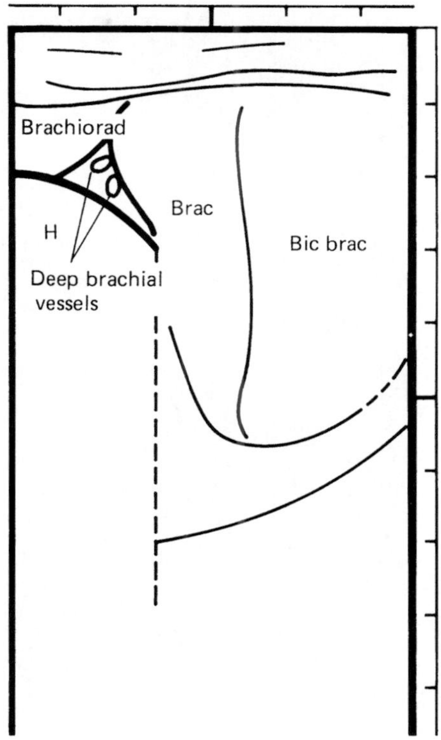
Brachiorad
Brac
Bic brac
H
Deep brachial vessels

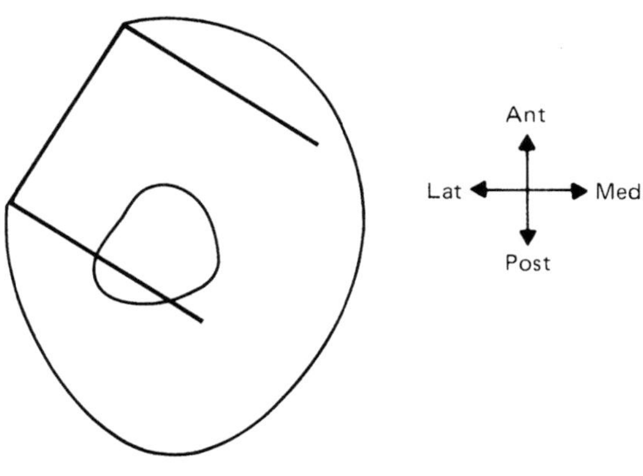
Ant
Lat
Med
Post

Arm, Lower Third

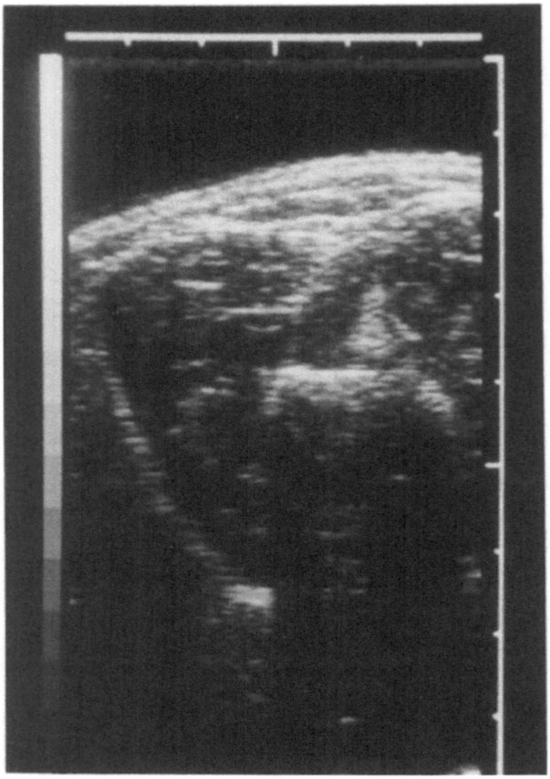

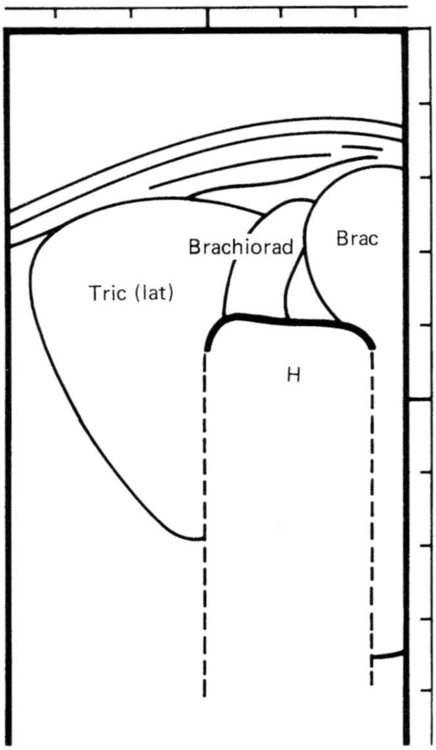

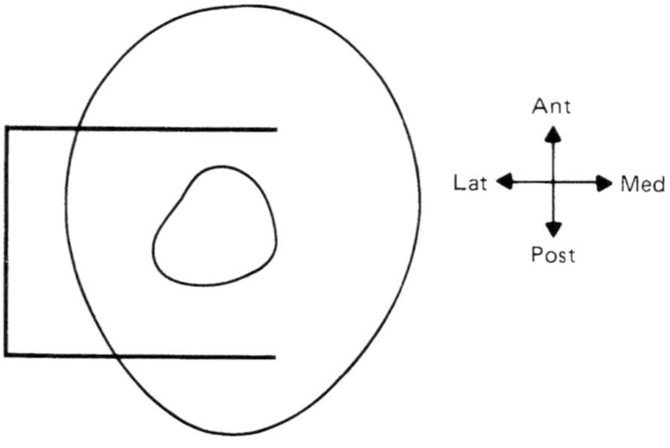

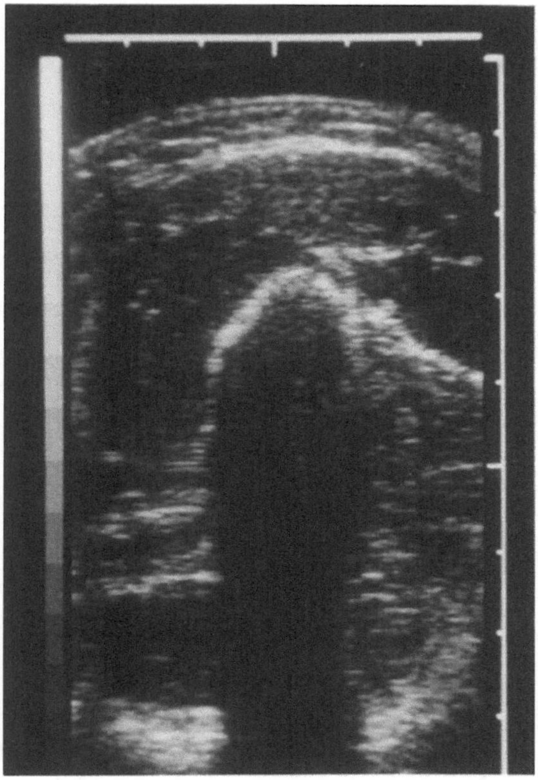

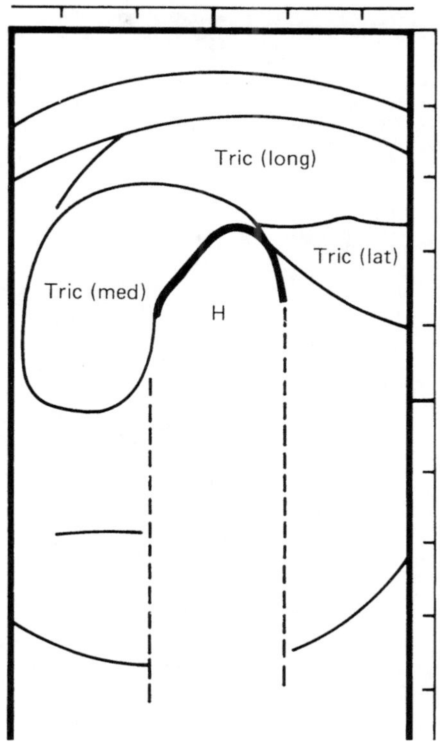
Tric (long)
Tric (lat)
Tric (med)
H

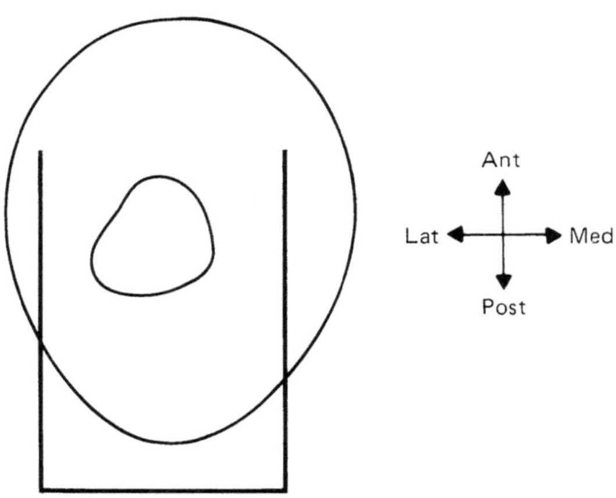
Ant
Lat
Med
Post

Arm, Lower Third

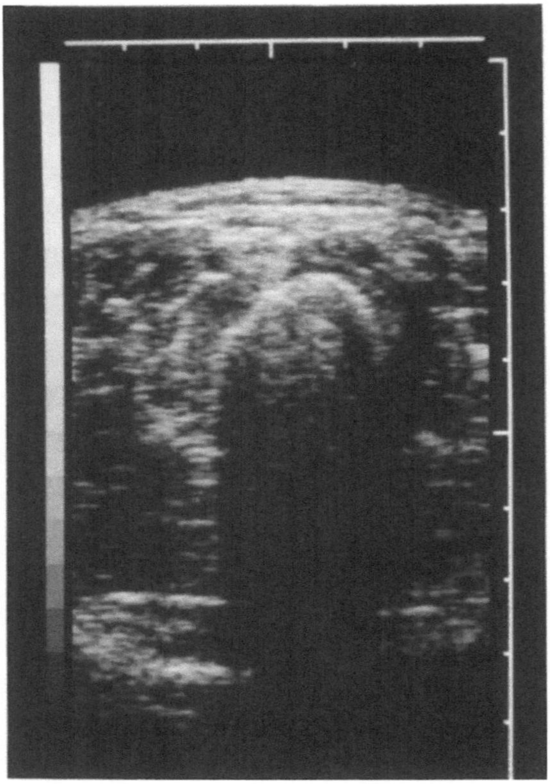

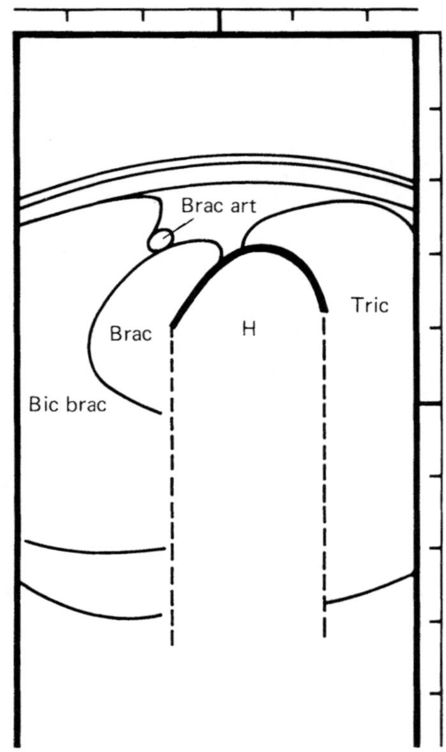

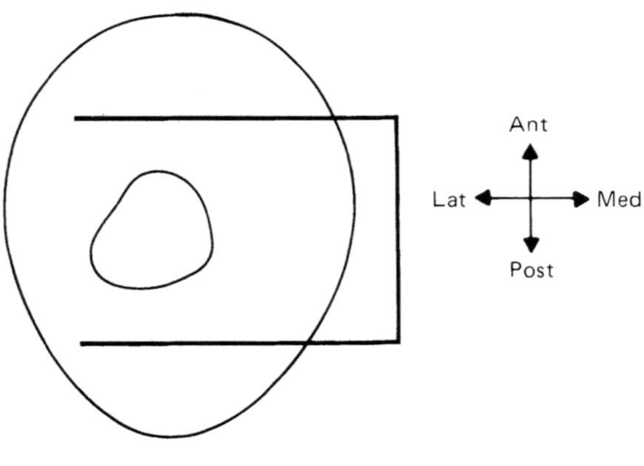

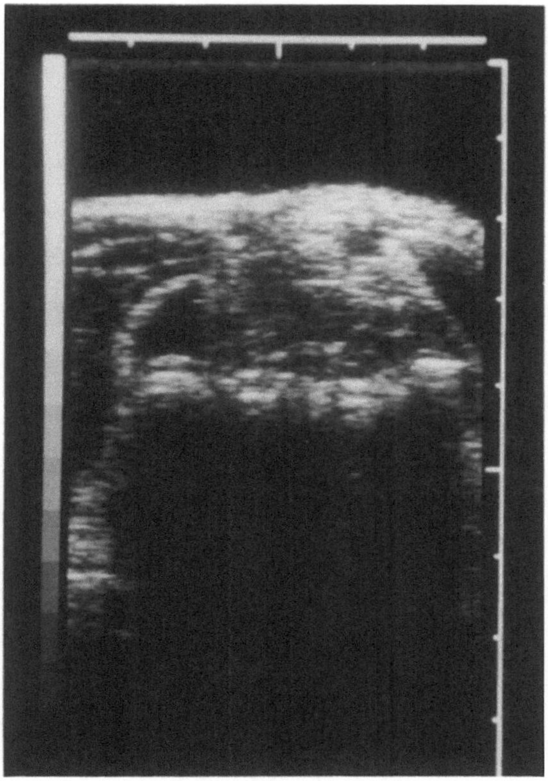

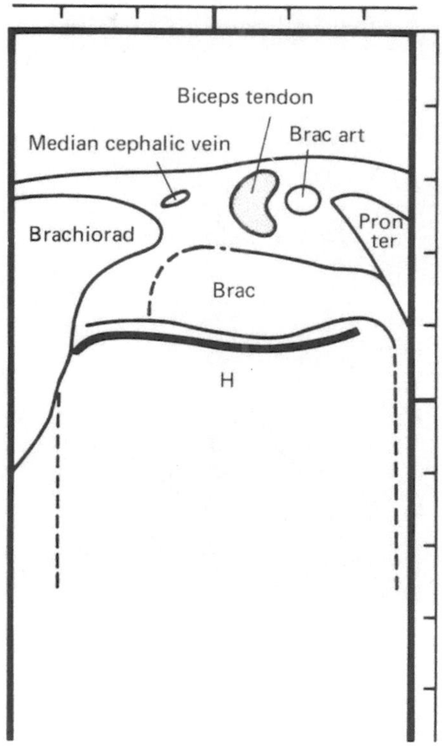
Biceps tendon
Median cephalic vein
Brac art
Brachiorad
Pron ter
Brac
H

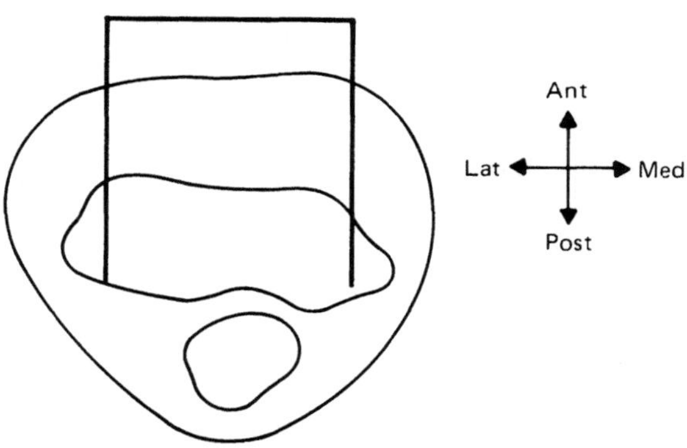
Ant
Lat
Med
Post

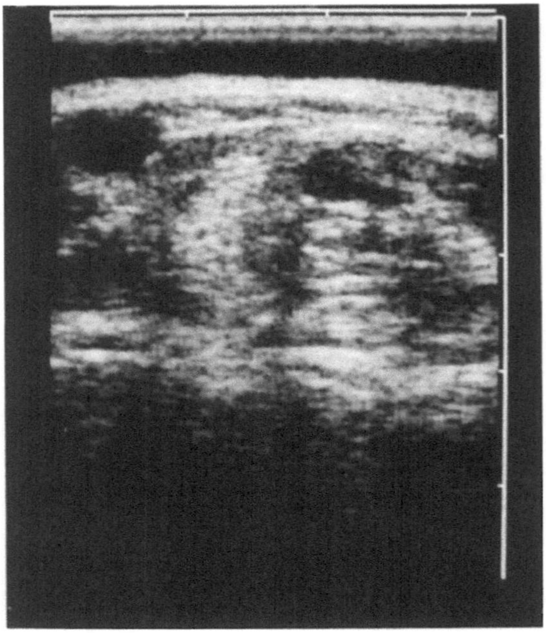

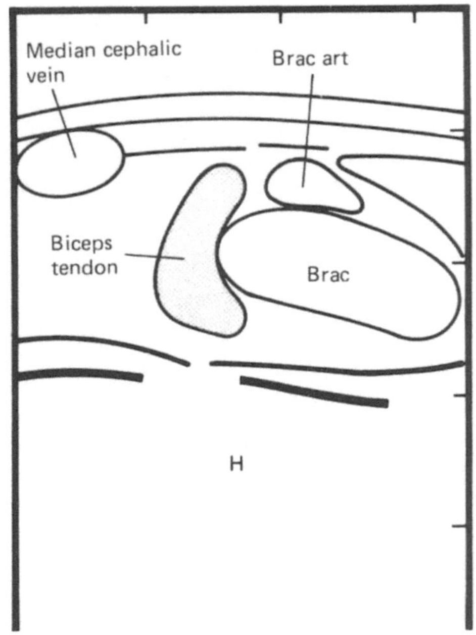
Median cephalic vein
Brac art
Biceps tendon
Brac
H

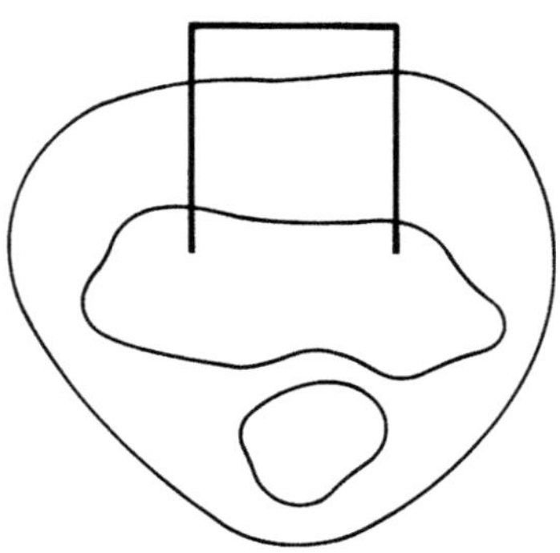

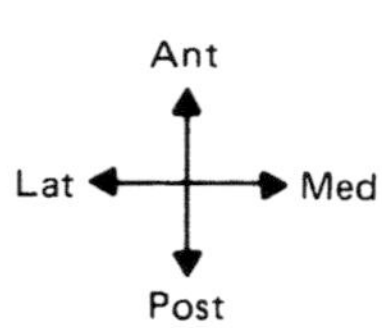
Ant
Lat
Med
Post

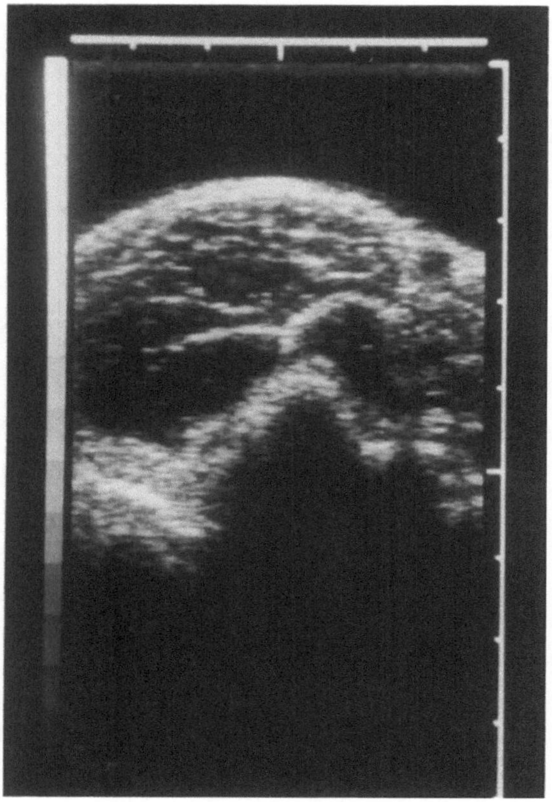

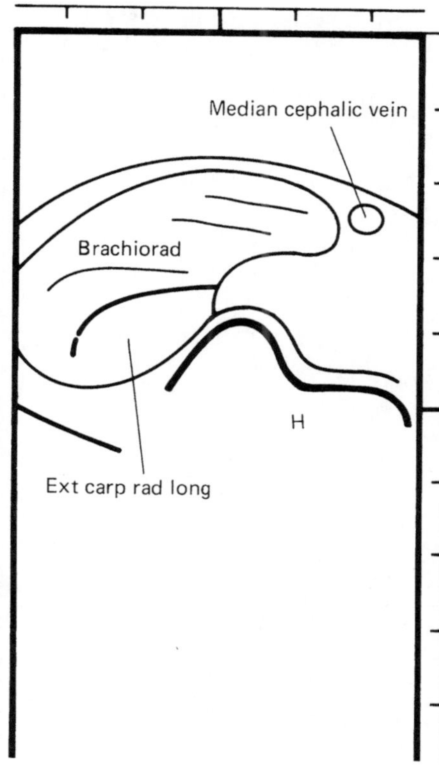
Median cephalic vein
Brachiorad
H
Ext carp rad long

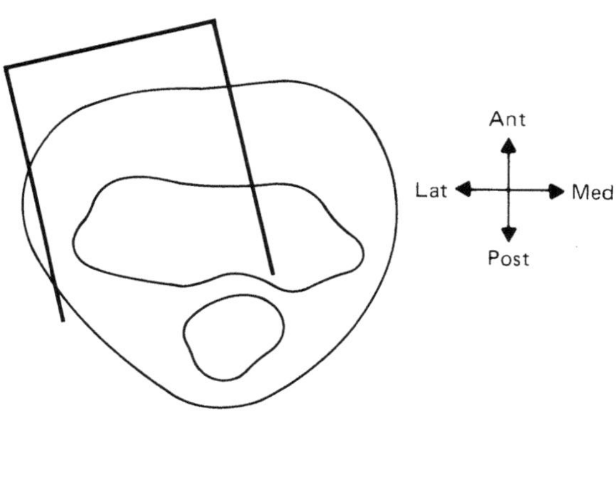
Ant
Lat
Med
Post

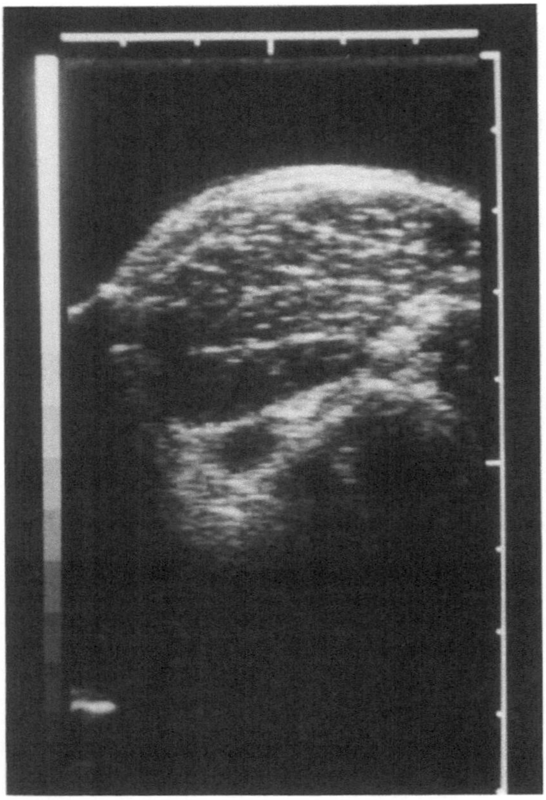

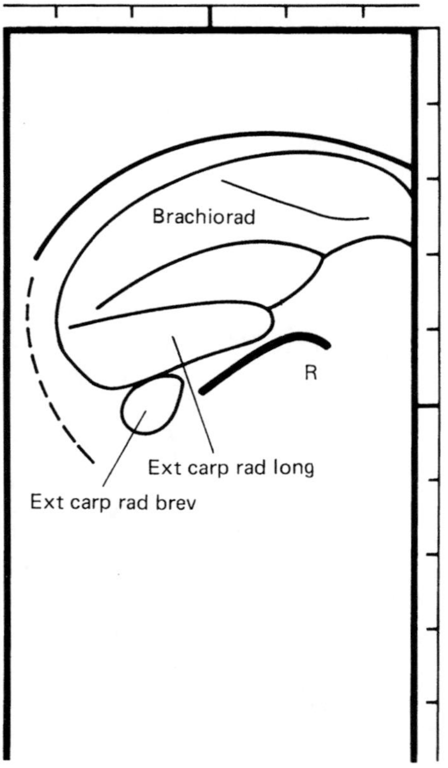
Brachiorad
R
Ext carp rad long
Ext carp rad brev

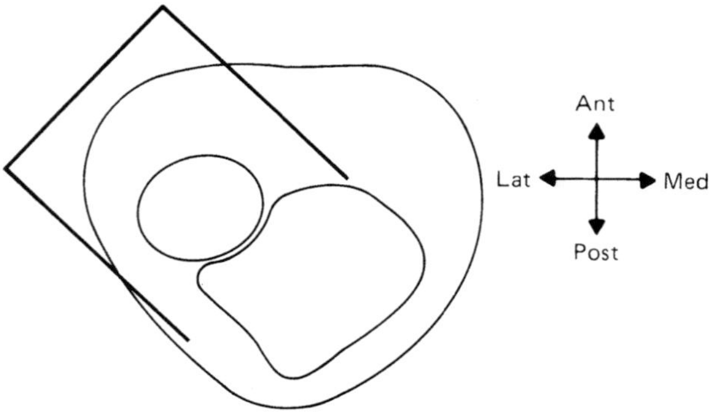
Ant
Lat
Med
Post

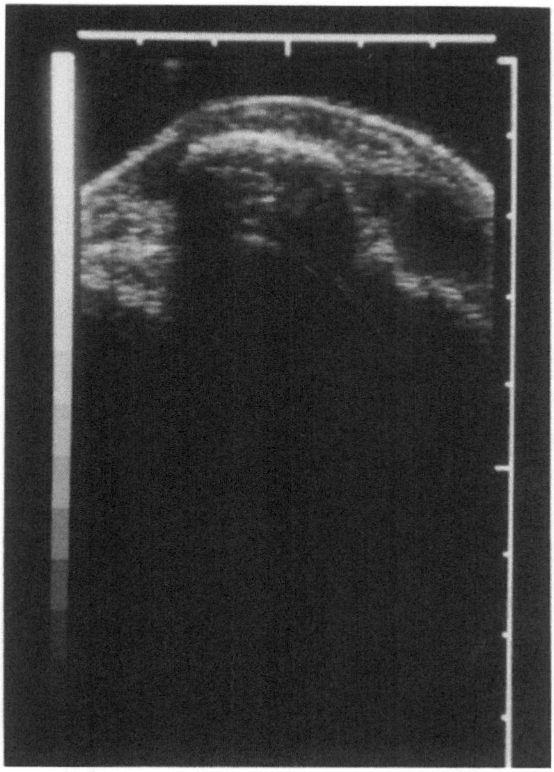

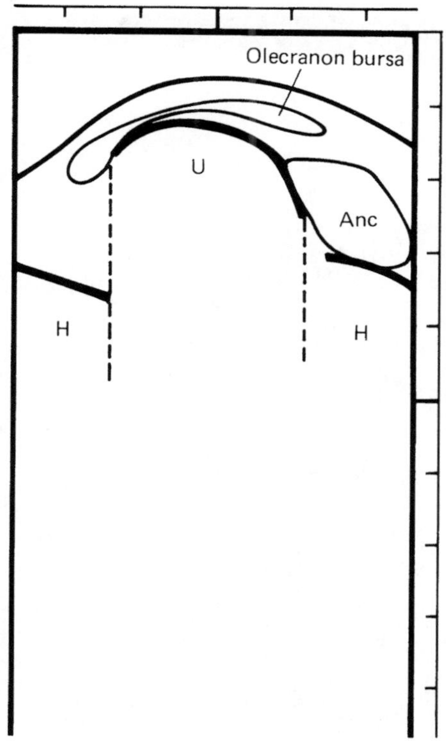
Olecranon bursa
U
Anc
H
H

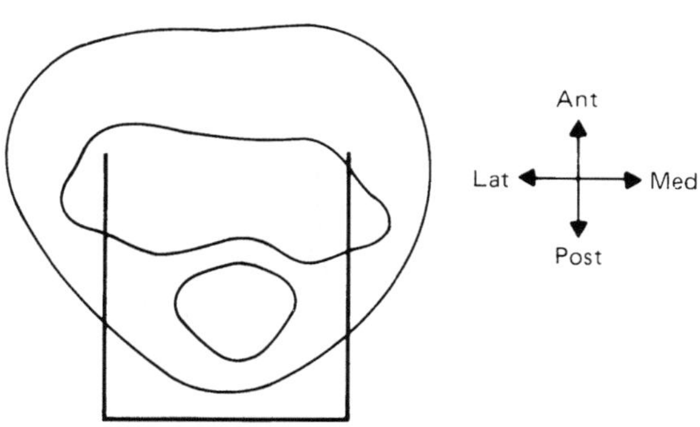
Ant
Lat
Med
Post

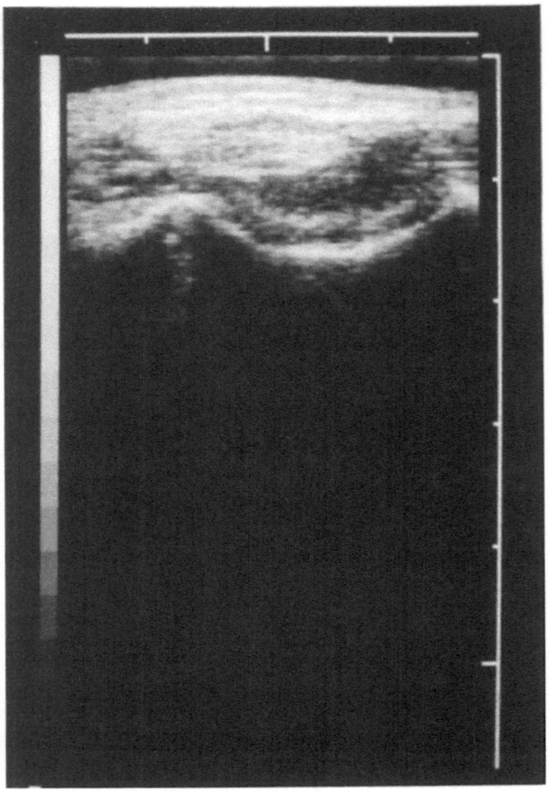

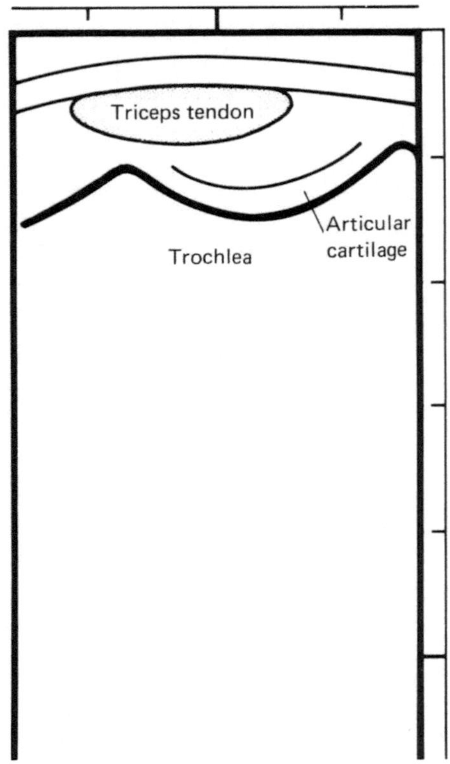
Triceps tendon
Articular cartilage
Trochlea

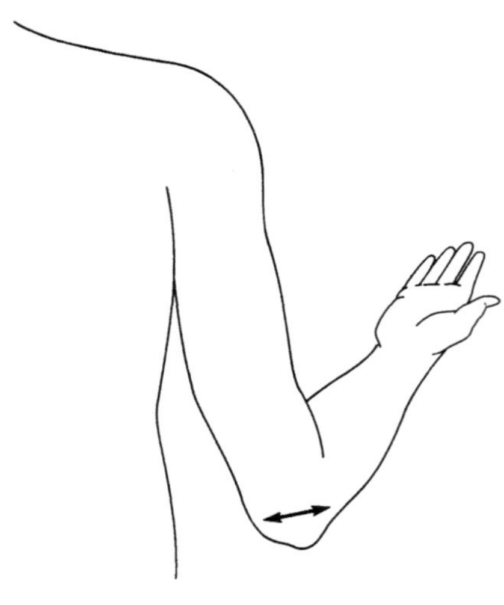

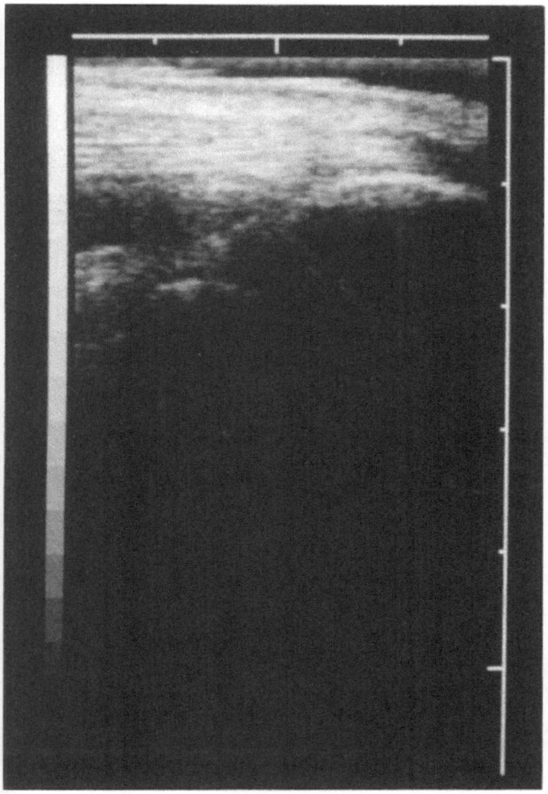

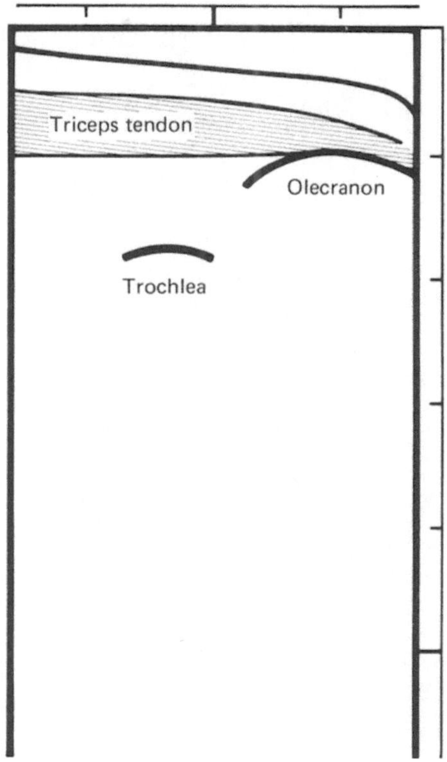
Triceps tendon
Olecranon
Trochlea

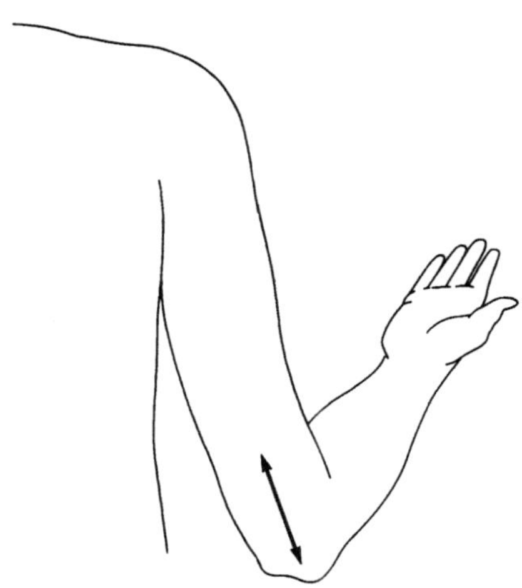

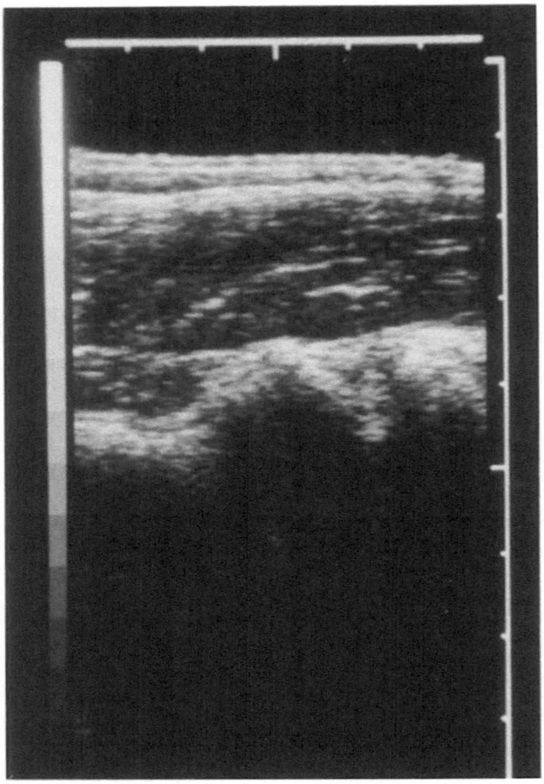

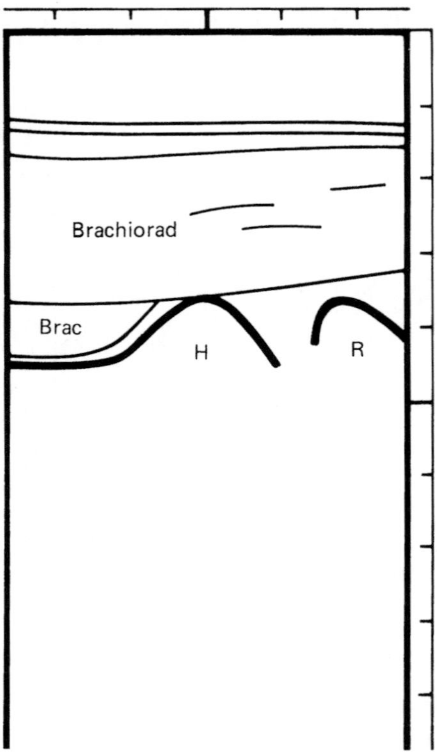
Brachiorad
Brac
H
R

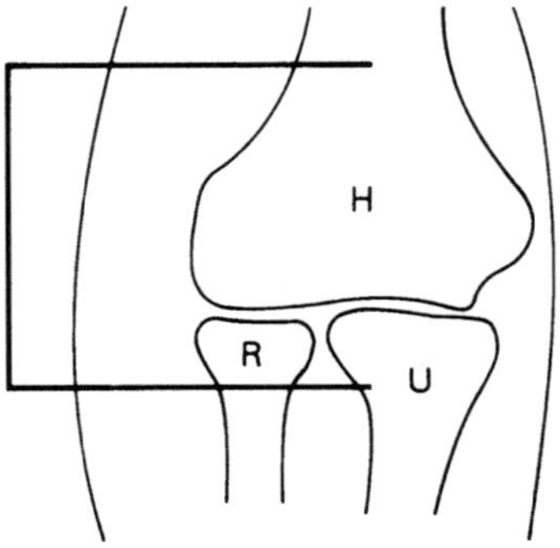
H
R
U

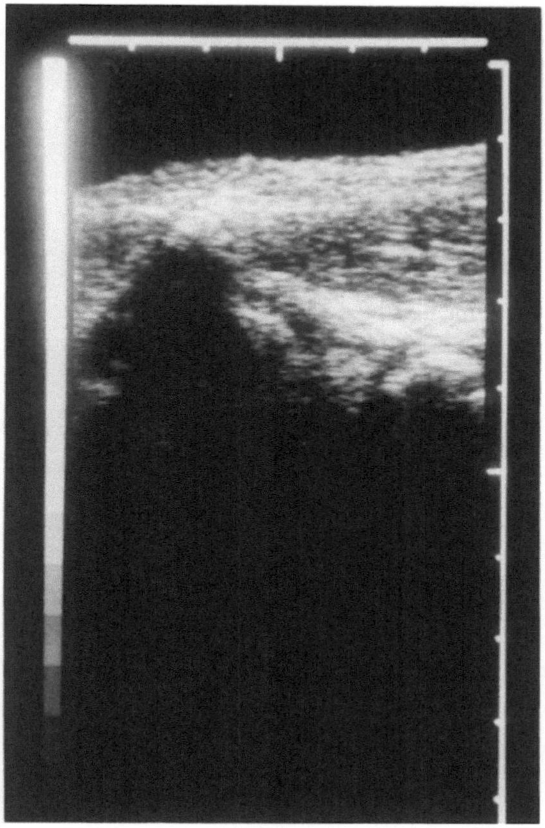

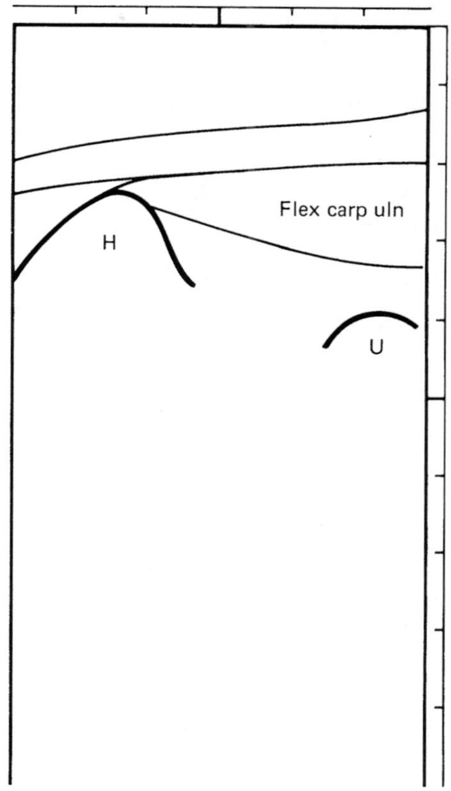
Flex carp uln
H
U

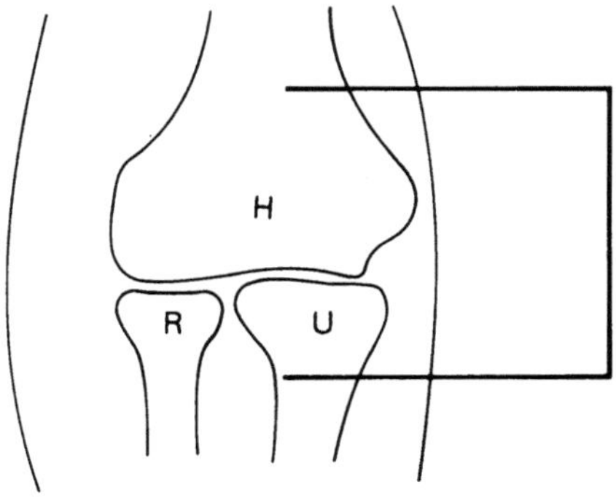
H
R
U

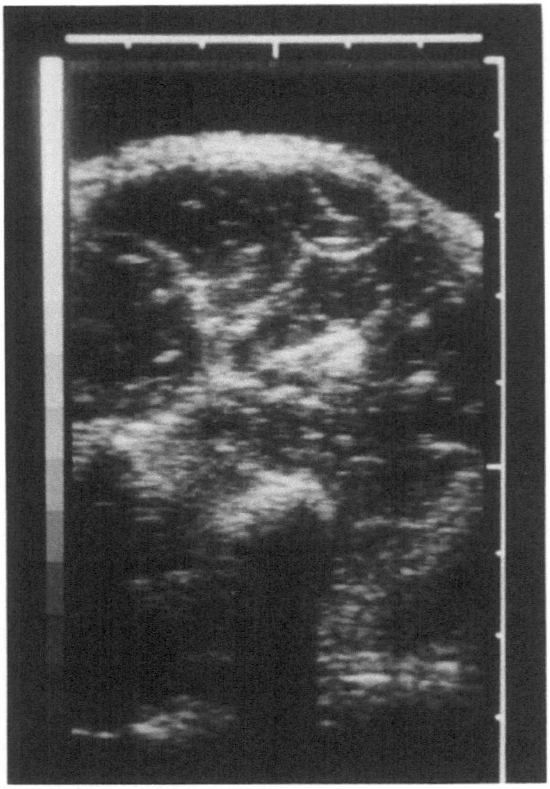

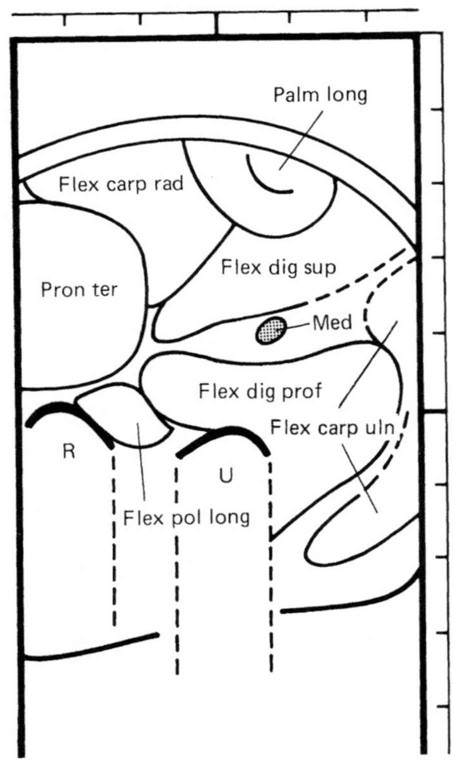
Palm long
Flex carp rad
Pron ter
Flex dig sup
Med
Flex dig prof
Flex carp uln
R
U
Flex pol long

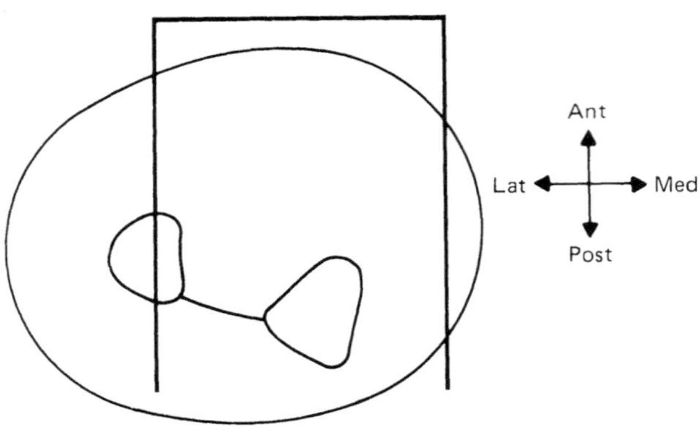
Ant
Lat
Med
Post

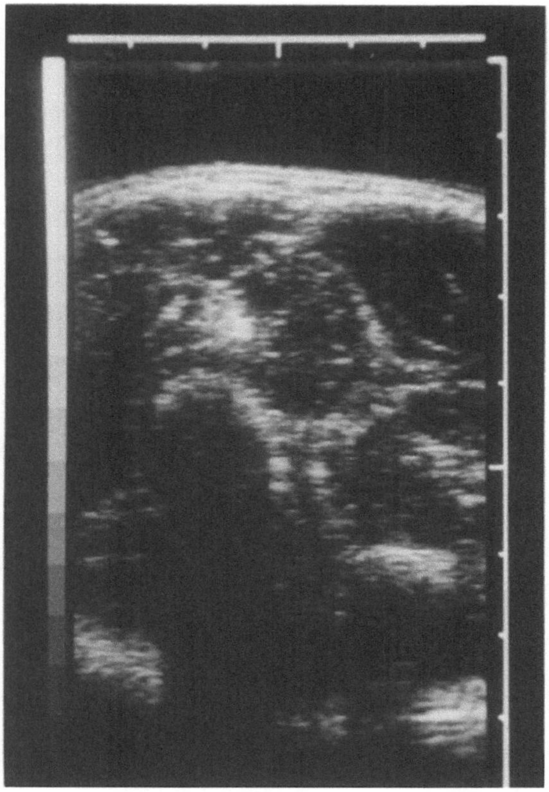

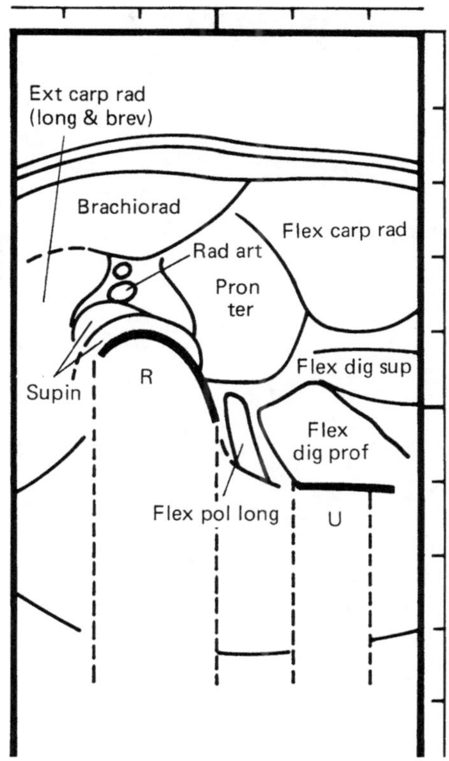
Ext carp rad
(long & brev)
Brachiorad
Flex carp rad
Rad art
Pron
ter
Supin
R
Flex dig sup
Flex
dig prof
Flex pol long
U

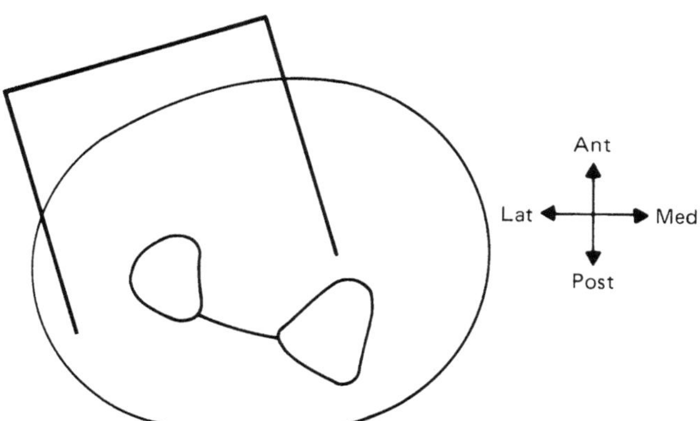
Ant
Lat
Med
Post

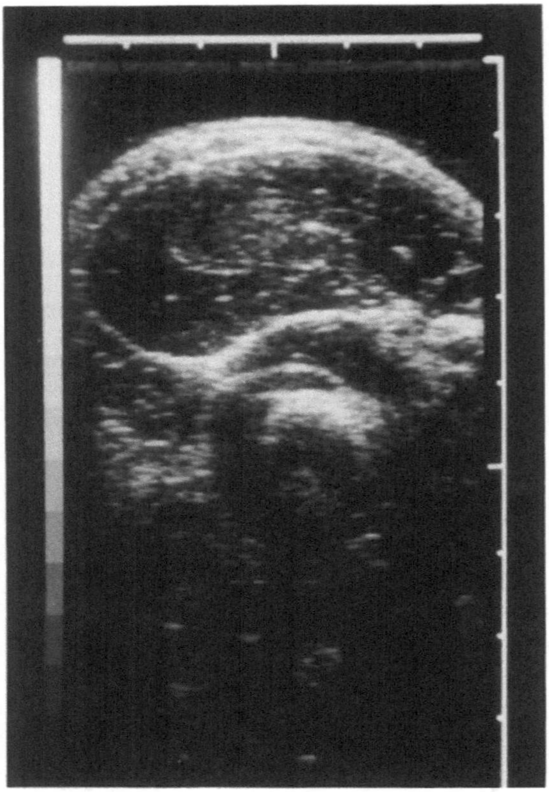

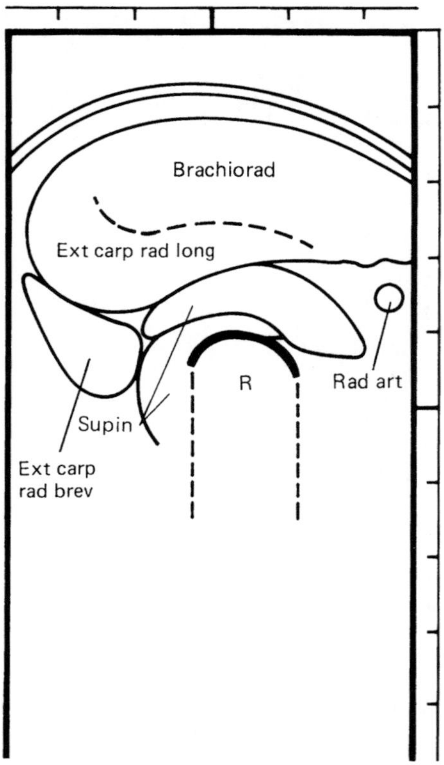
Brachiorad
Ext carp rad long
R
Rad art
Supin
Ext carp
rad brev

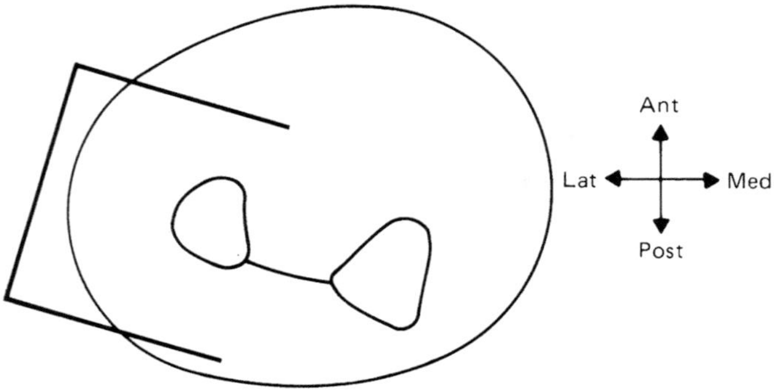
Ant
Lat
Med
Post

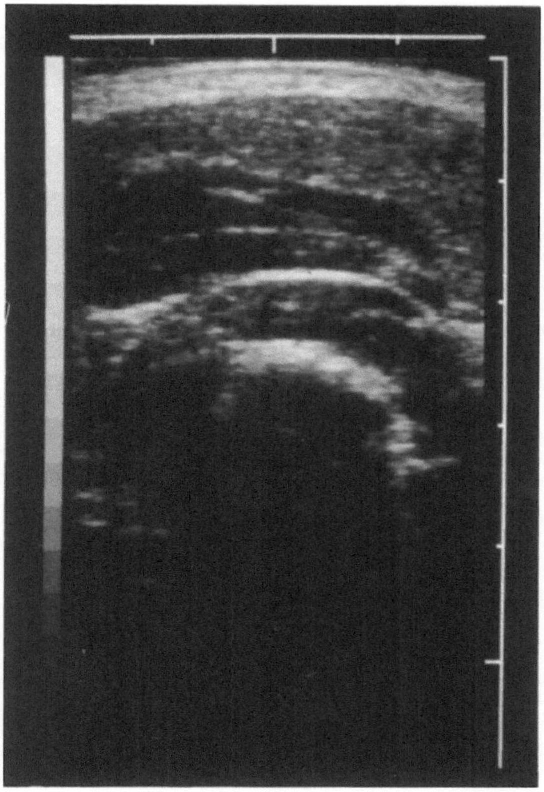

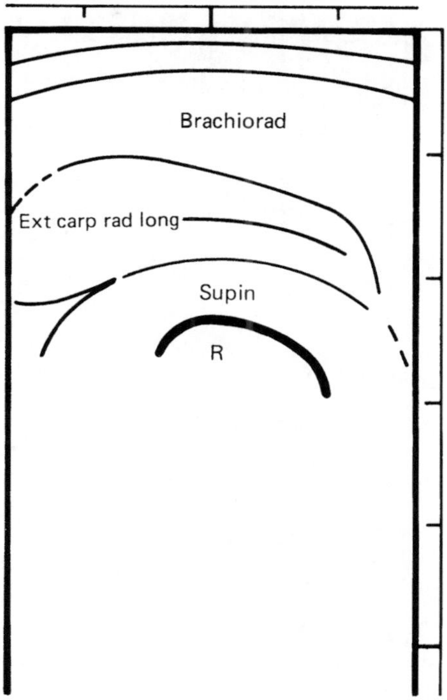
Brachiorad
Ext carp rad long
Supin
R

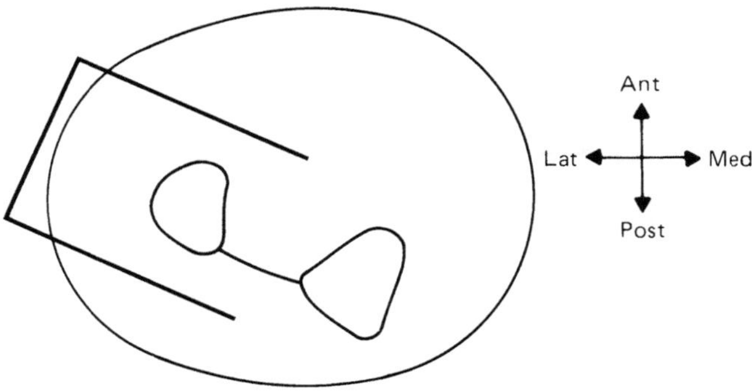
Ant
Lat
Med
Post

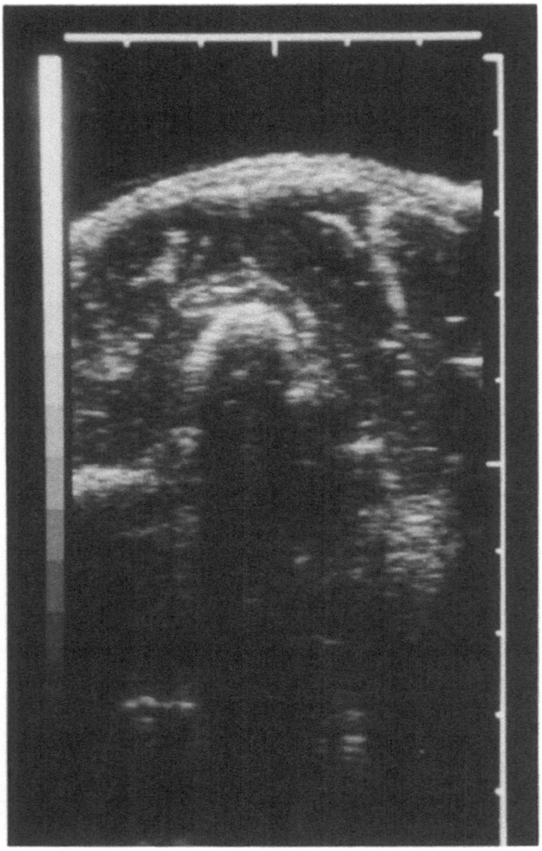

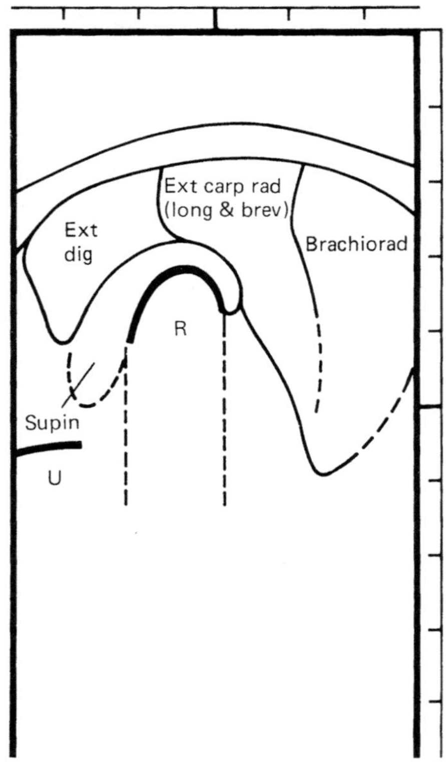
Ext carp rad
(long & brev)
Ext
dig
Brachiorad
R
Supin
U

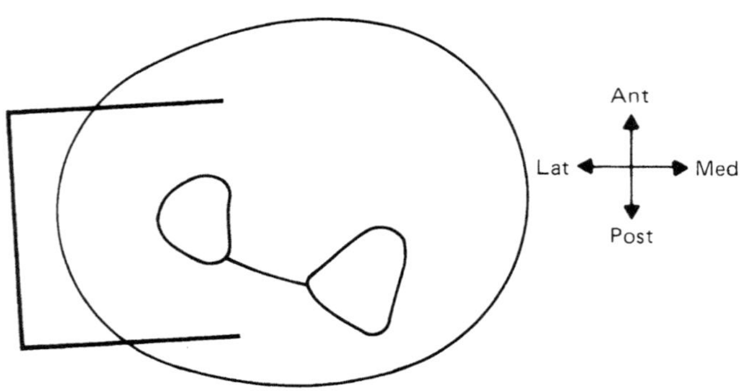
Ant
Lat
Med
Post

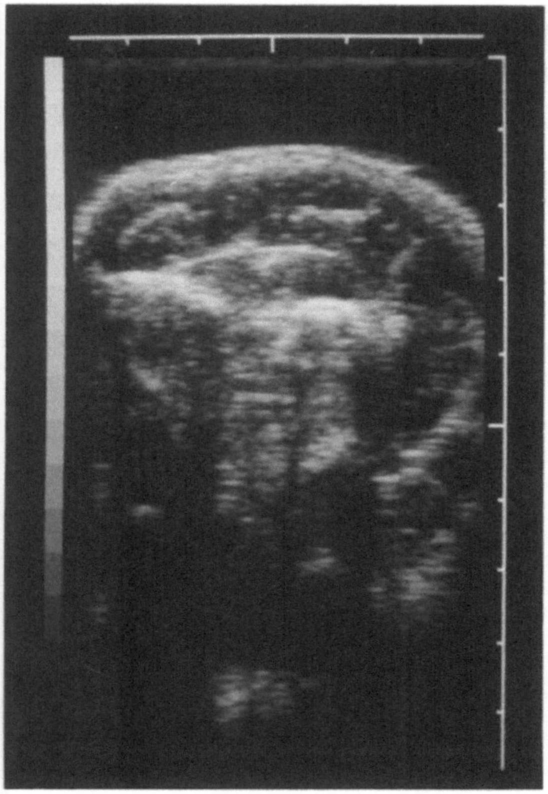

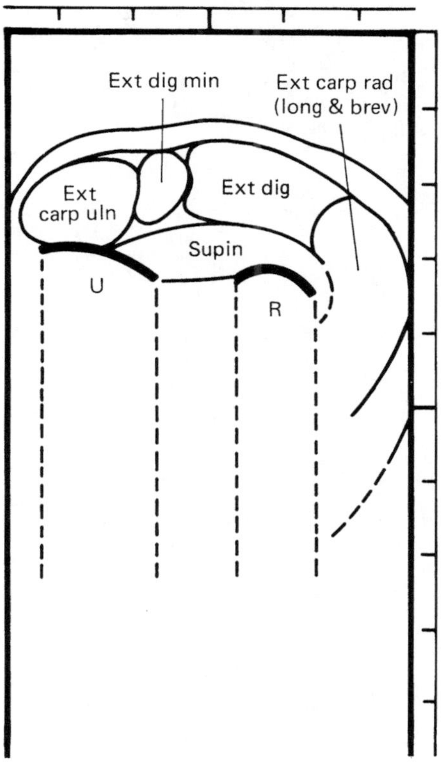
Ext dig min
Ext carp rad (long & brev)
Ext carp uln
Ext dig
Supin
U
R

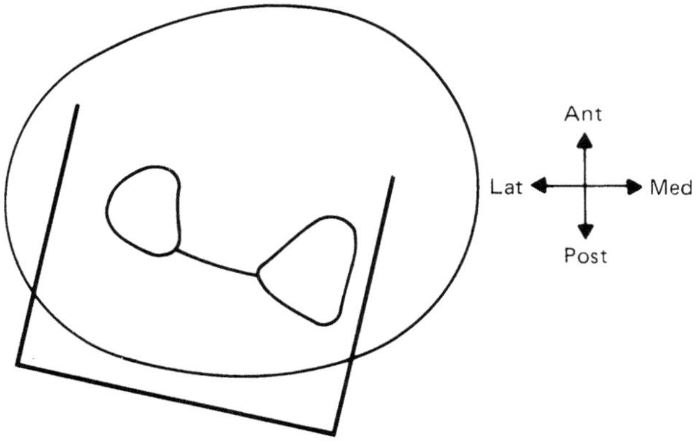
Ant
Lat
Med
Post

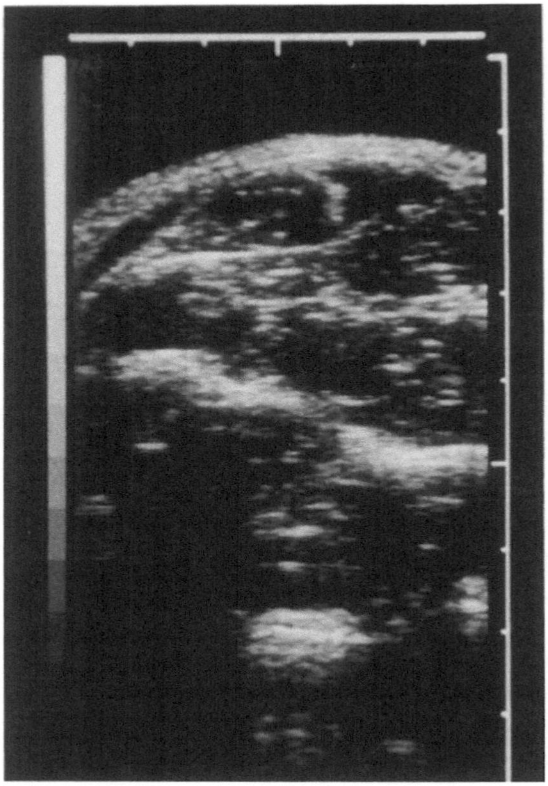

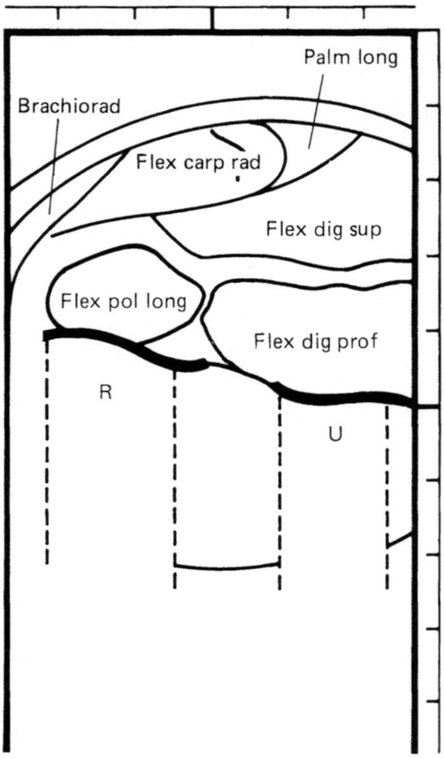
Palm long
Brachiorad
Flex carp rad
Flex dig sup
Flex pol long
Flex dig prof
R
U

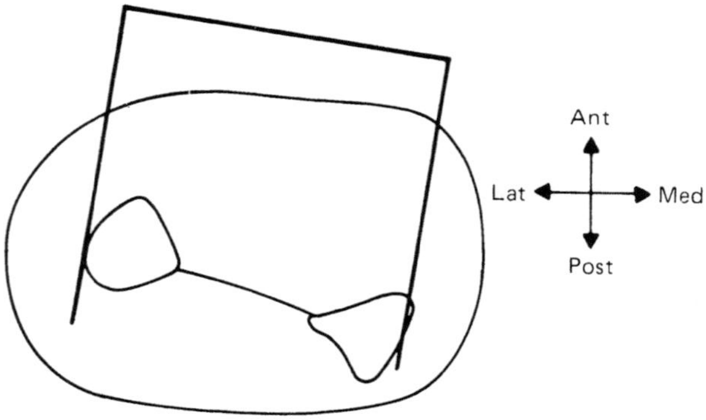
Ant
Lat
Med
Post

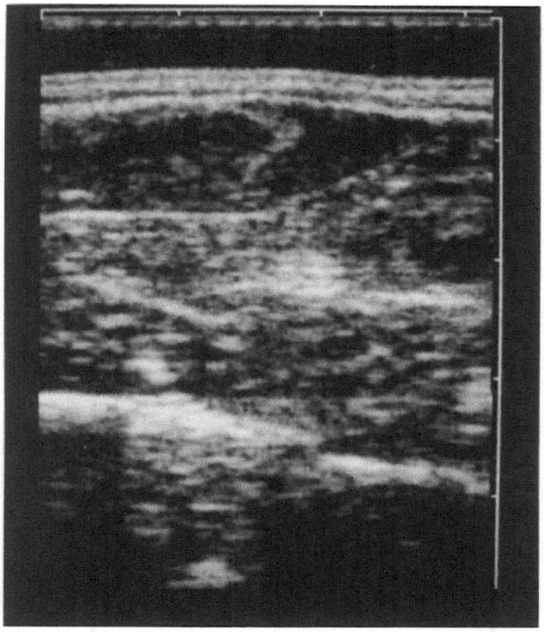

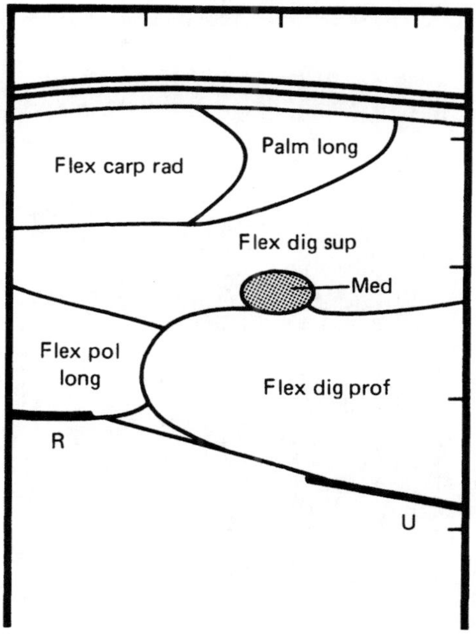
Flex carp rad
Palm long
Flex dig sup
Med
Flex pol long
Flex dig prof
R
U

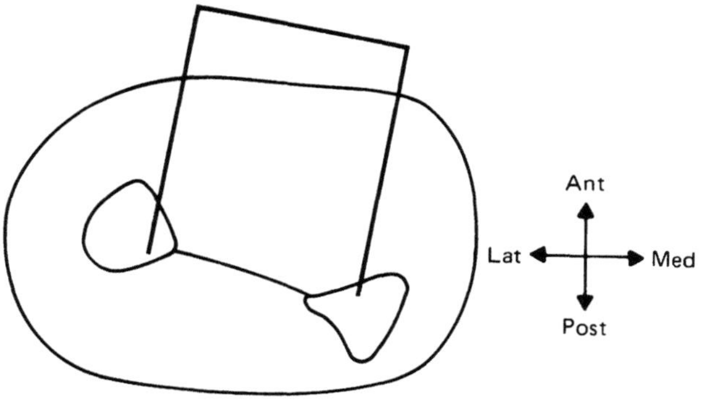
Ant
Lat
Med
Post

Forearm, Midthird

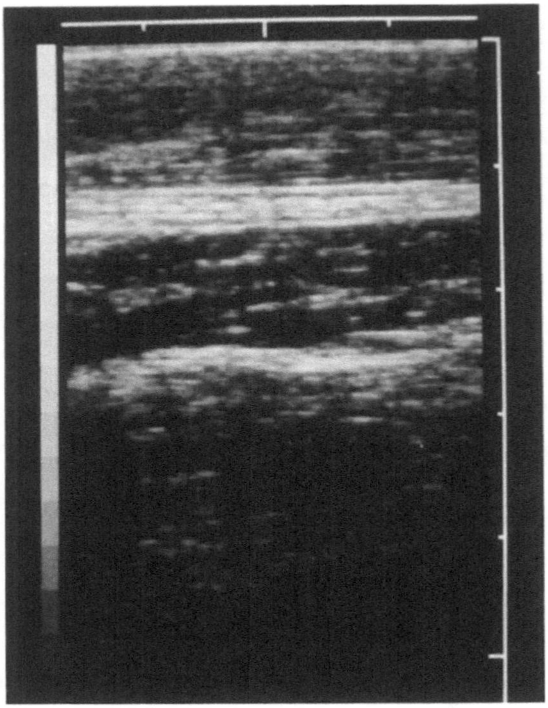

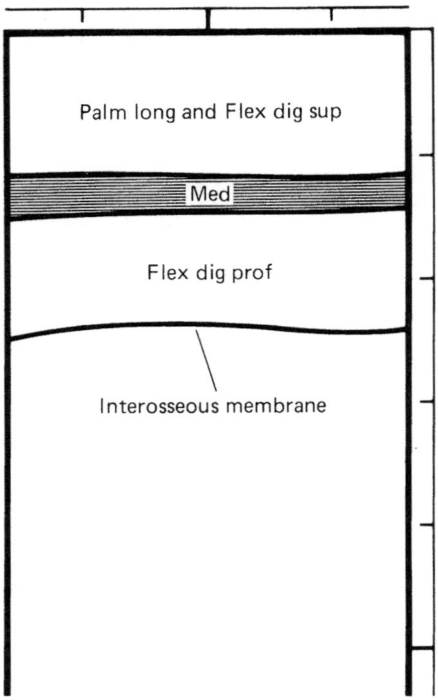

(Reprinted with permission from Fornage BD: Peripheral Nerves of the Extremities: Imaging with US. Radiology 1988;167:179–182

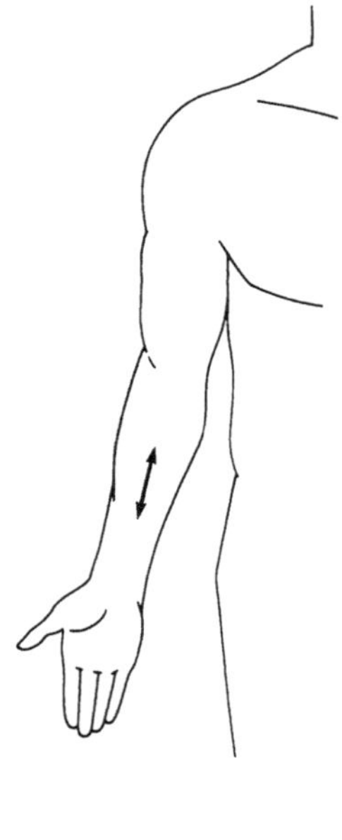

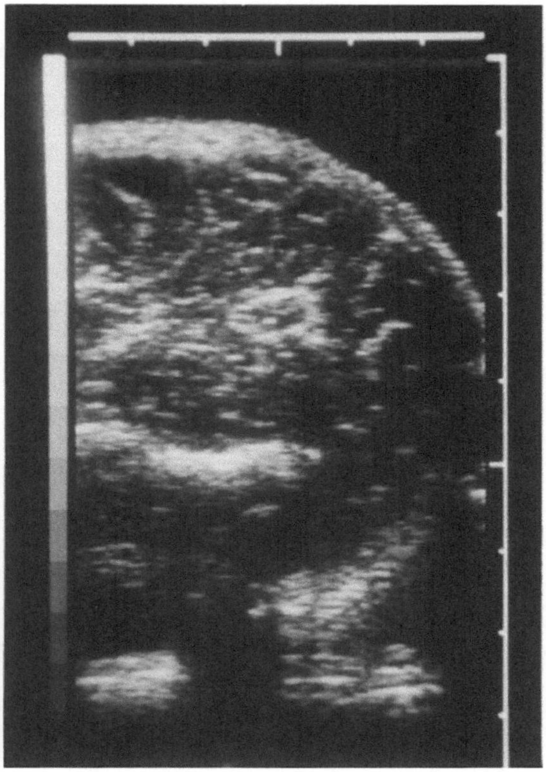

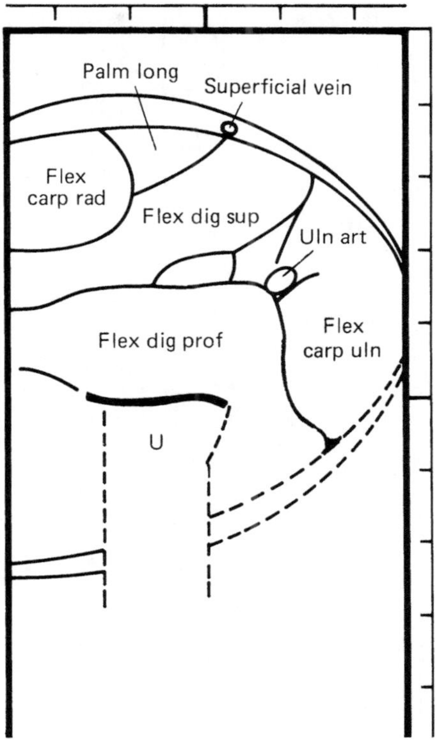
Palm long
Superficial vein
Flex carp rad
Flex dig sup
Uln art
Flex dig prof
Flex carp uln
U

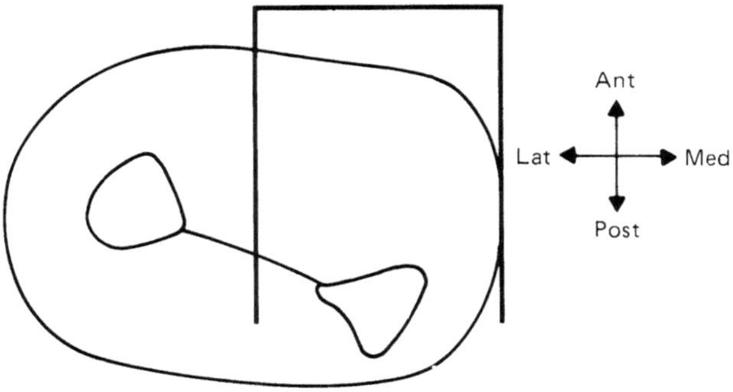
Ant
Lat
Med
Post

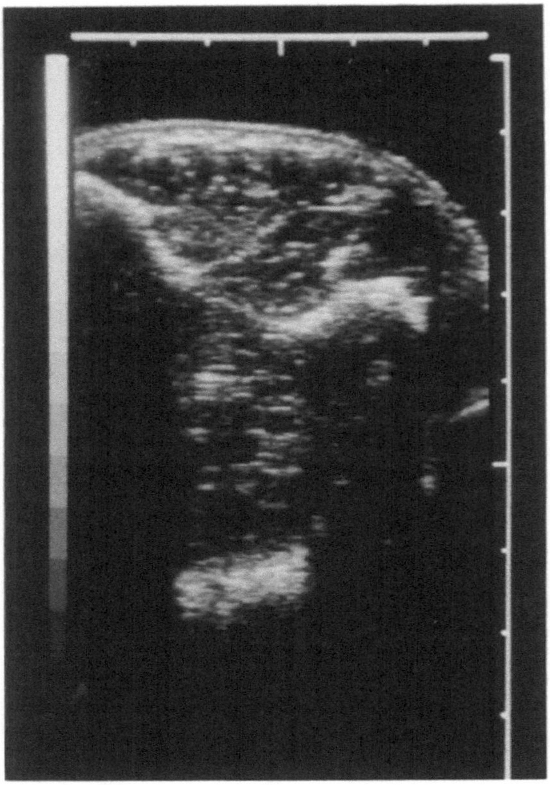

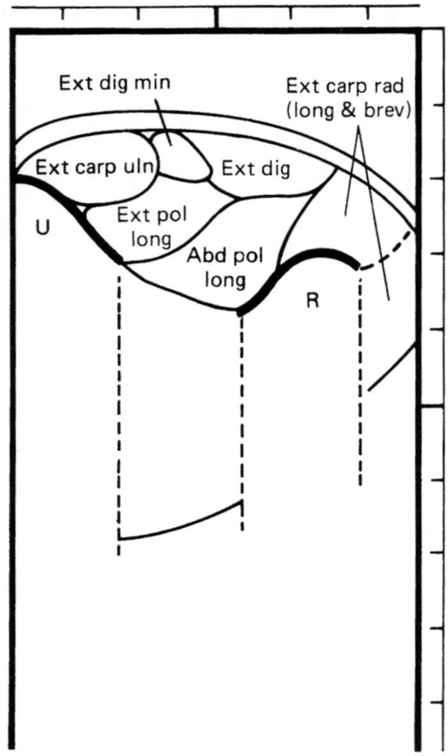
Ext dig min
Ext carp rad (long & brev)
Ext carp uln
Ext dig
U
Ext pol long
Abd pol long
R

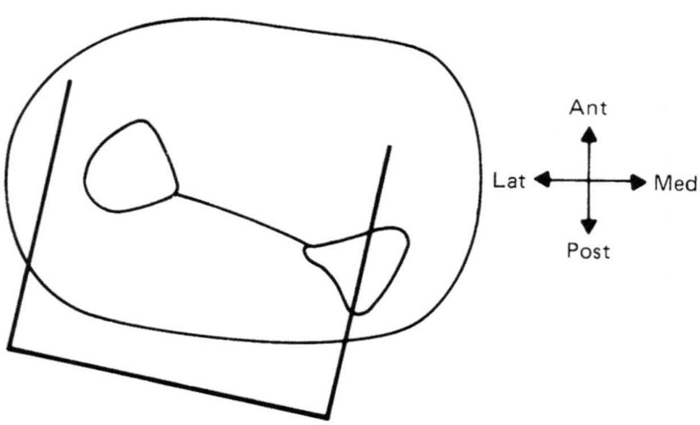
Ant
Lat
Med
Post

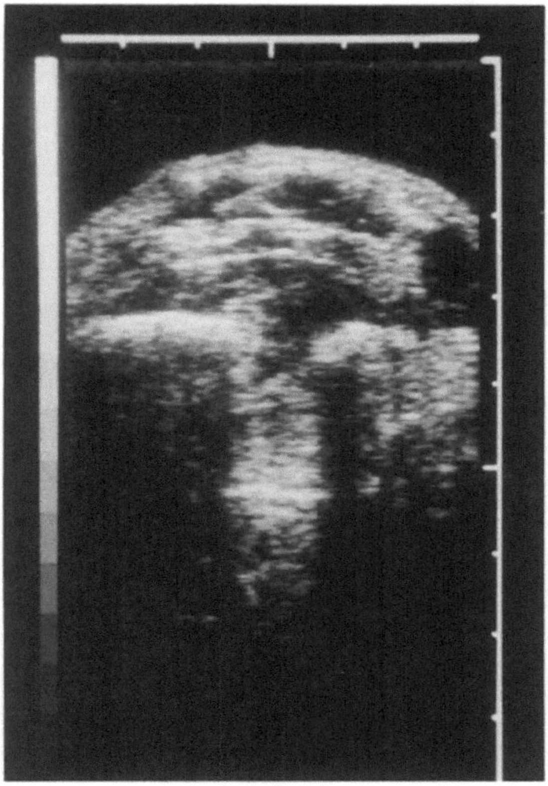

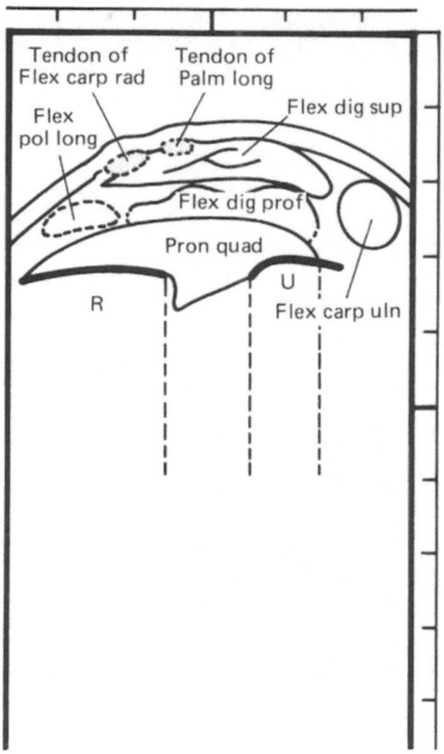
Tendon of
Flex carp rad
Tendon of
Palm long
Flex dig sup
Flex
pol long
Flex dig prof
Pron quad
U
R
Flex carp uln

Ant
Lat
Med
Post

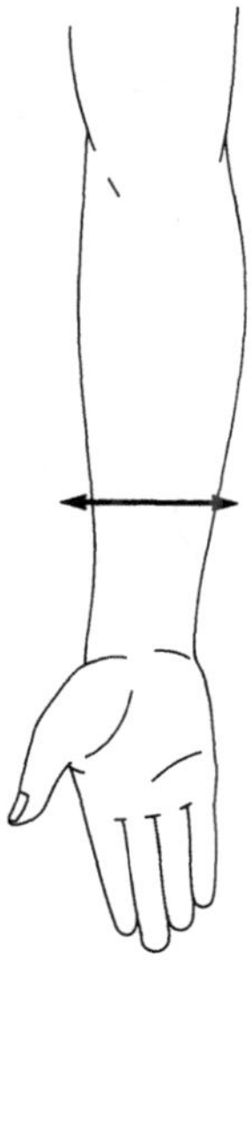

Forearm, Lower Third

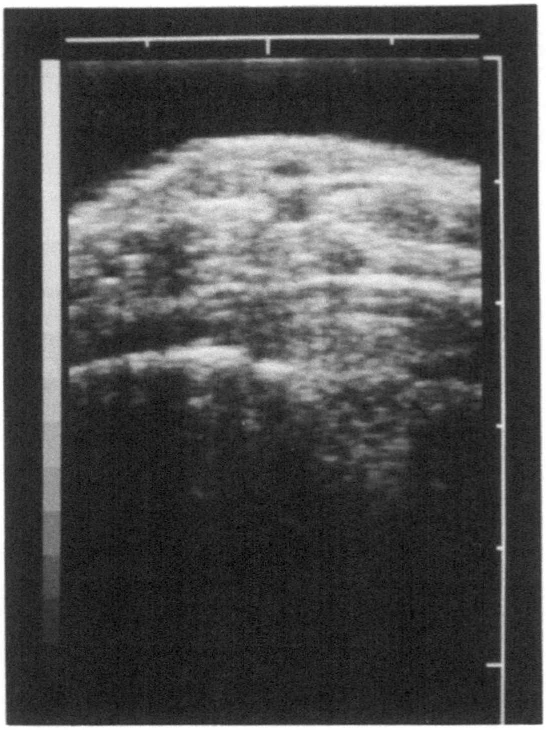

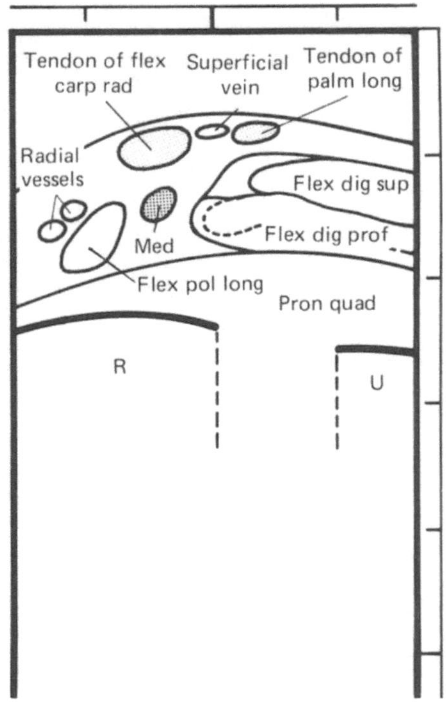

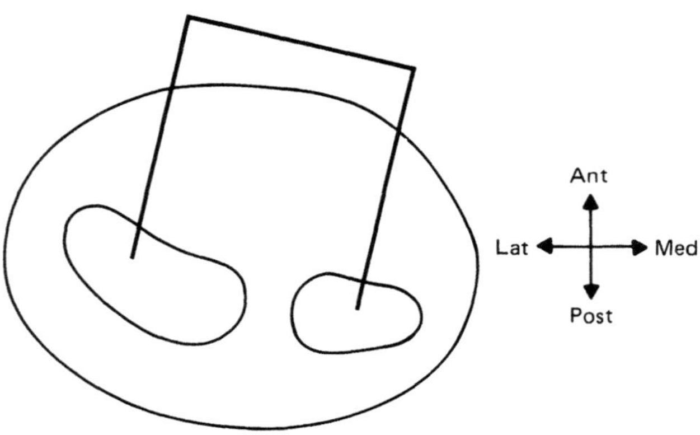

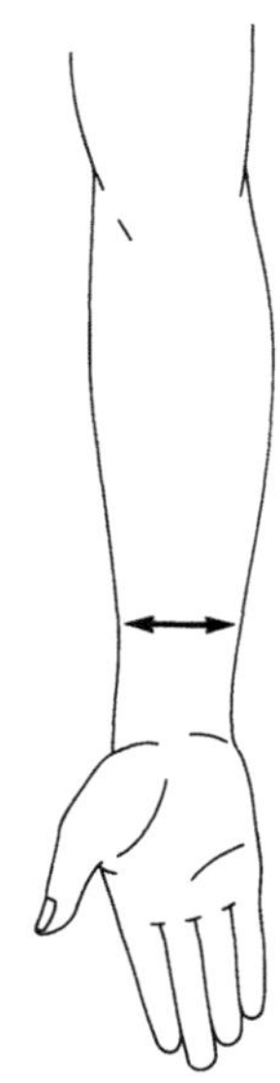

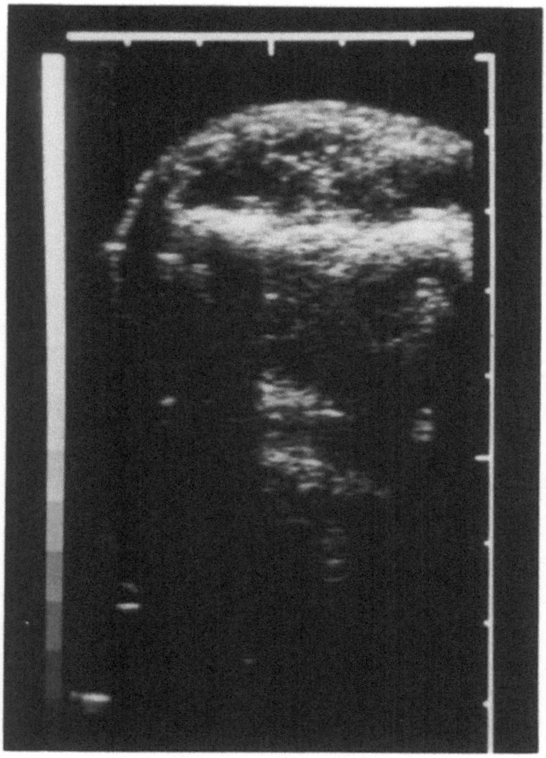

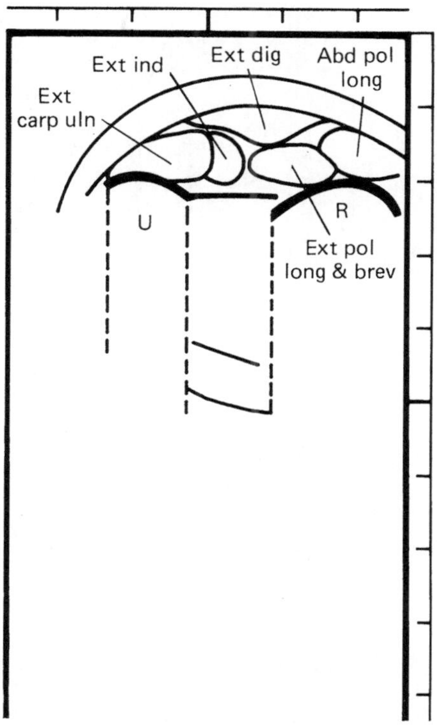
Ext ind
Ext dig
Abd pol long
Ext carp uln
U
R
Ext pol long & brev

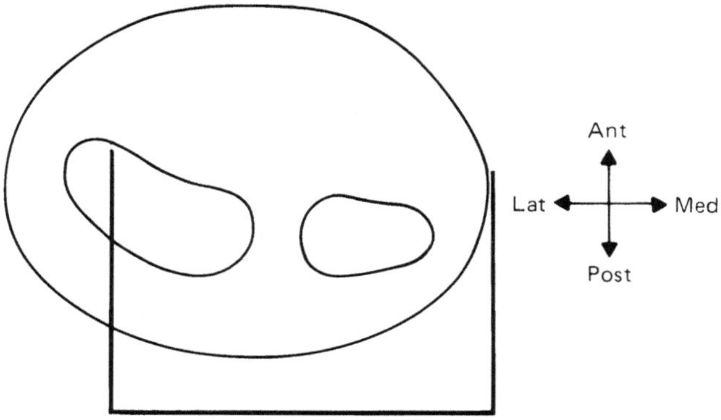
Ant
Lat
Med
Post

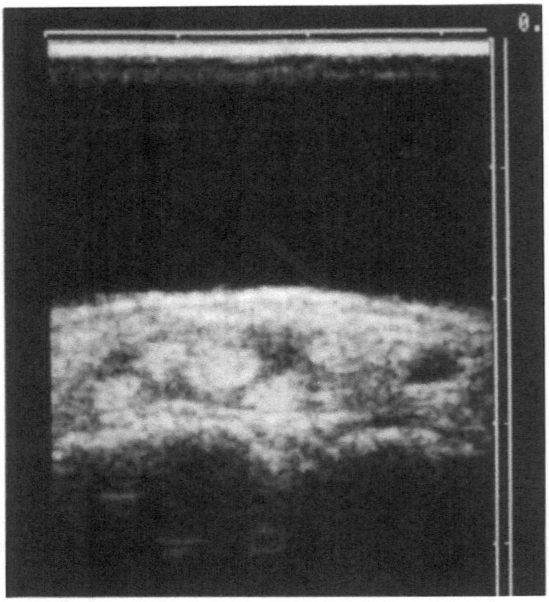

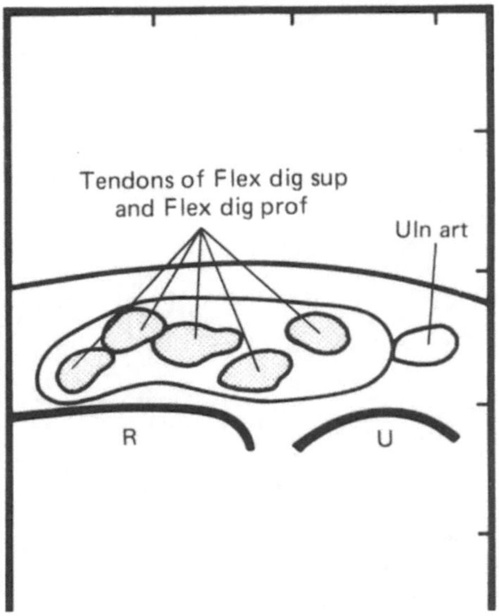

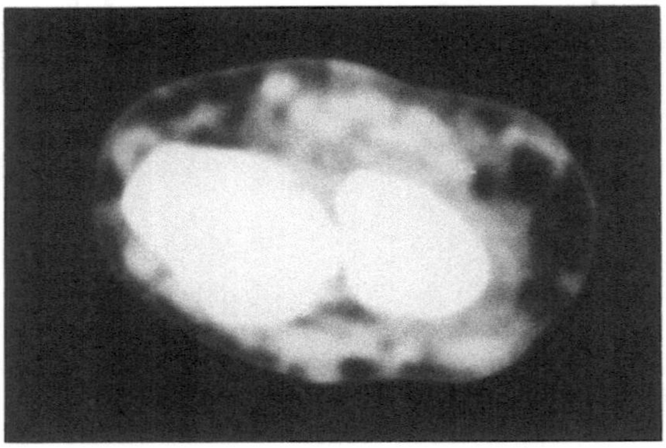

a. Sonogram
b. CT scan performed at same level

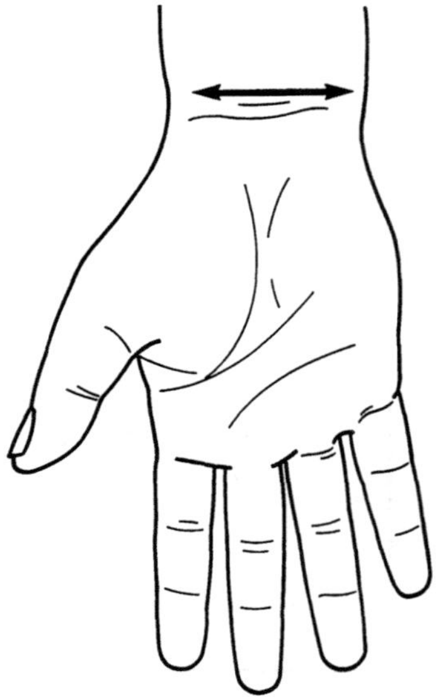

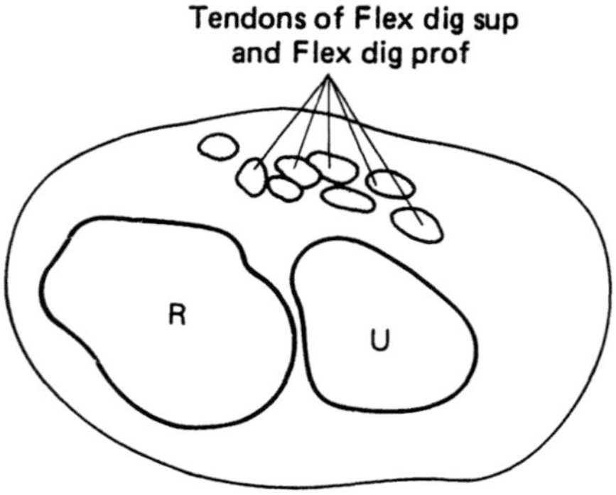

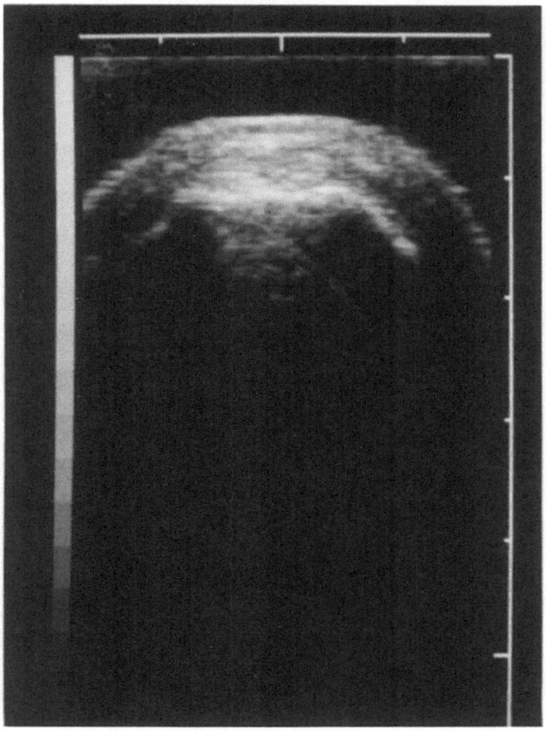

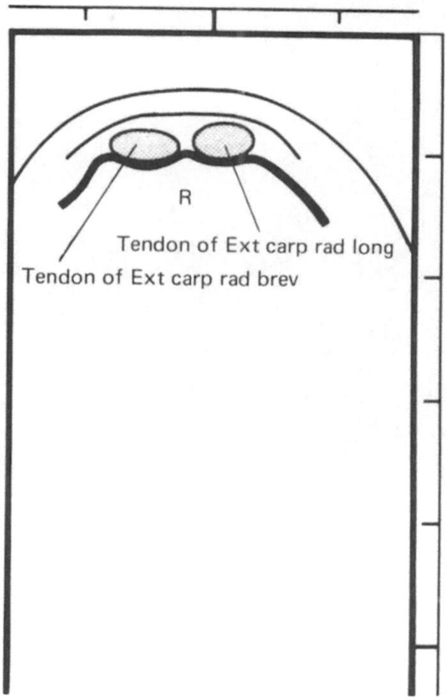
R
Tendon of Ext carp rad long
Tendon of Ext carp rad brev

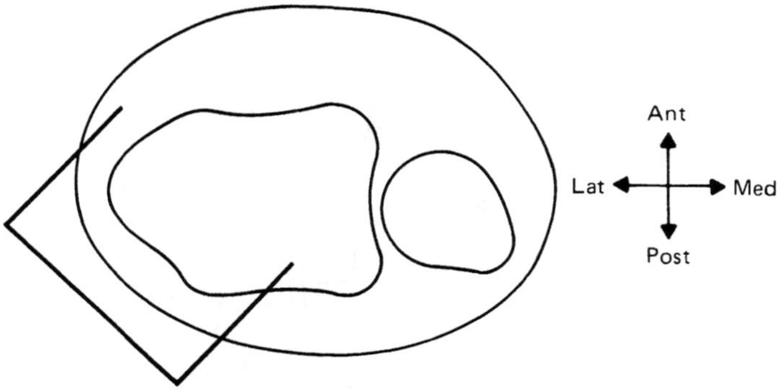
Ant
Lat
Med
Post

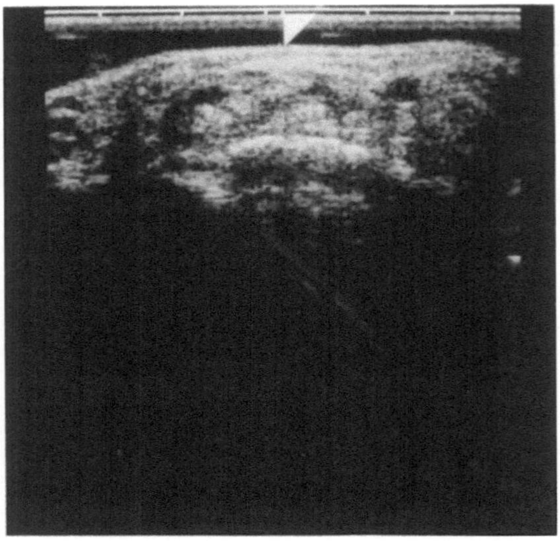

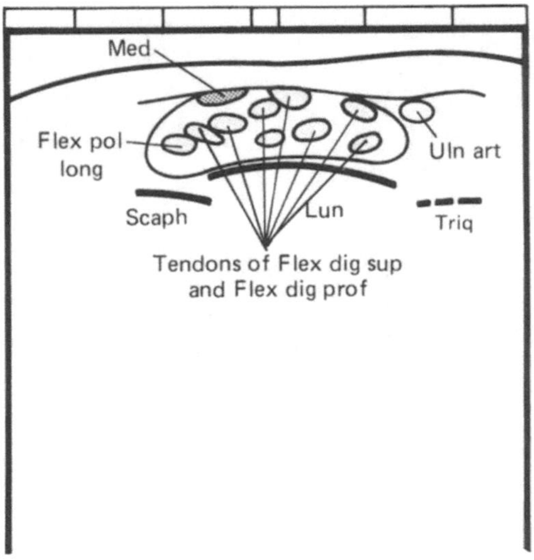
Med
Flex pol long
Uln art
Scaph
Lun
Triq
Tendons of Flex dig sup and Flex dig prof

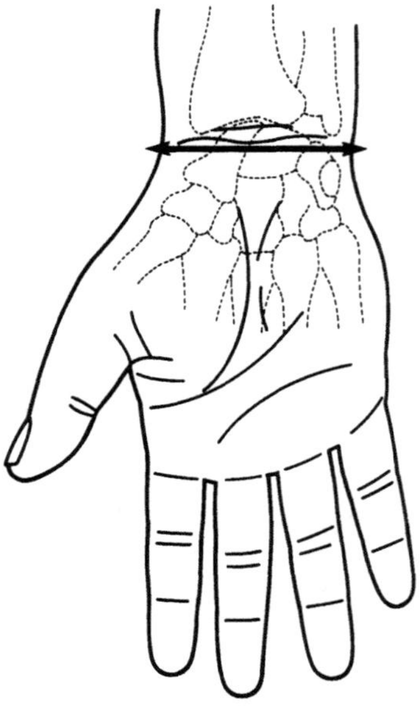

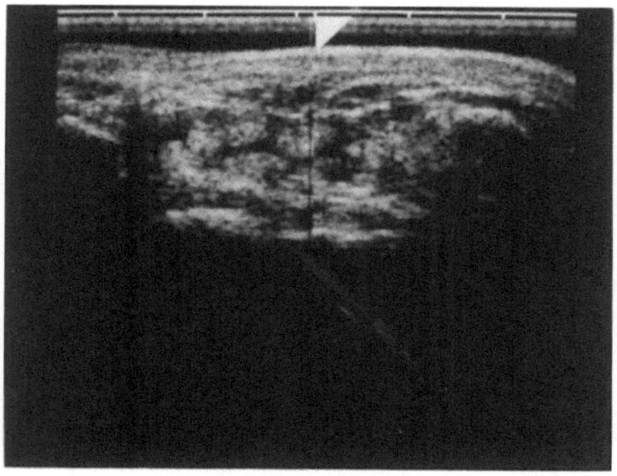

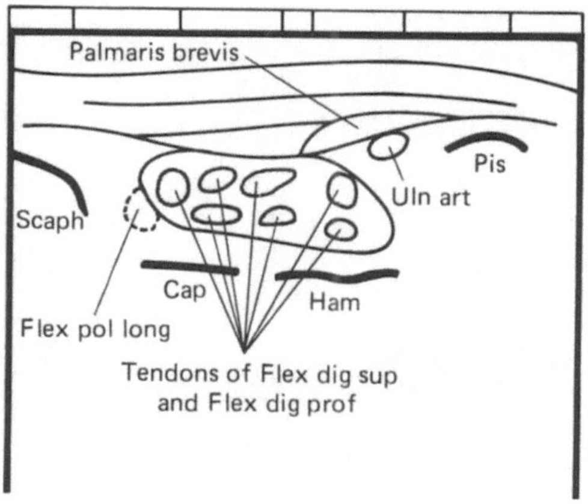

(Reprinted with permission from Fornage BD, Rifkin MD: Ultrasound examination of tendons. Radiol. Clin. North Am., 1988;26(1):87–107.)

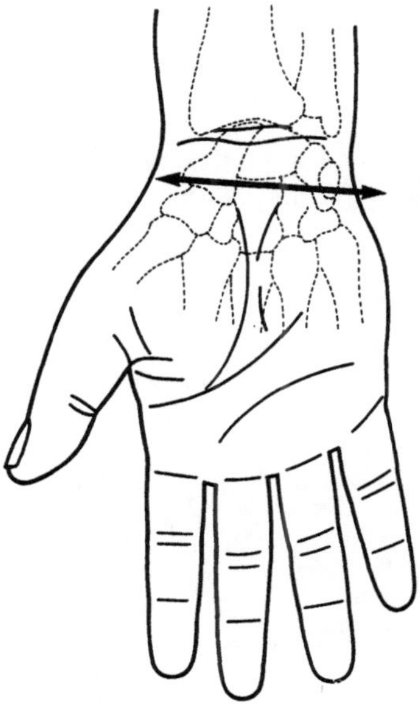

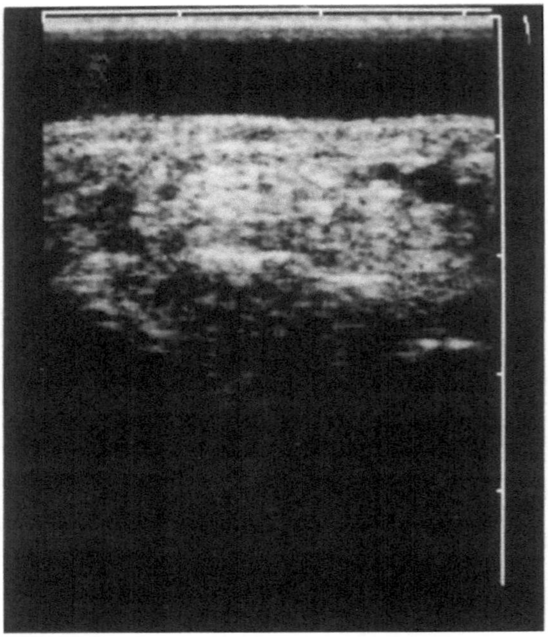

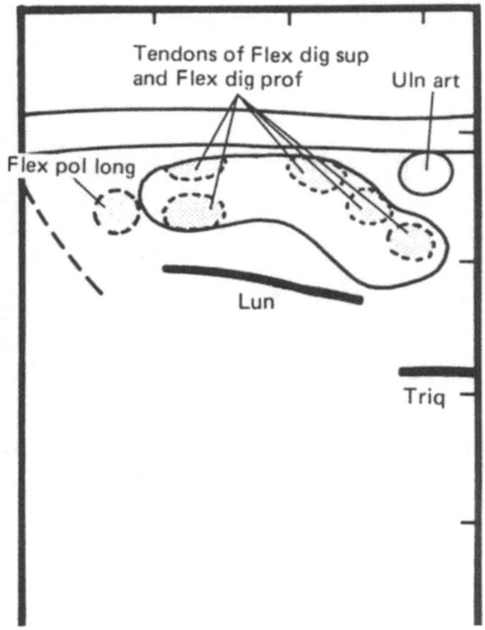
Tendons of Flex dig sup
and Flex dig prof
Uln art
Flex pol long
Lun
Triq

a

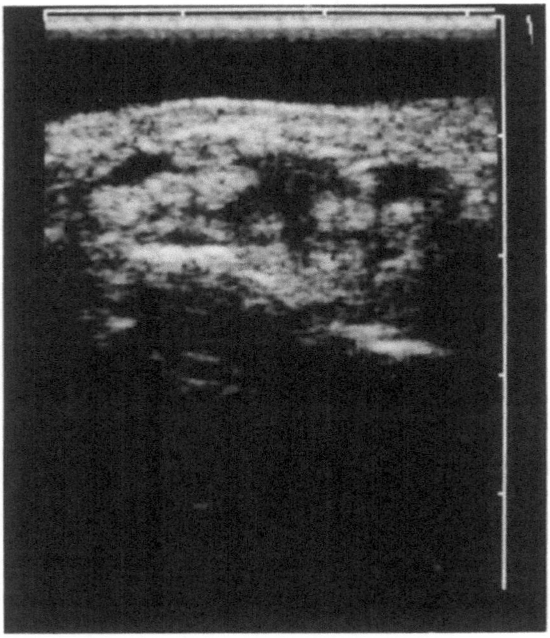

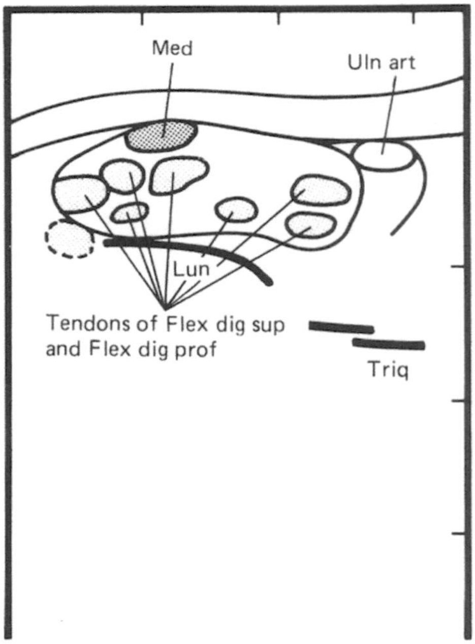
Med
Uln art
Lun
Tendons of Flex dig sup
and Flex dig prof
Triq

b

c

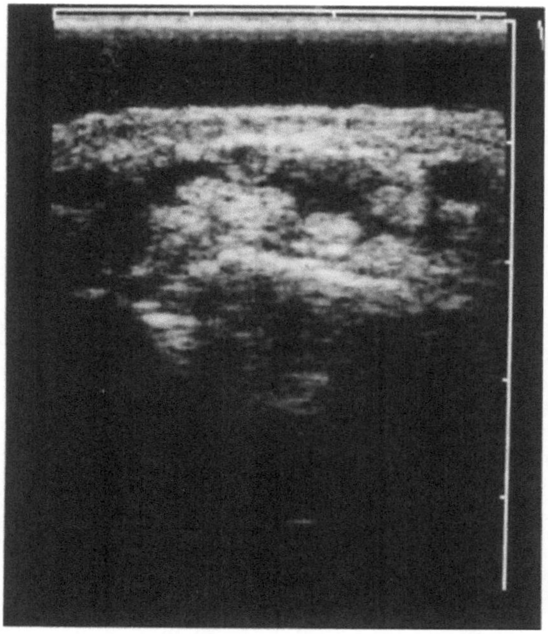

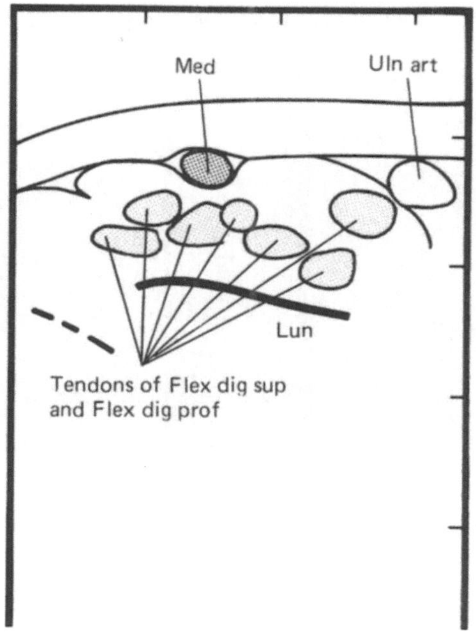

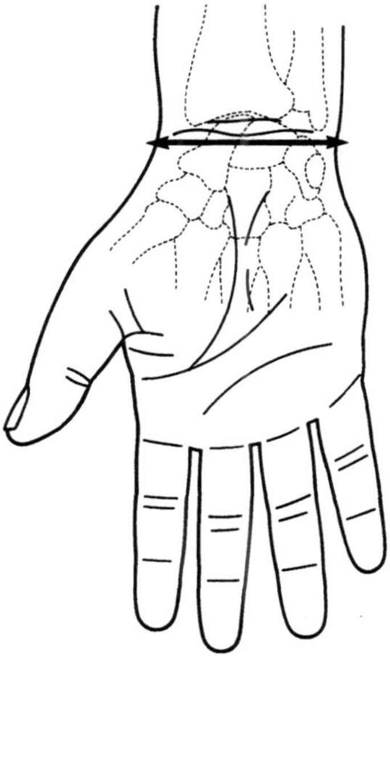

a. At rest

b and c. During flexion of the fingers
and wrist

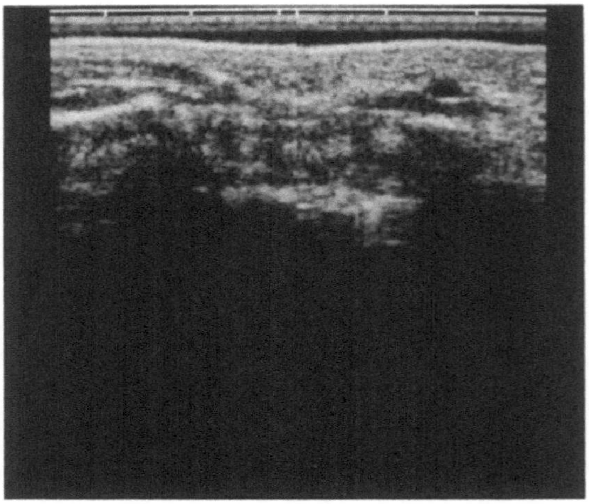

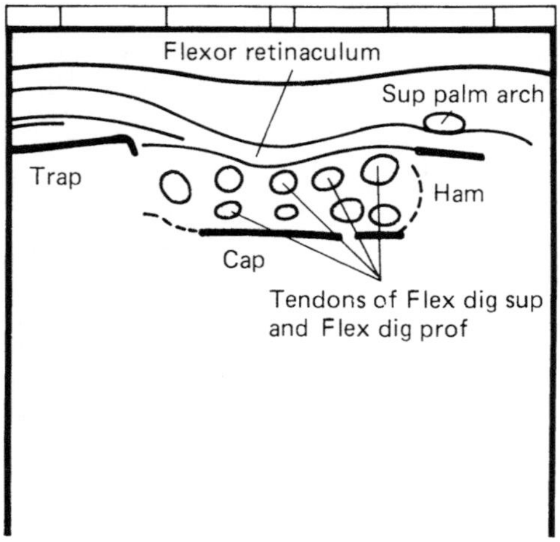
Flexor retinaculum
Sup palm arch
Trap
Ham
Cap
Tendons of Flex dig sup
and Flex dig prof

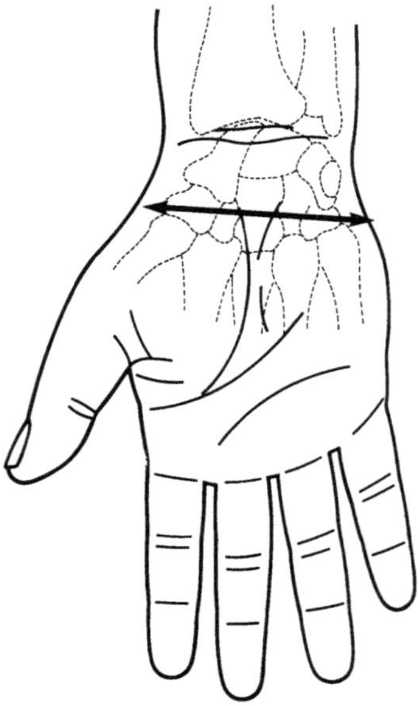

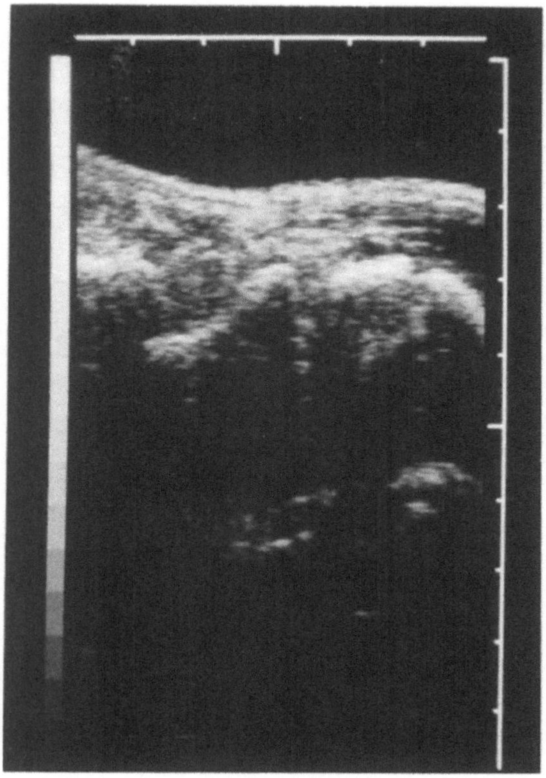

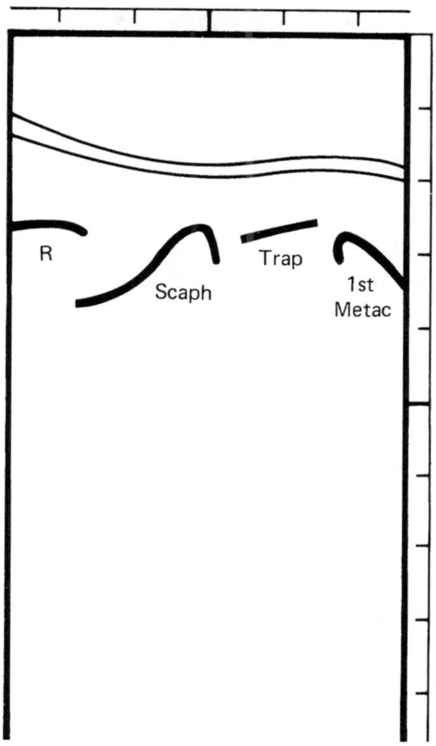
R
Scaph
Trap
1st
Metac

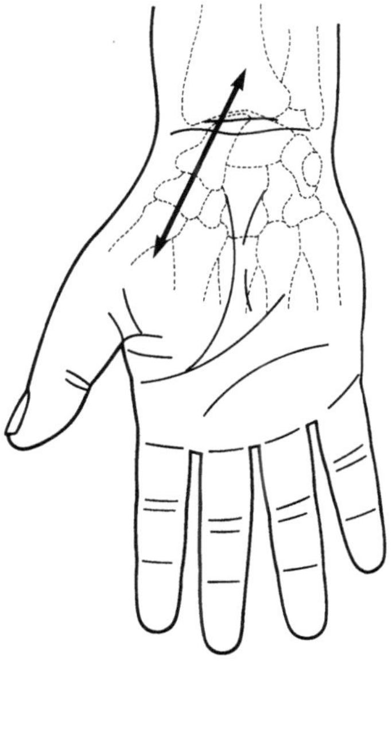

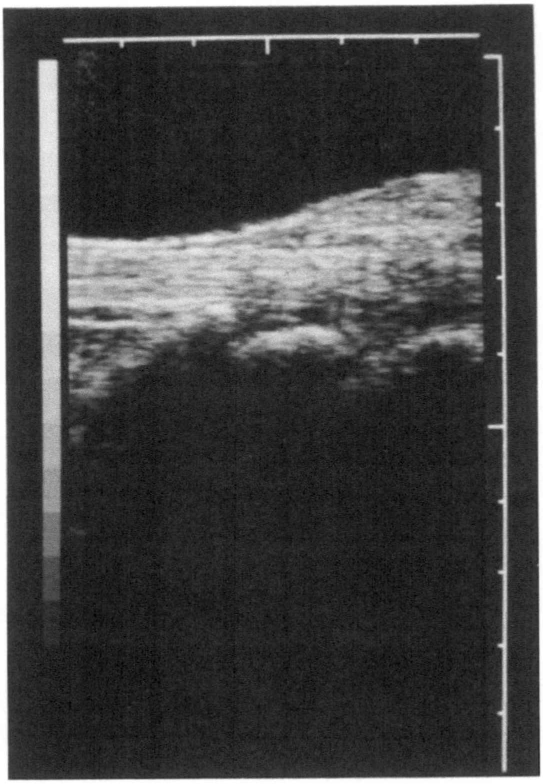

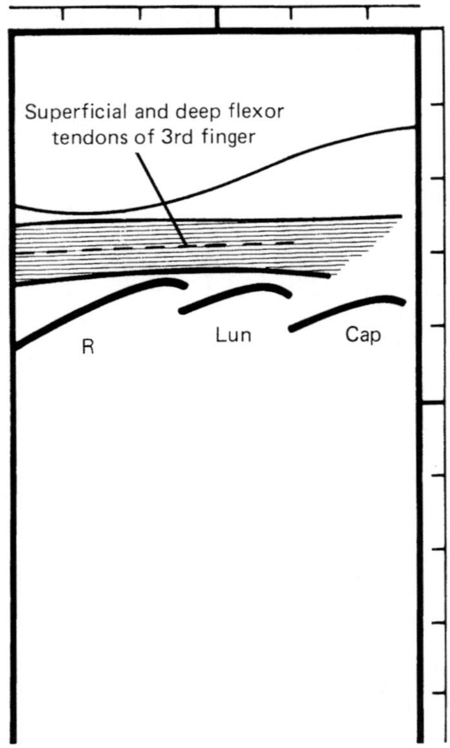
Superficial and deep flexor tendons of 3rd finger
R
Lun
Cap

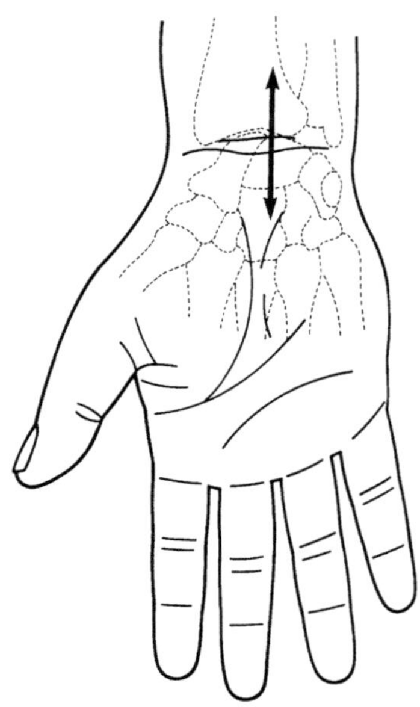

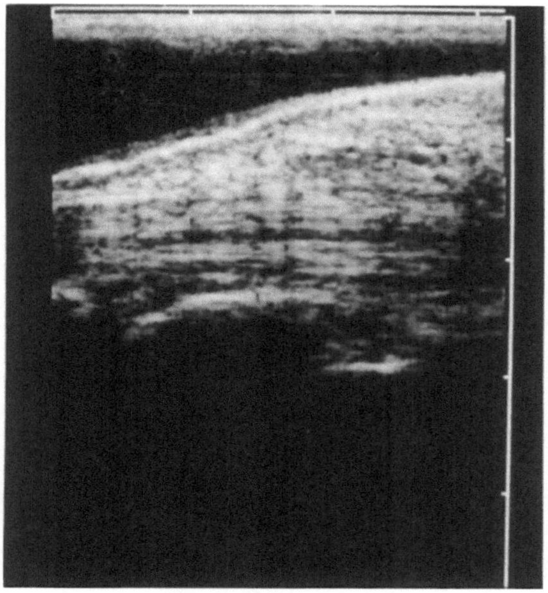

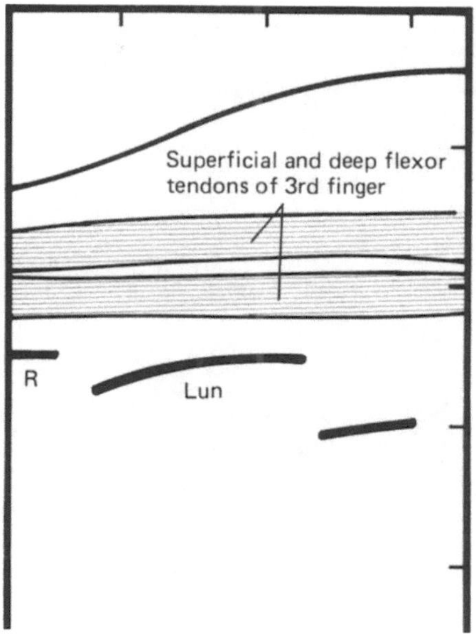
Superficial and deep flexor
tendons of 3rd finger
R
Lun

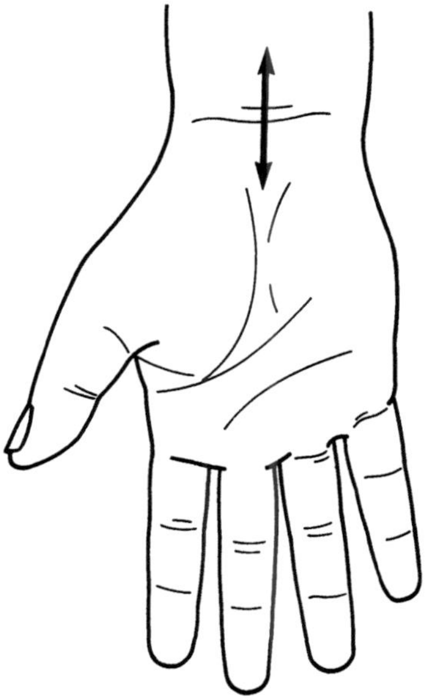

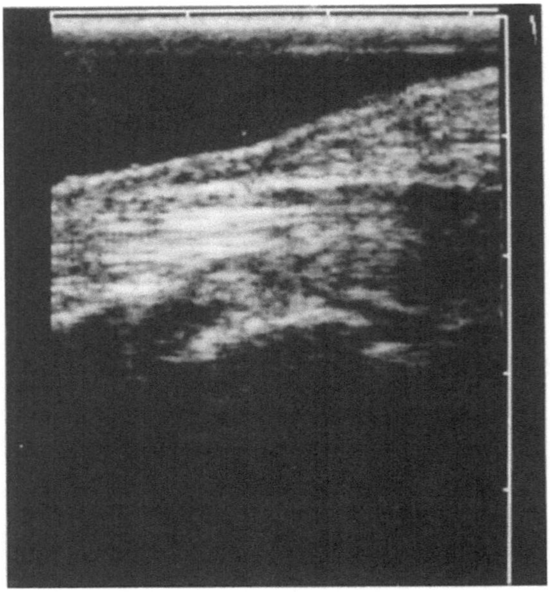

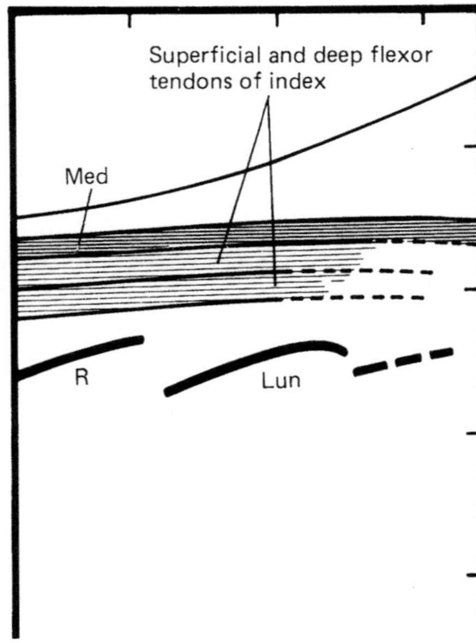
Superficial and deep flexor tendons of index
Med
R
Lun

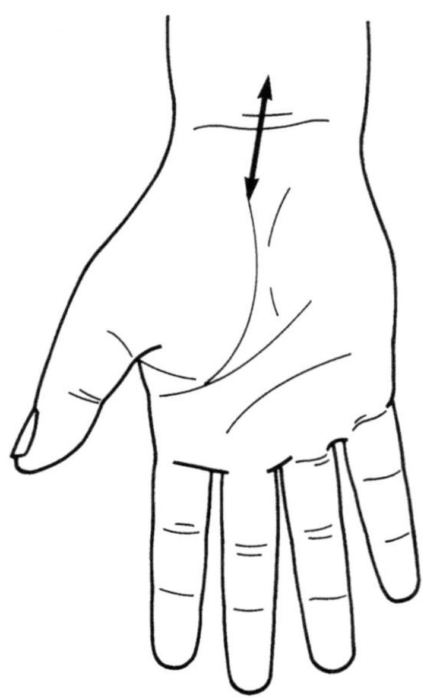

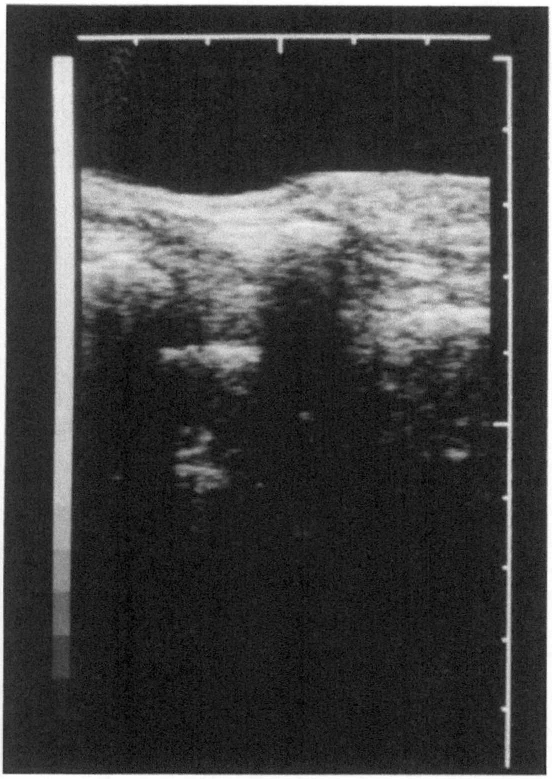

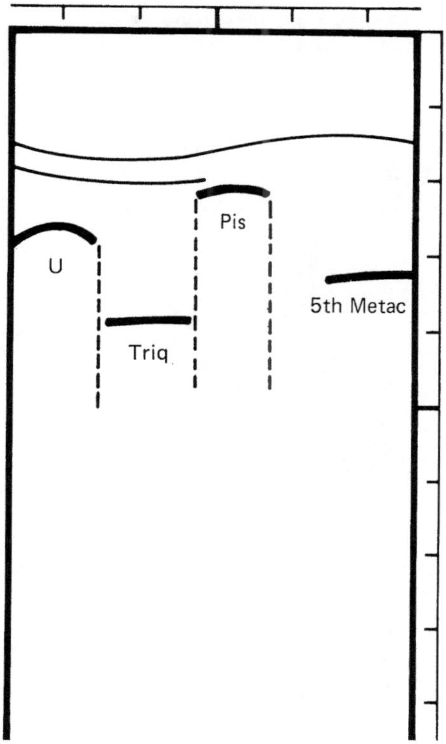
Pis
U
5th Metac
Triq

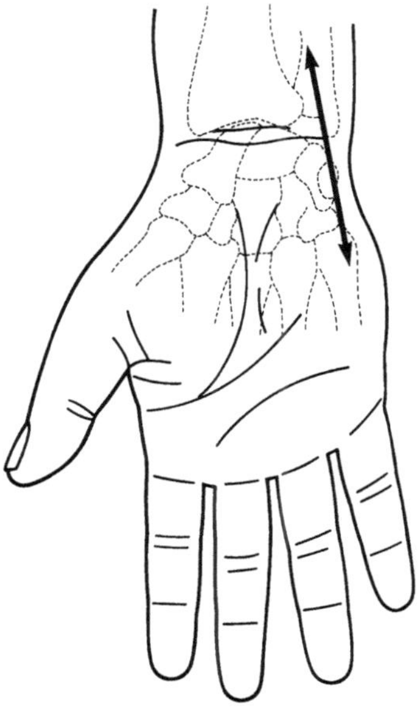

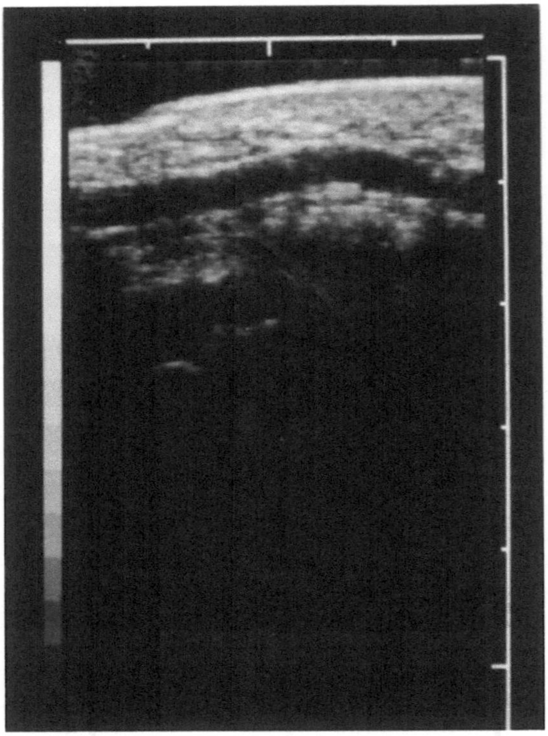

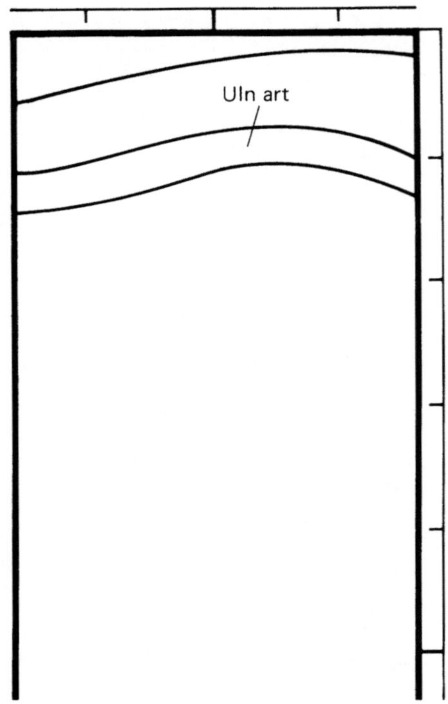
Uln art

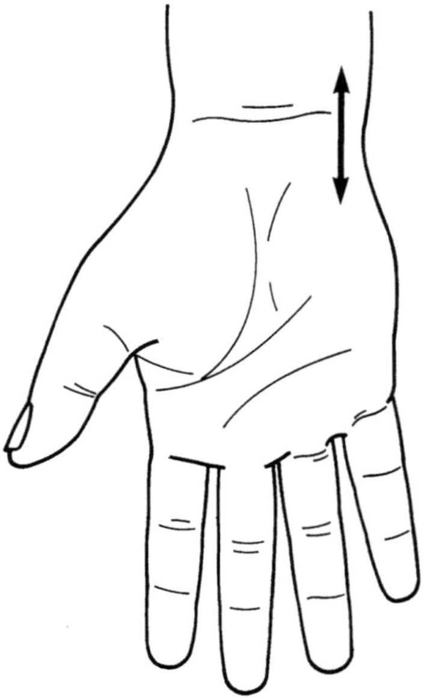

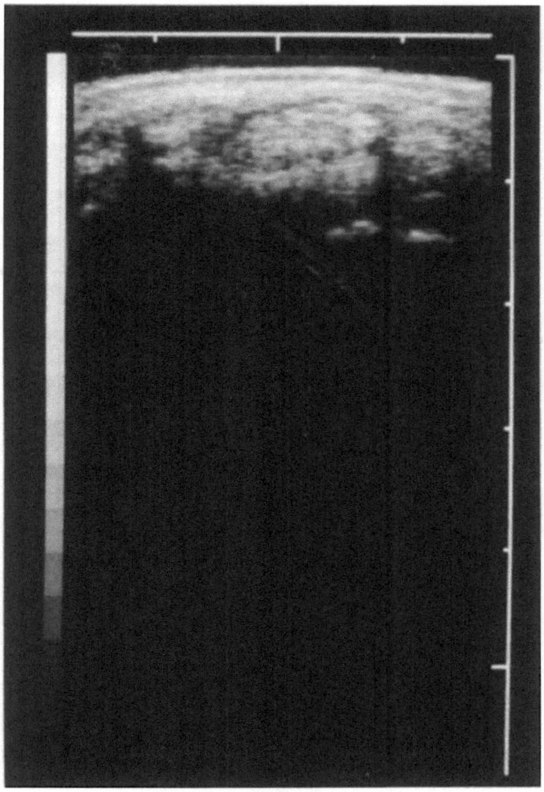

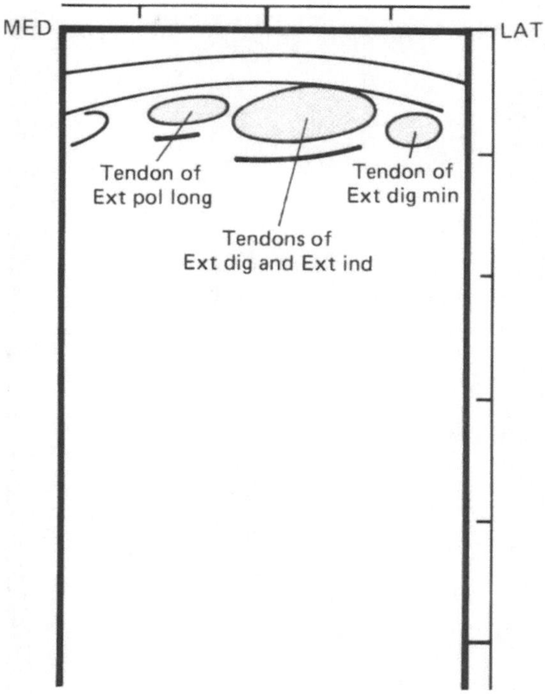
MED
LAT
Tendon of
Ext pol long
Tendon of
Ext dig min
Tendons of
Ext dig and Ext ind

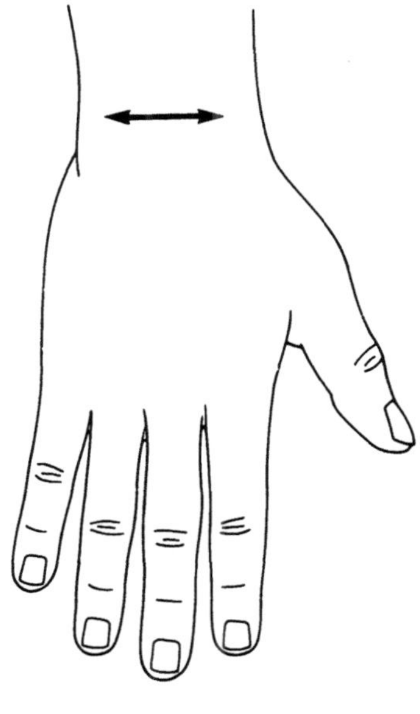

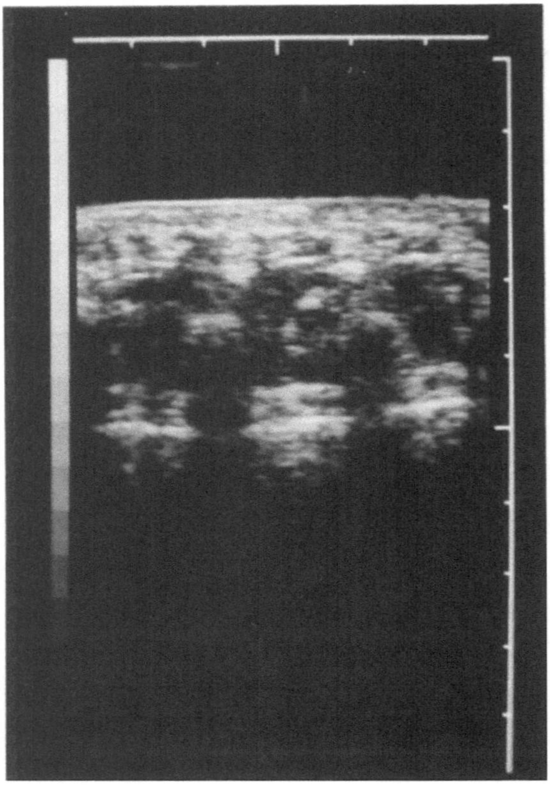

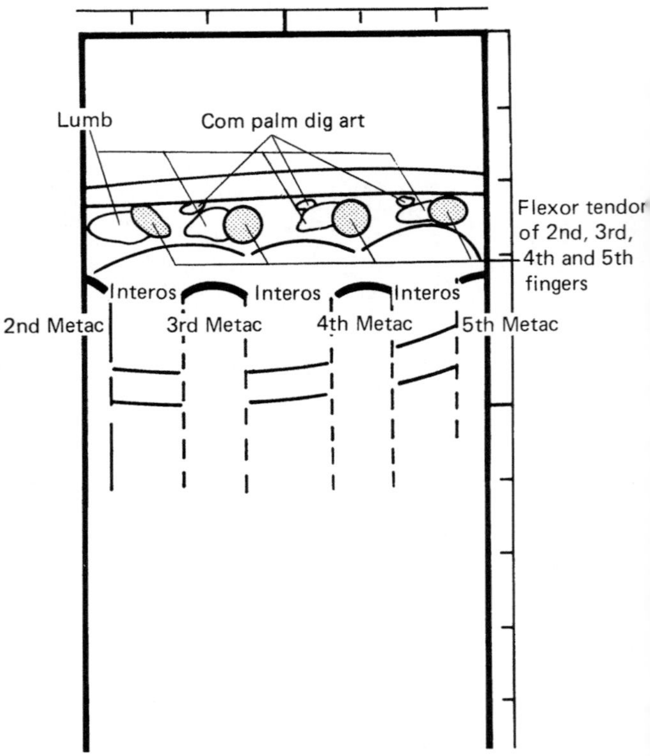

Lumb
Com palm dig art
Flexor tendon of 2nd, 3rd, 4th and 5th fingers
Interos
Interos
Interos
2nd Metac
3rd Metac
4th Metac
5th Metac

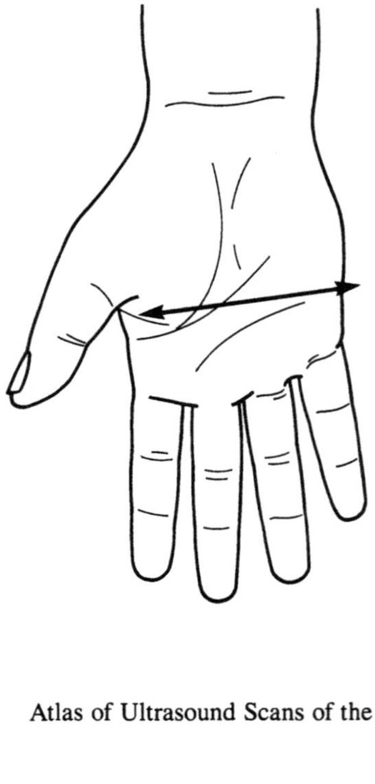

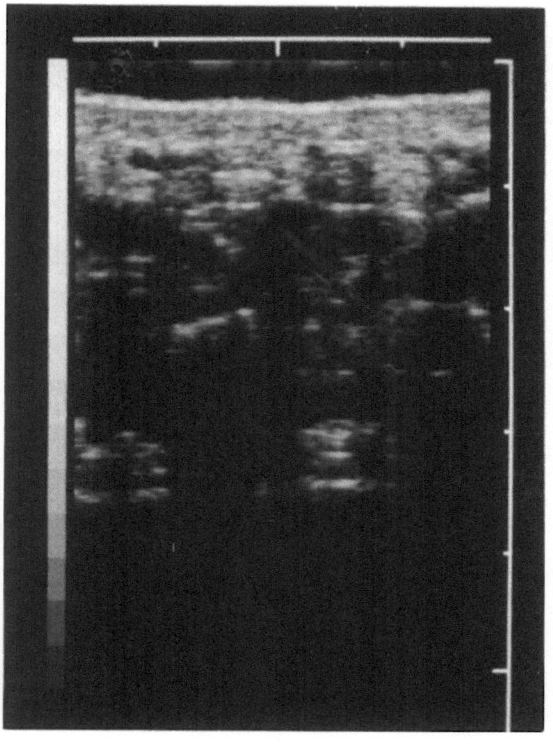

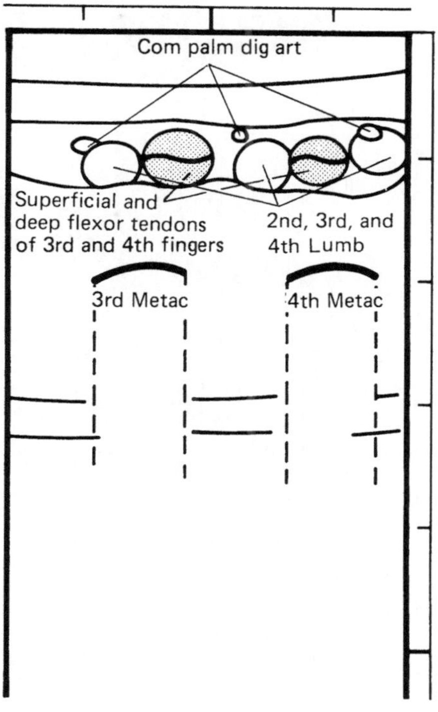
Com palm dig art
Superficial and deep flexor tendons of 3rd and 4th fingers
2nd, 3rd, and 4th Lumb
3rd Metac
4th Metac

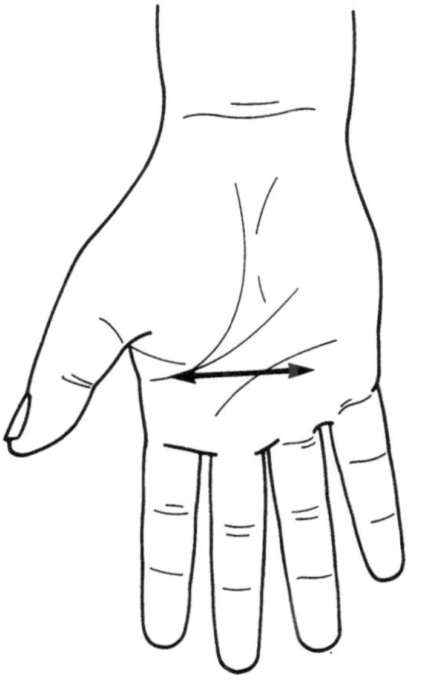

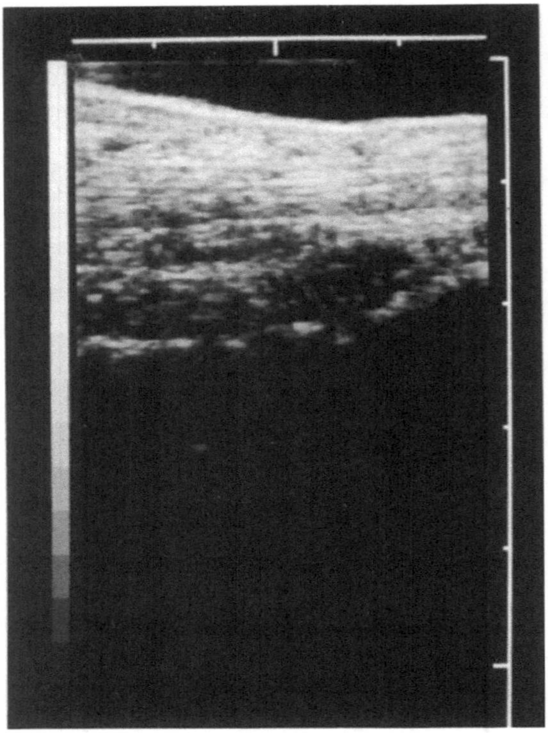

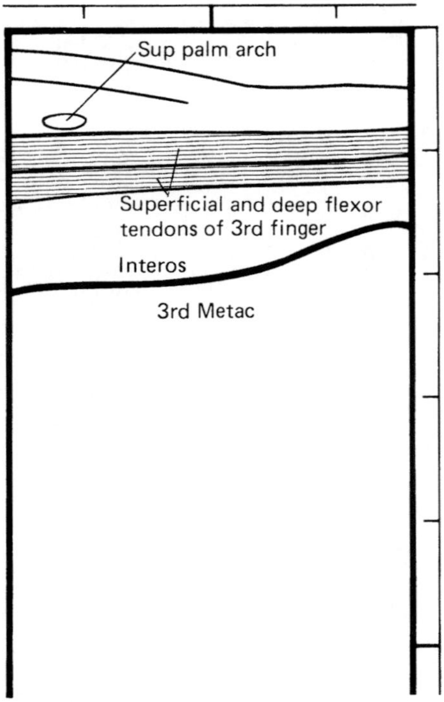
Sup palm arch
Superficial and deep flexor tendons of 3rd finger
Interos
3rd Metac

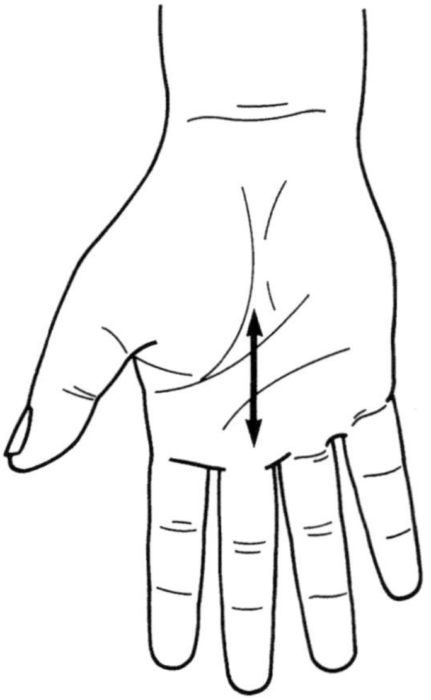

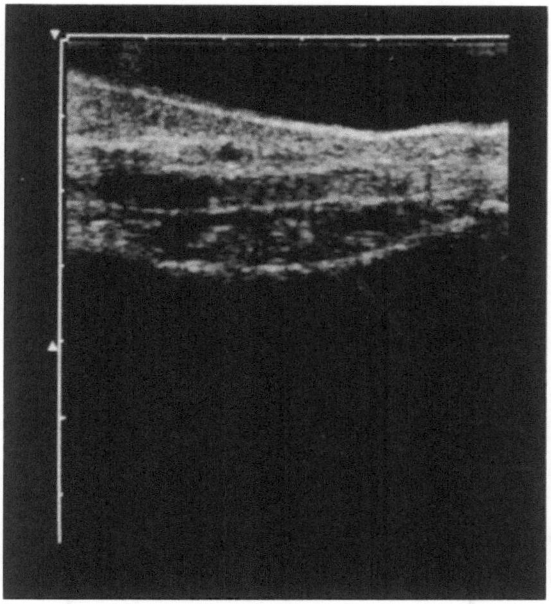

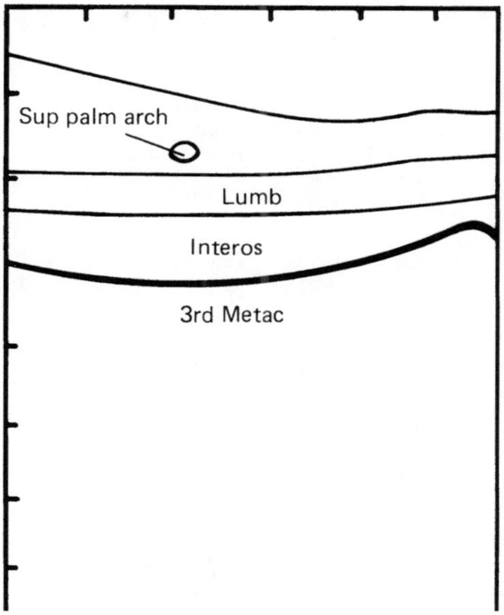
Sup palm arch
Lumb
Interos
3rd Metac

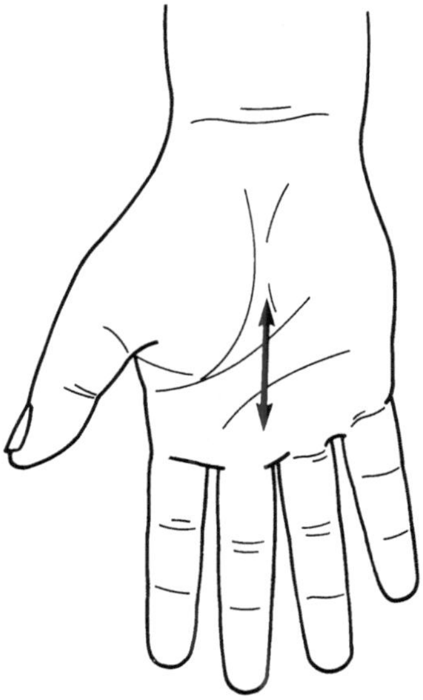

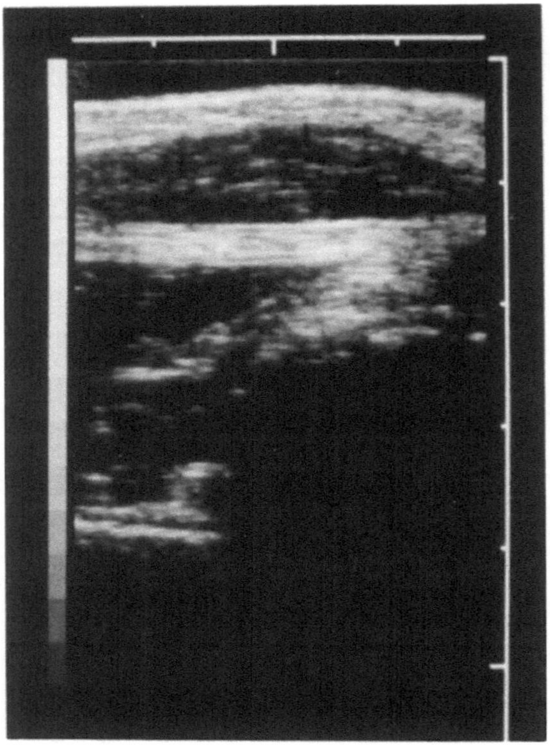

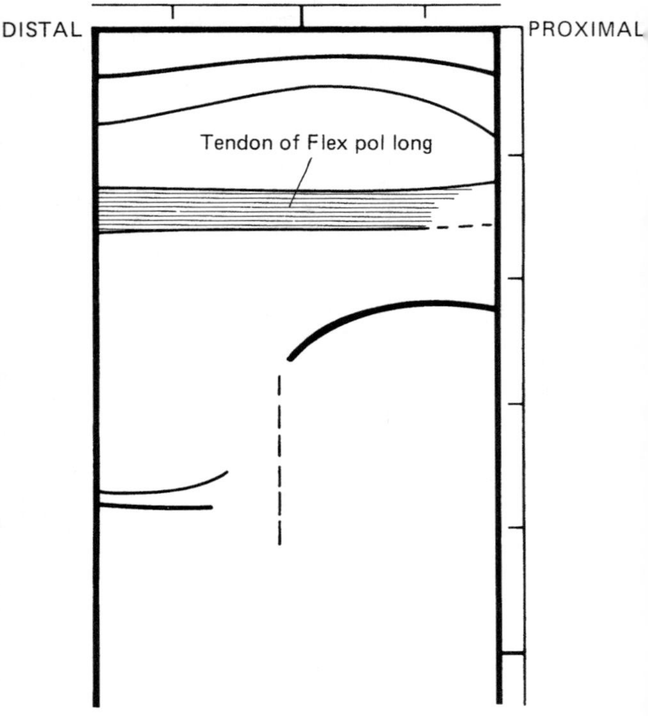
DISTAL
PROXIMAL
Tendon of Flex pol long

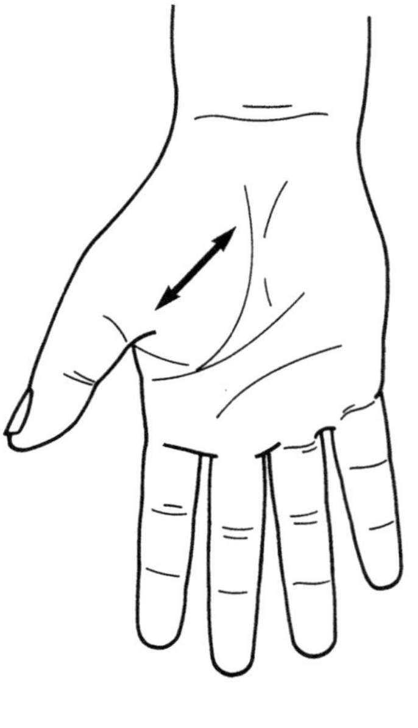

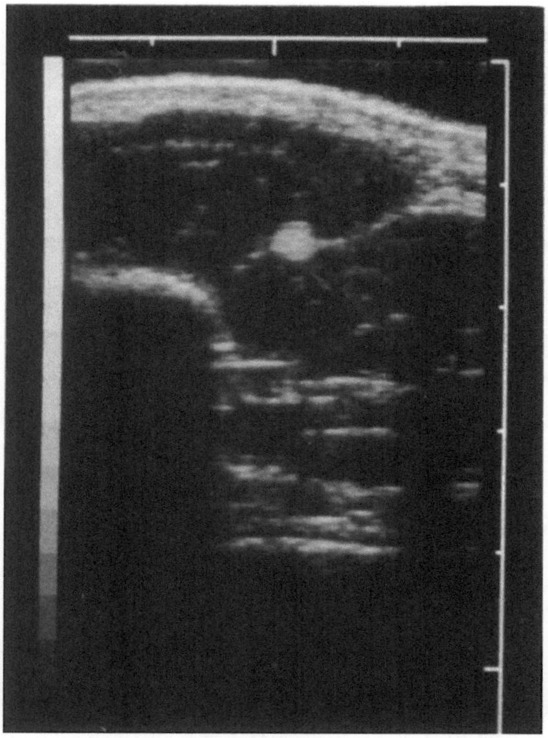

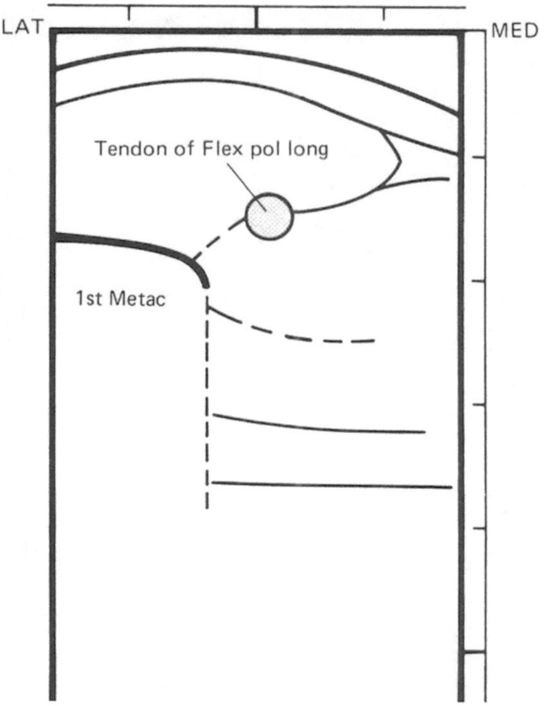
LAT
MED
Tendon of Flex pol long
1st Metac

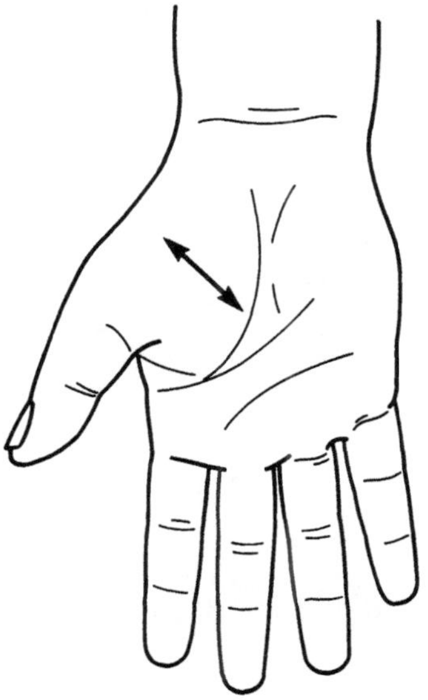

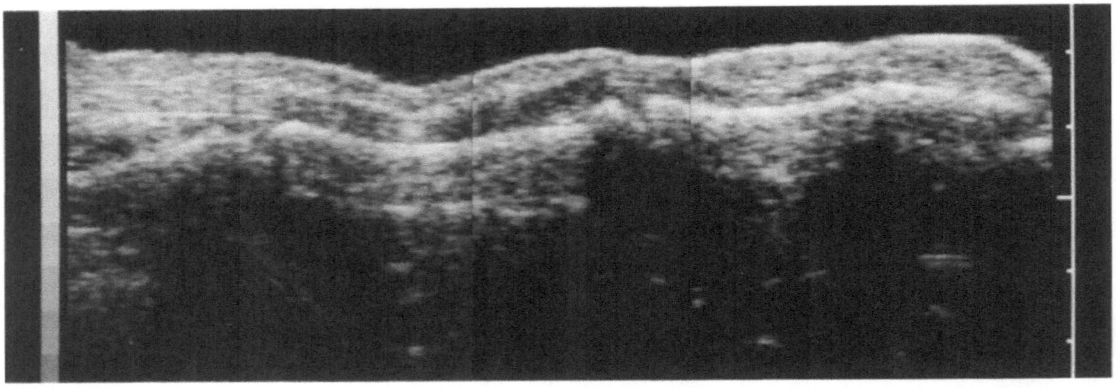
a

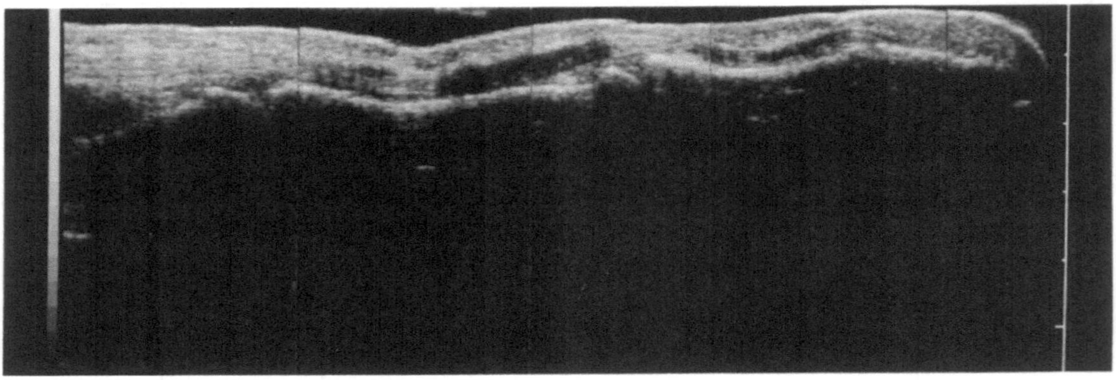
b

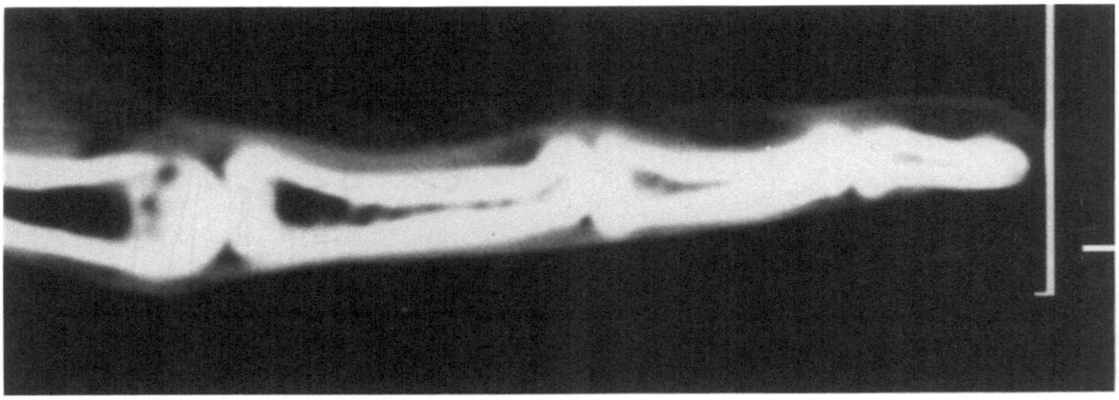
c

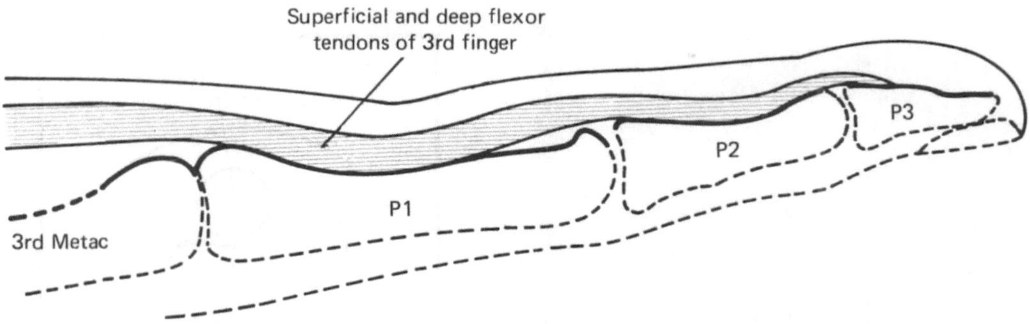

a. Montage of sonograms obtained with the use of a 5-MHz probe.

b. Montage of sonograms obtained with the use of a 7.5- MHz probe.

c. CT scan.

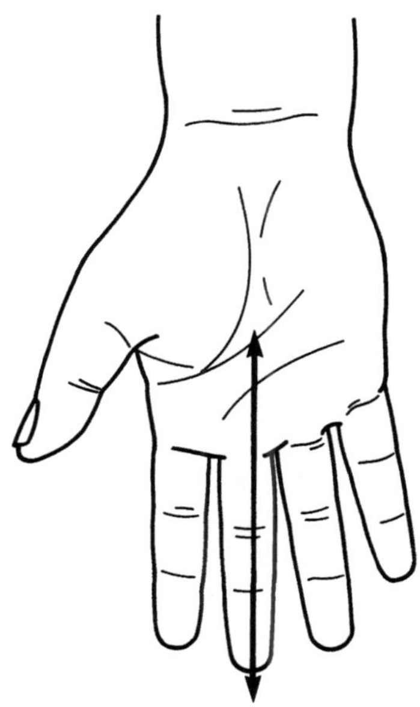

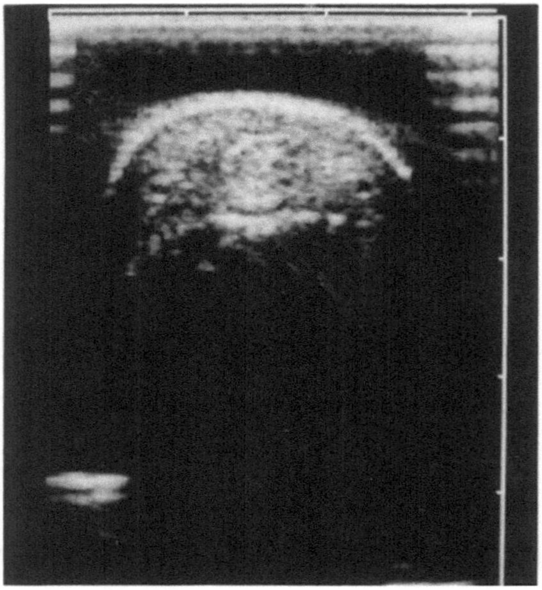

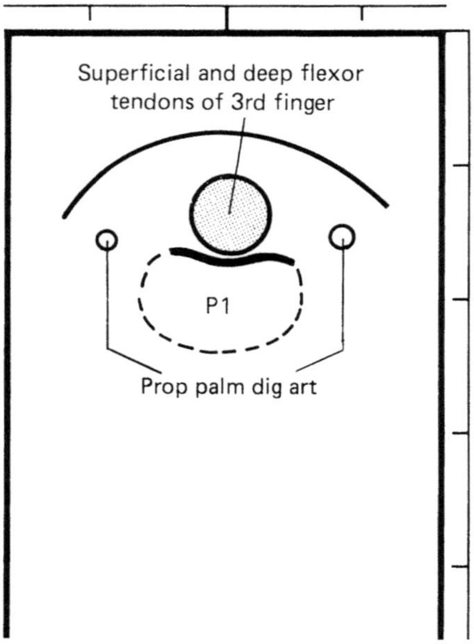
Superficial and deep flexor tendons of 3rd finger
P1
Prop palm dig art

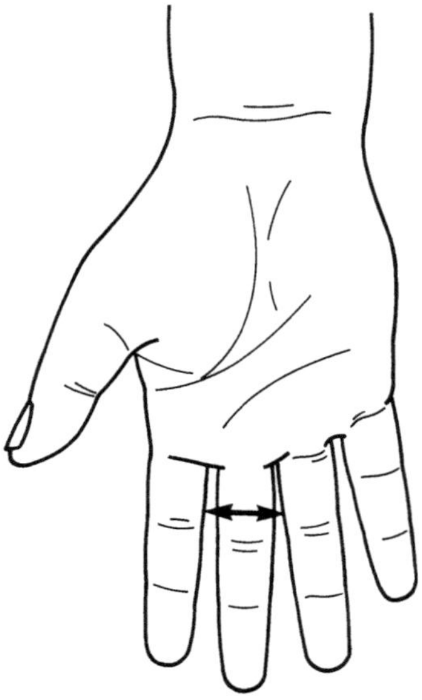

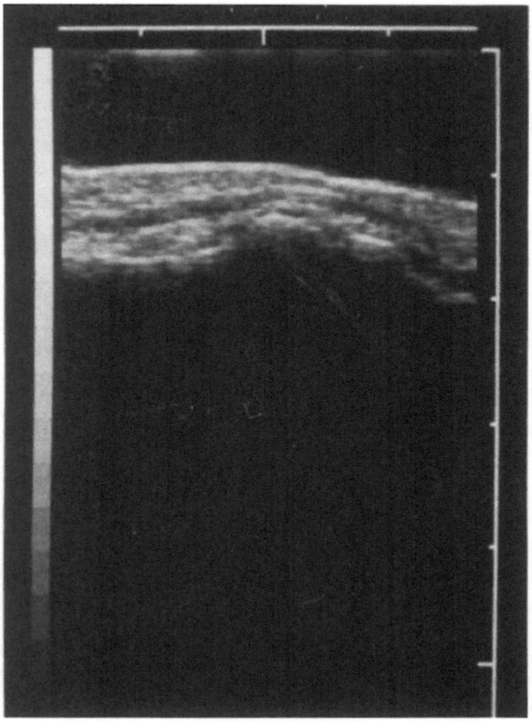

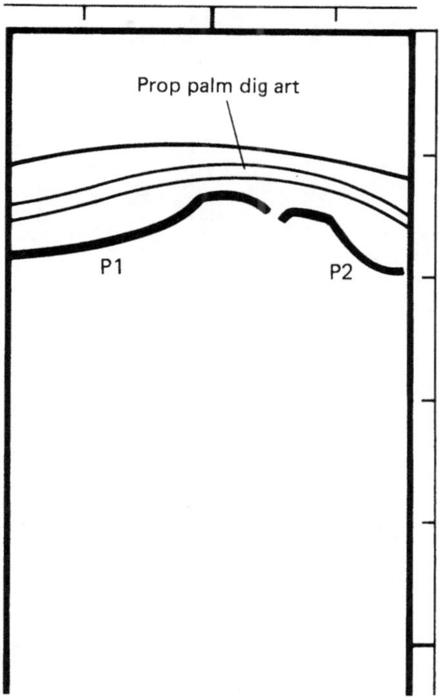
Prop palm dig art
P1
P2

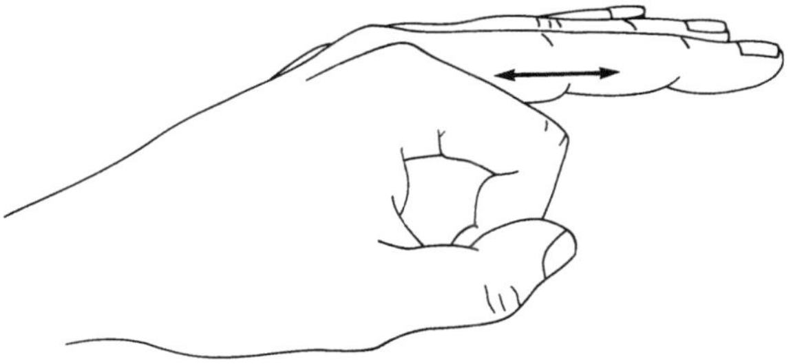

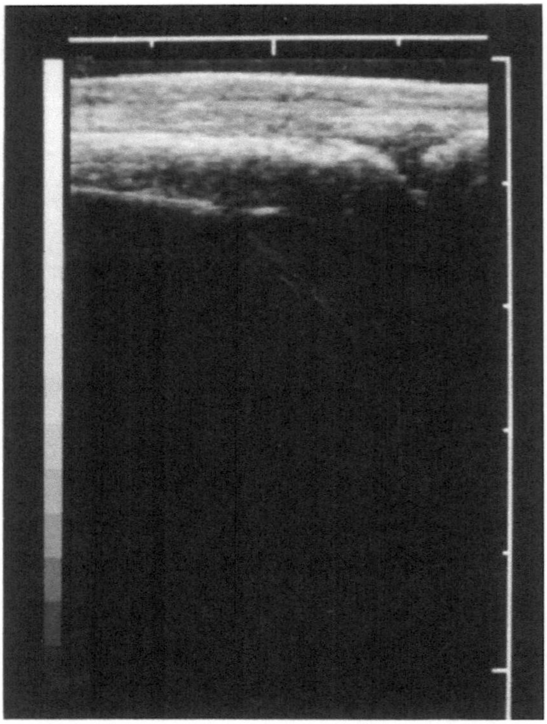

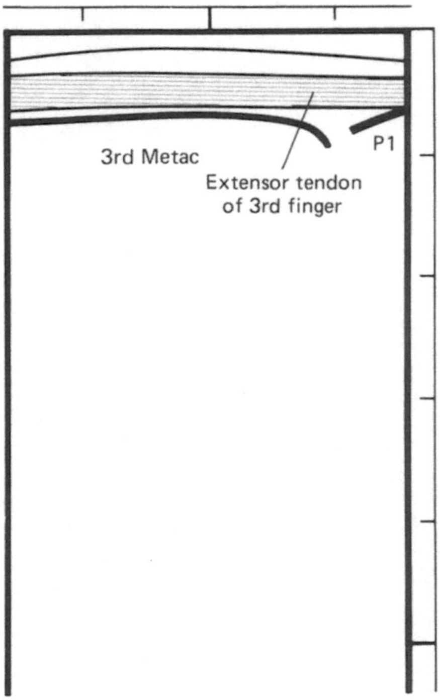
3rd Metac
P1
Extensor tendon
of 3rd finger

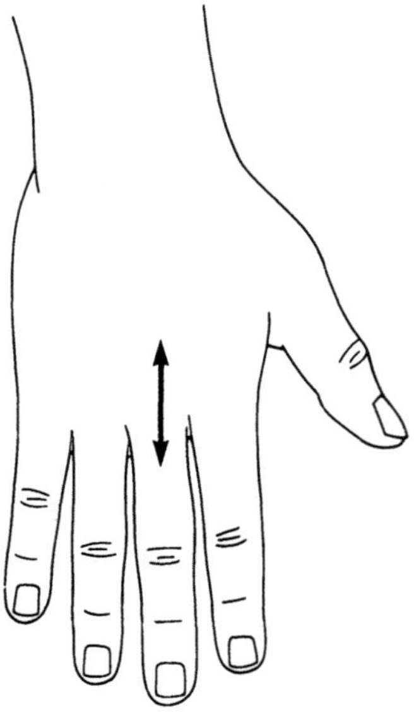

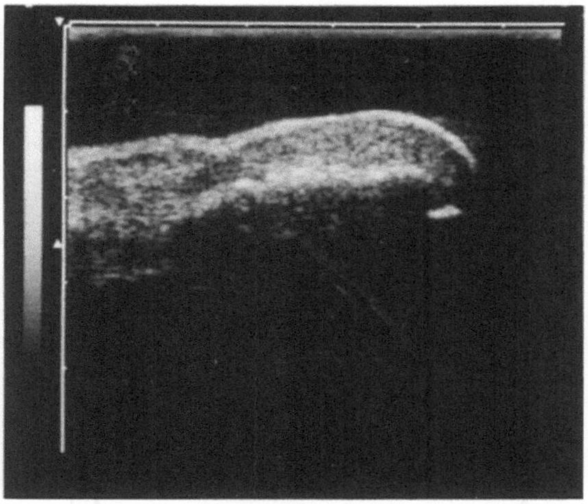

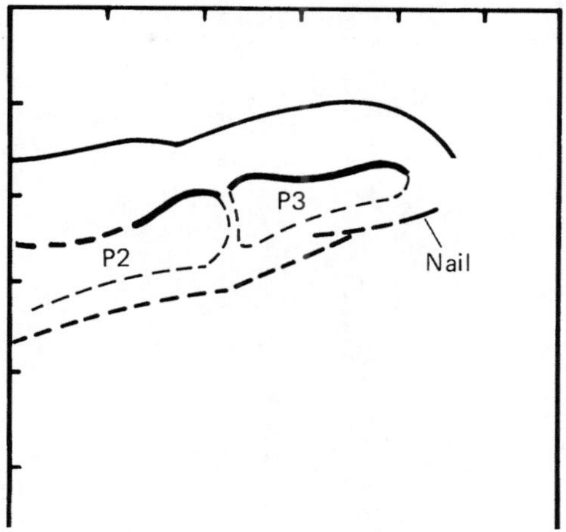
P3
P2
Nail

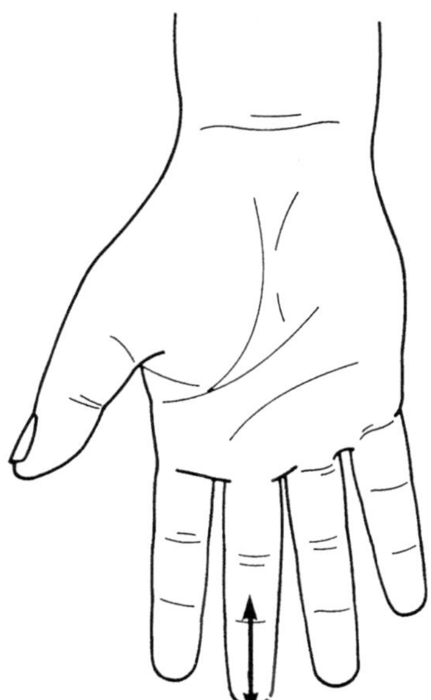

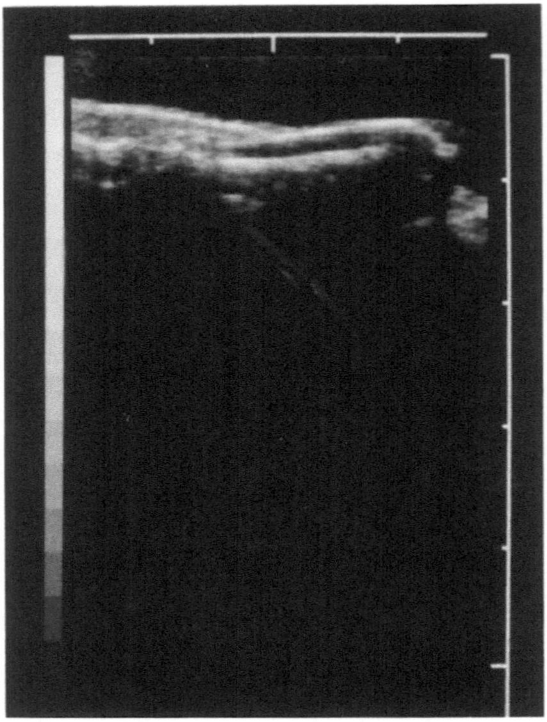

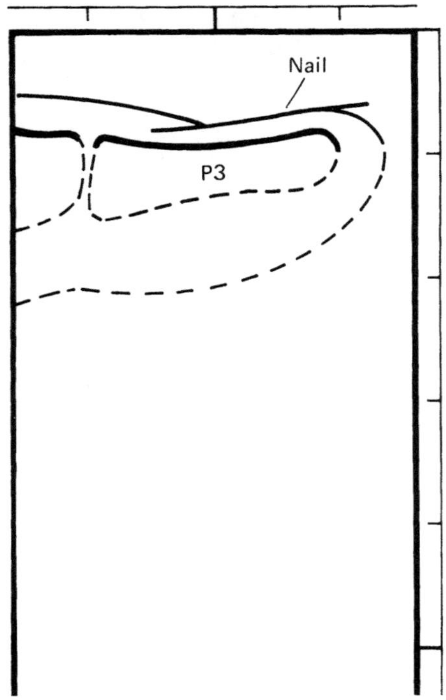
Nail
P3

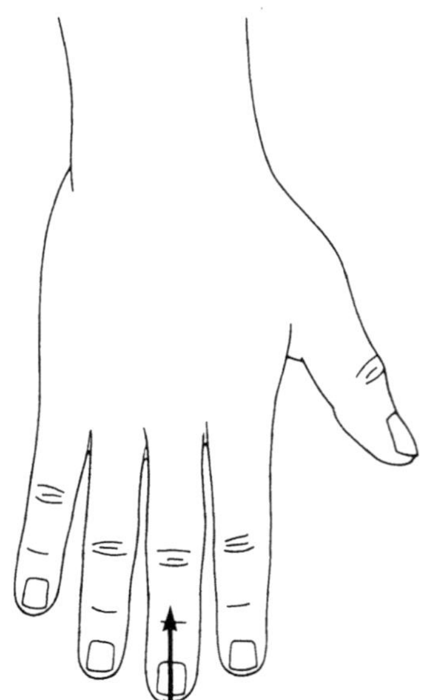

B. Lower Extremity

Thigh Upper Third

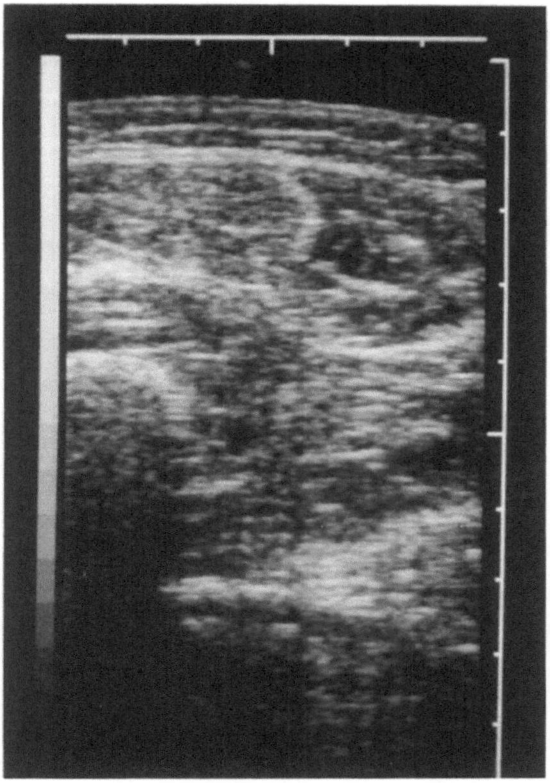

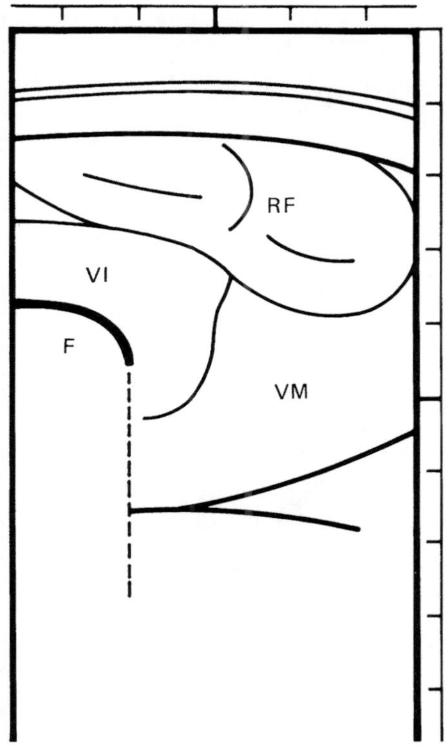

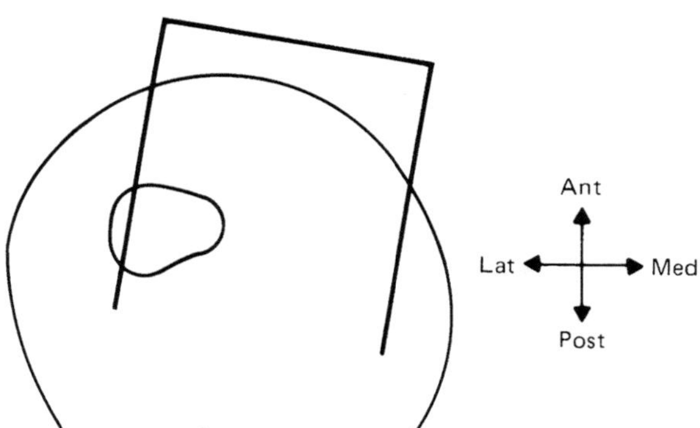

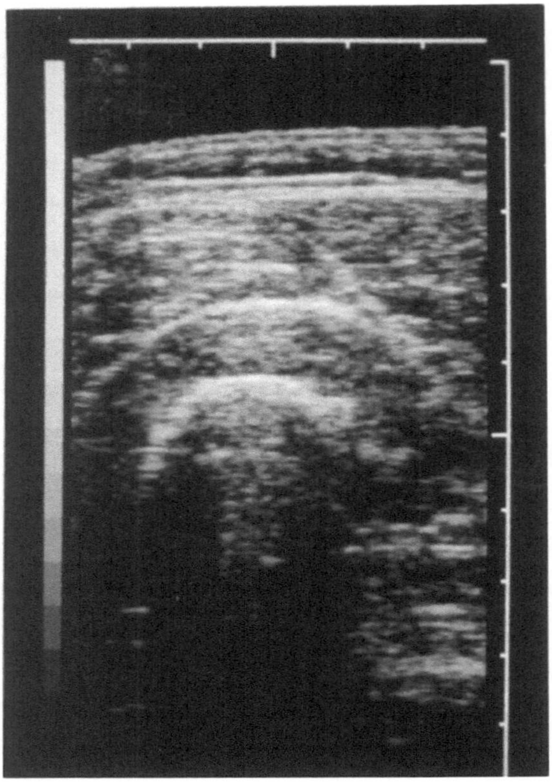

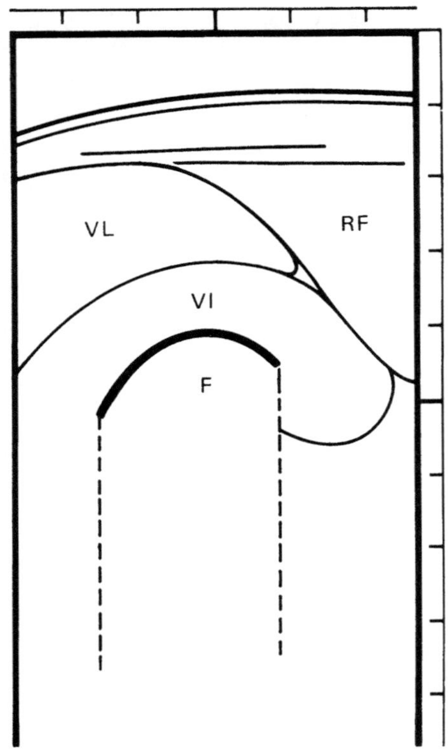
VL
RF
VI
F

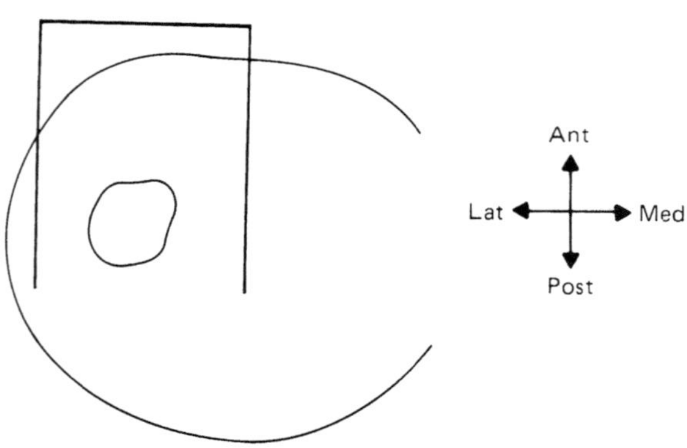
Ant
Lat
Med
Post

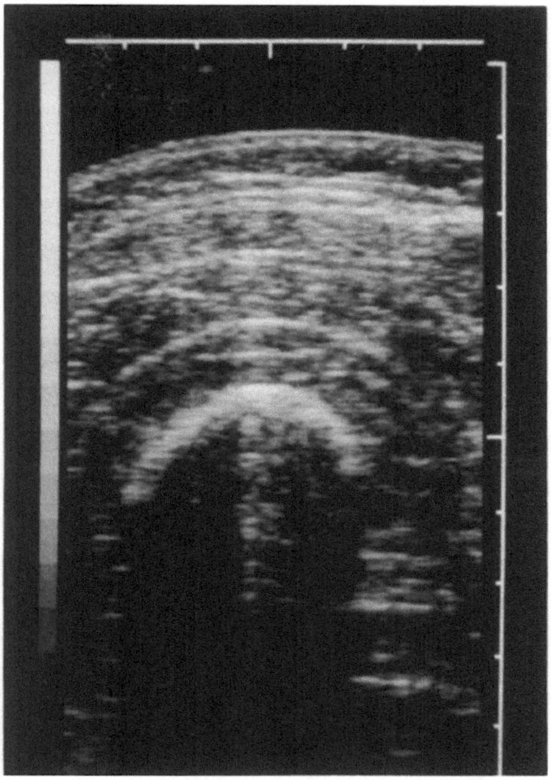

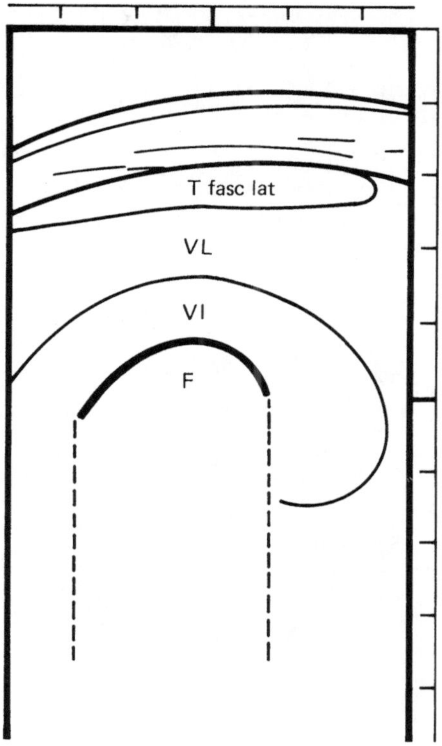
T fasc lat
VL
VI
F

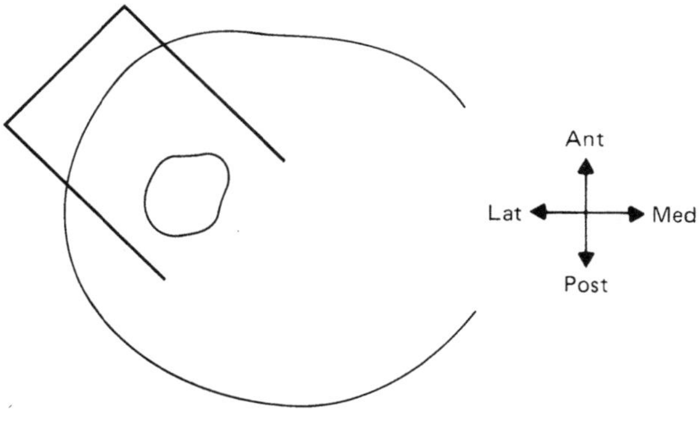
Ant
Lat
Med
Post

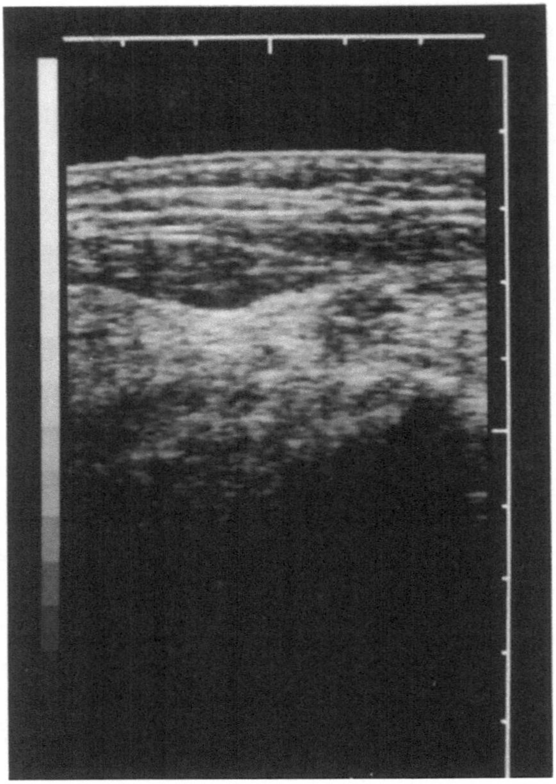

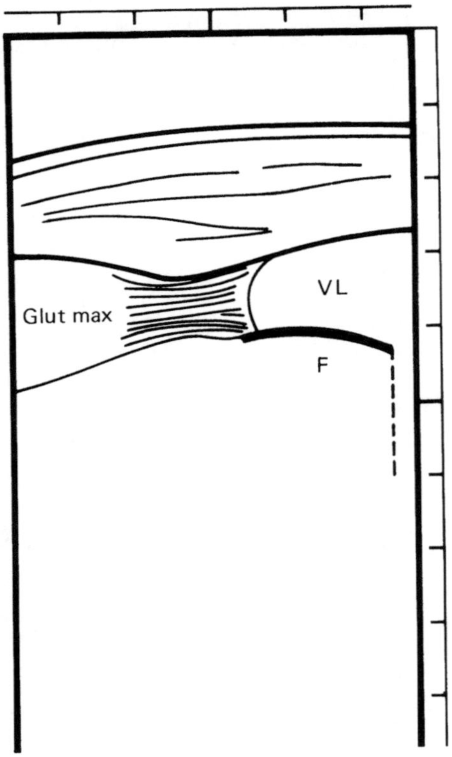
VL
Glut max
F

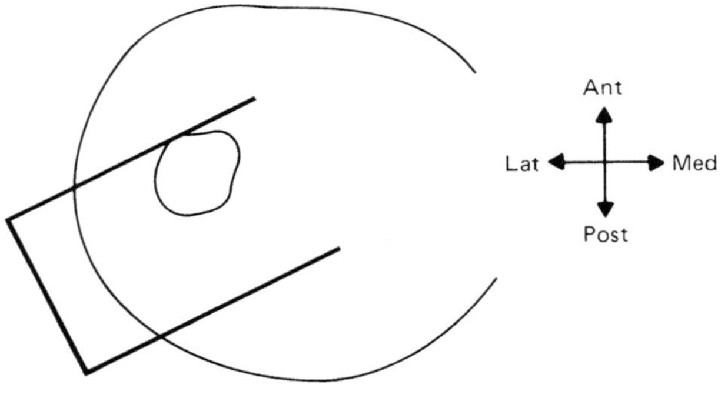
Ant
Lat
Med
Post

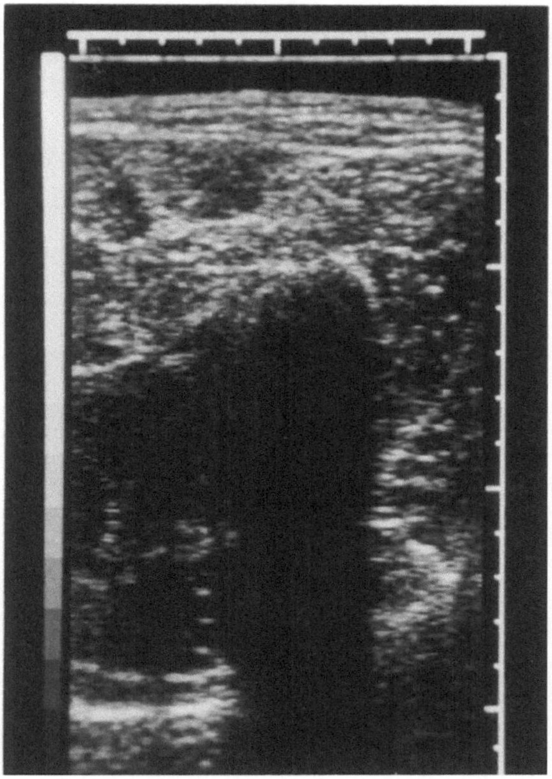

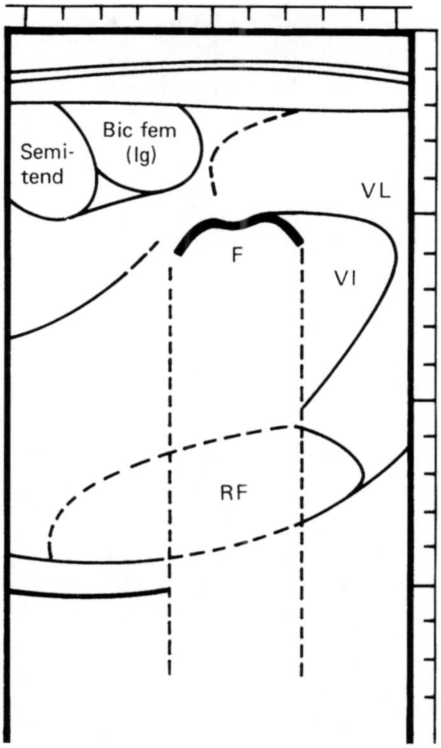
Semi-
tend
Bic fem
(lg)
VL
F
VI
RF

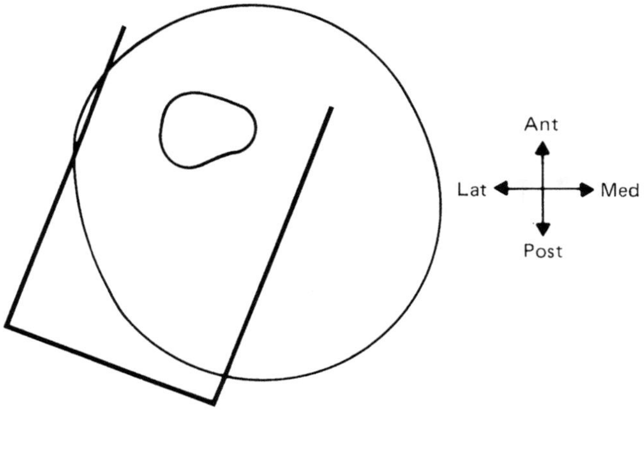
Ant
Lat
Med
Post

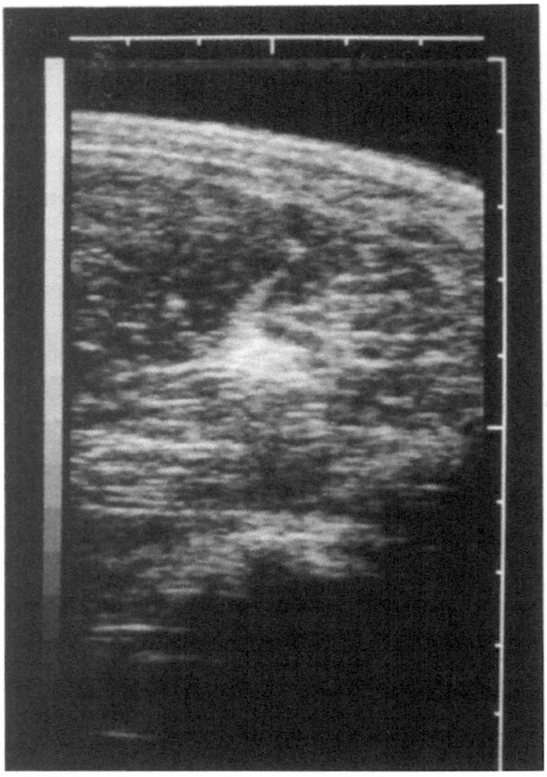

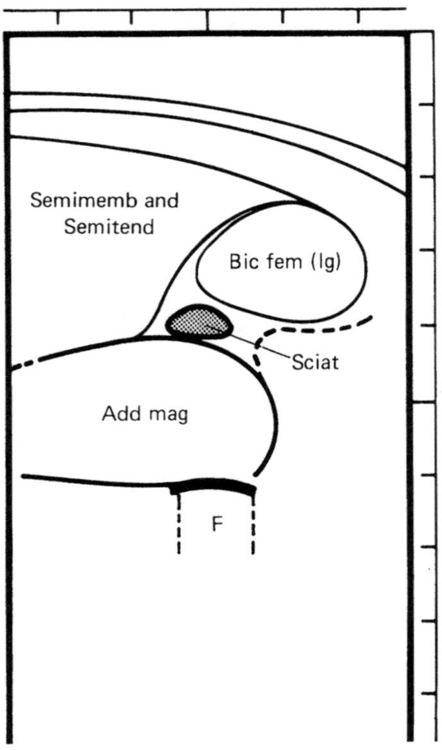
Semimemb and
Semitend
Bic fem (lg)
Sciat
Add mag
F

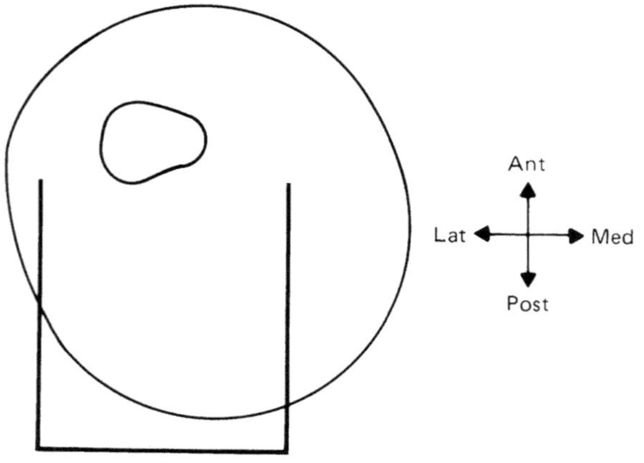
Ant
Lat
Med
Post

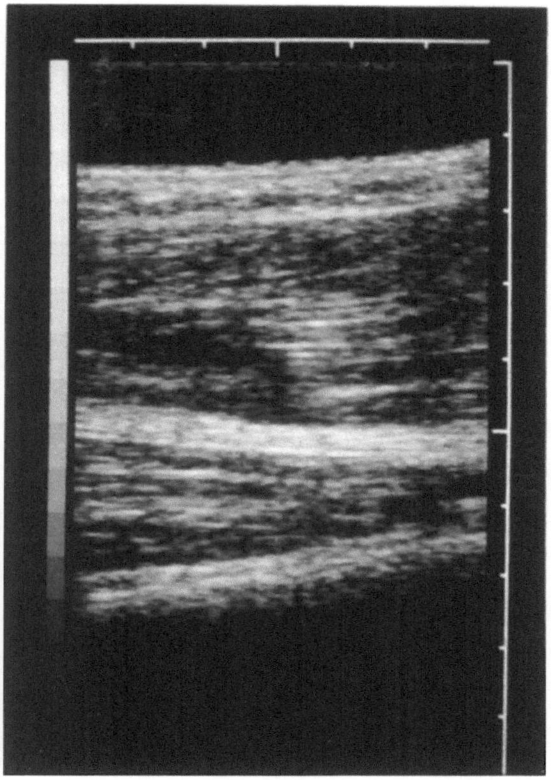

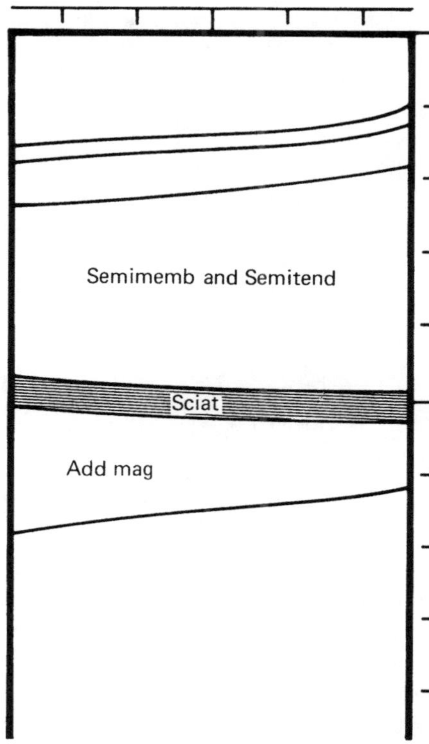
Semimemb and Semitend
Sciat
Add mag

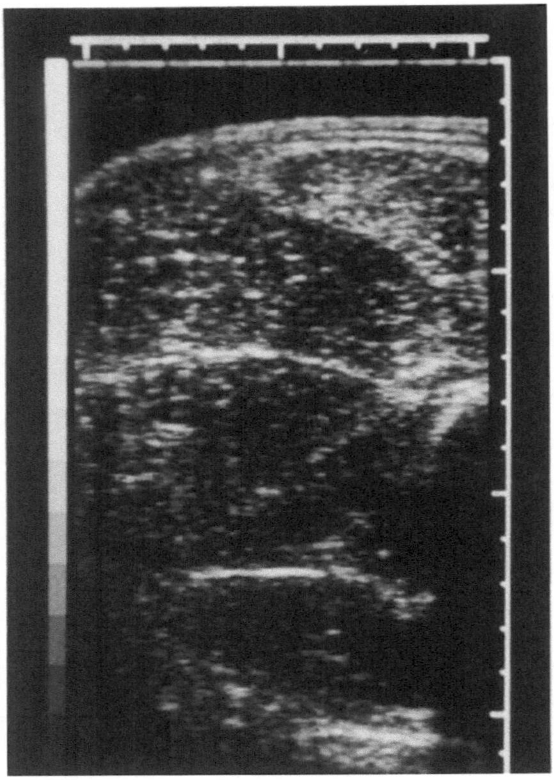

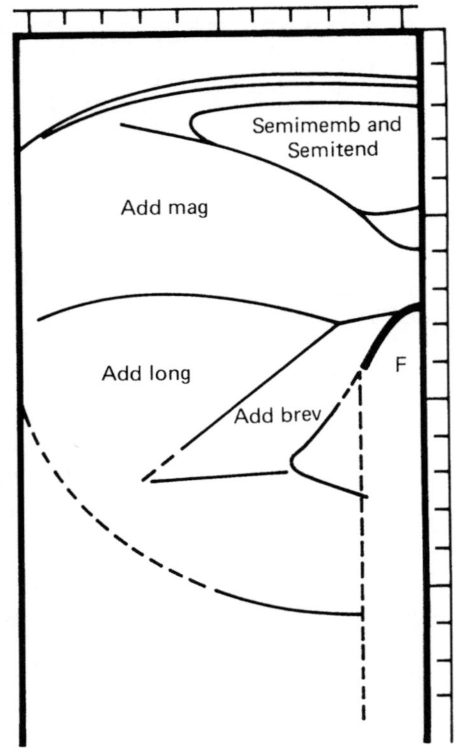
Semimemb and
Semitend
Add mag
Add long
F
Add brev

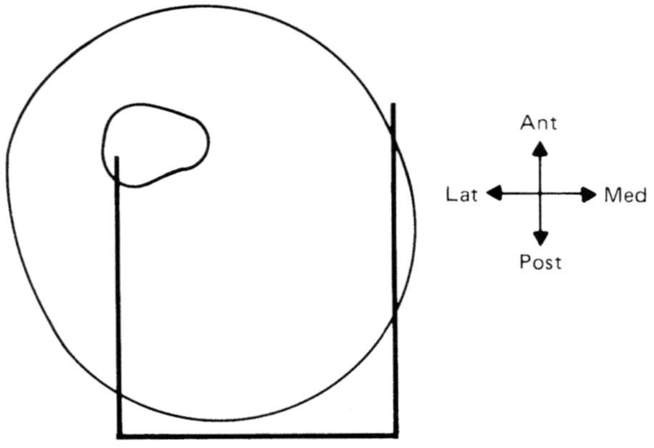
Ant
Lat
Med
Post

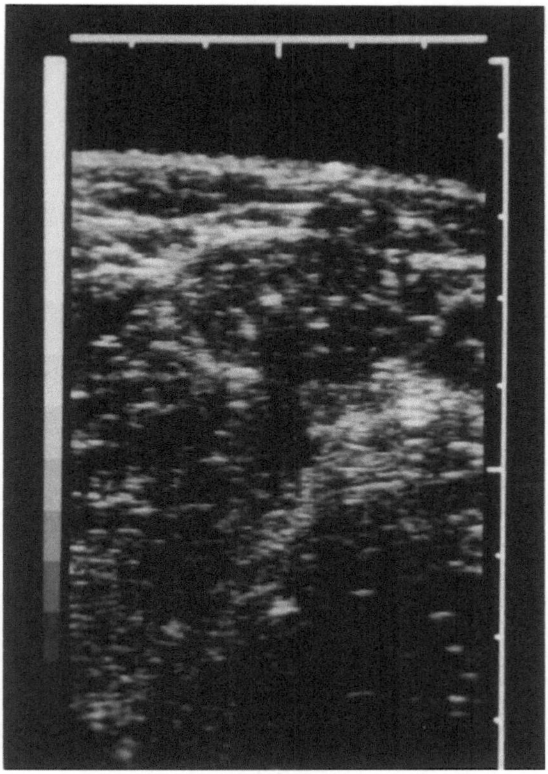

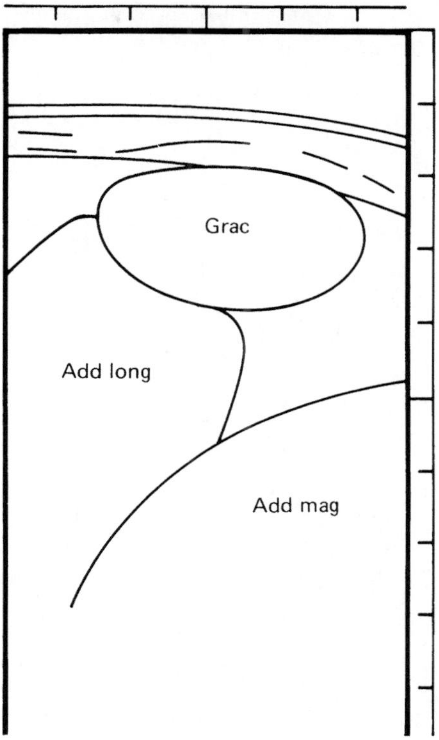
Grac
Add long
Add mag

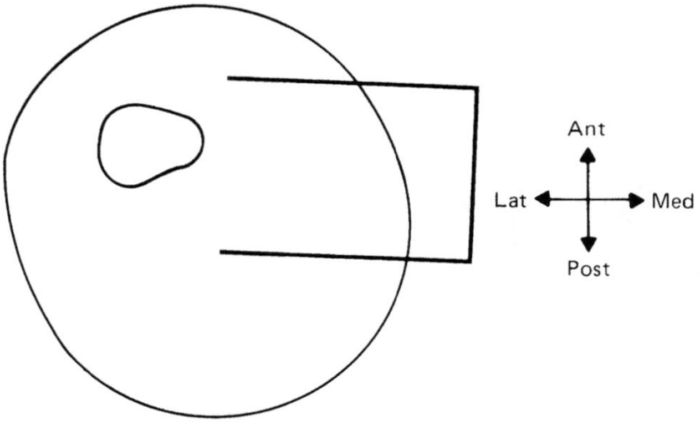
Ant
Lat
Med
Post

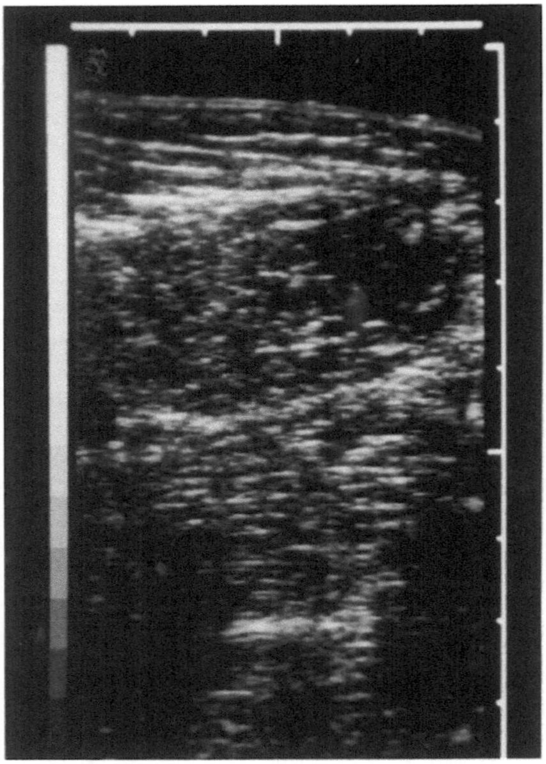

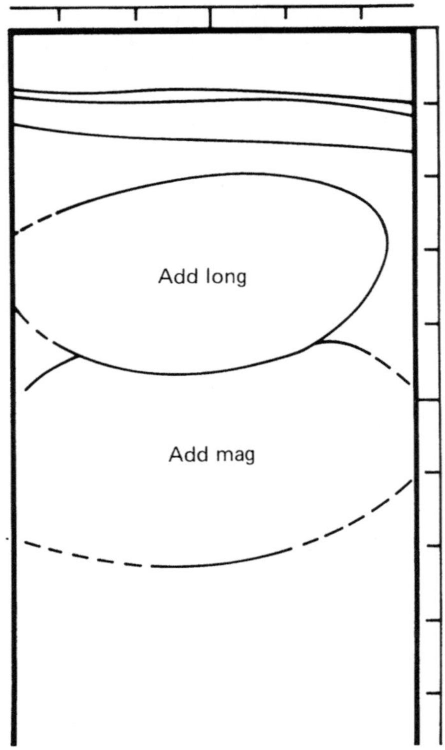
Add long
Add mag

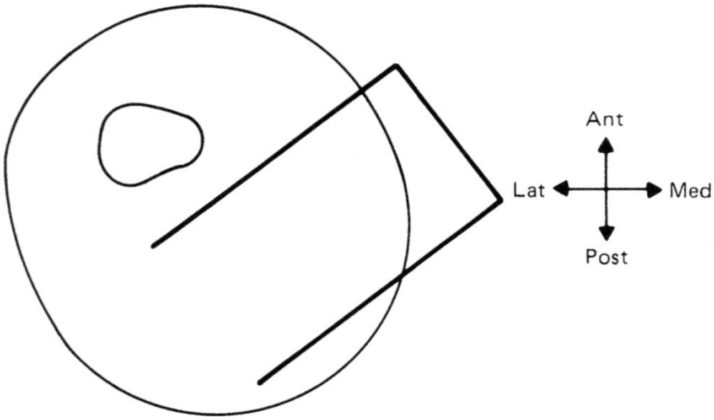
Ant
Lat
Med
Post

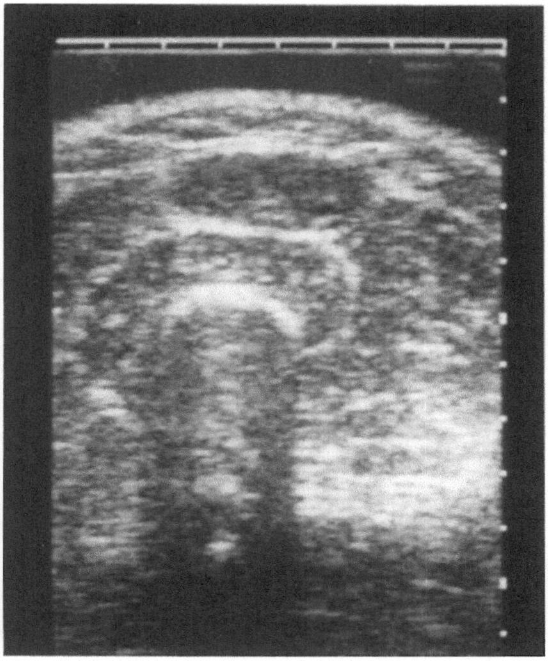

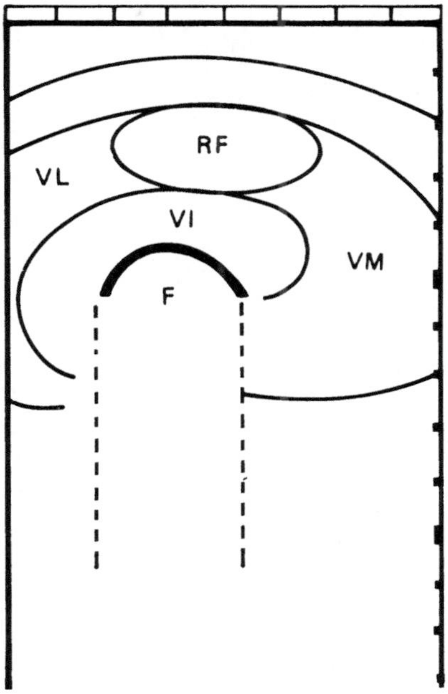
RF
VL
VI
VM
F

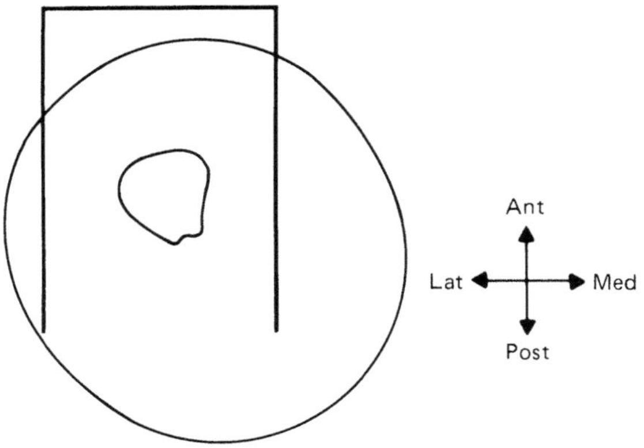
Ant
Lat
Med
Post

Thigh, Midthird

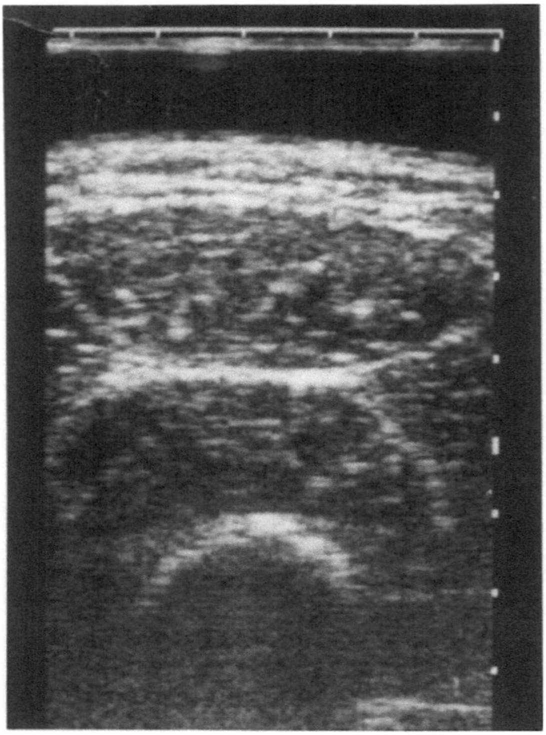

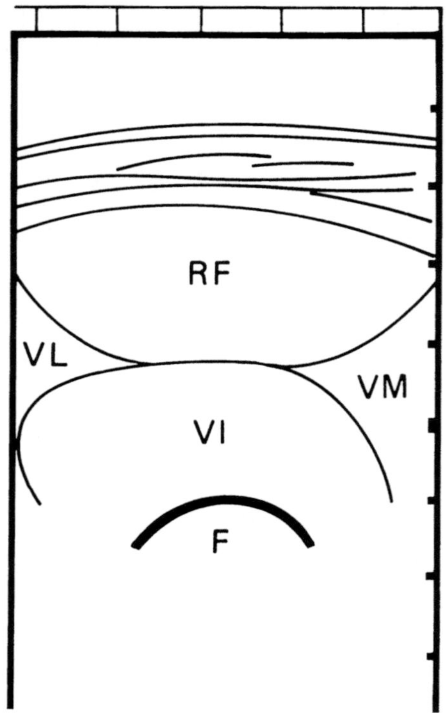

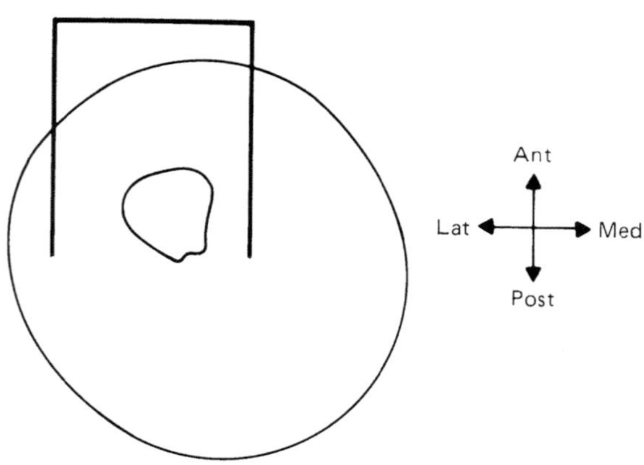

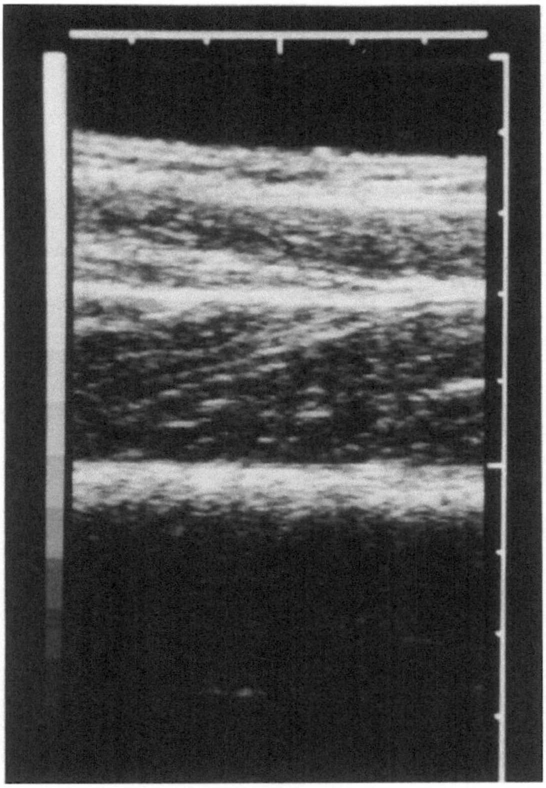

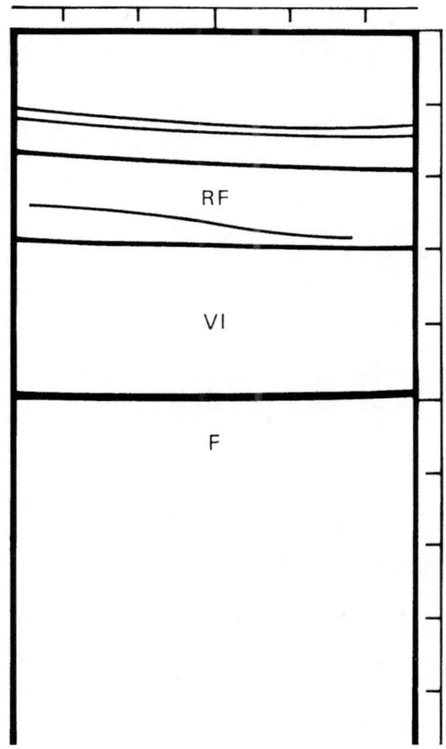
RF
VI
F

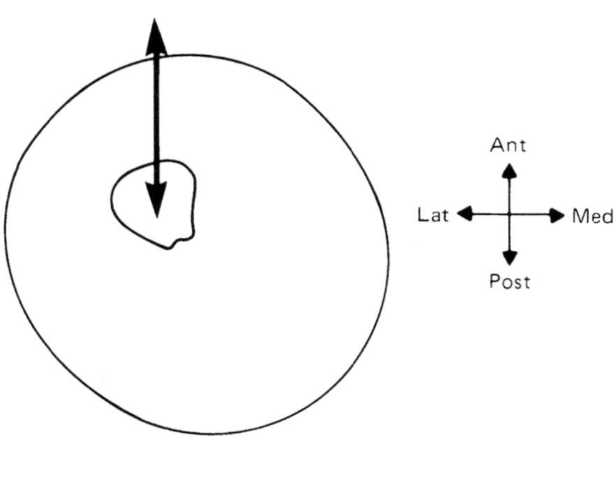
Ant
Lat
Med
Post

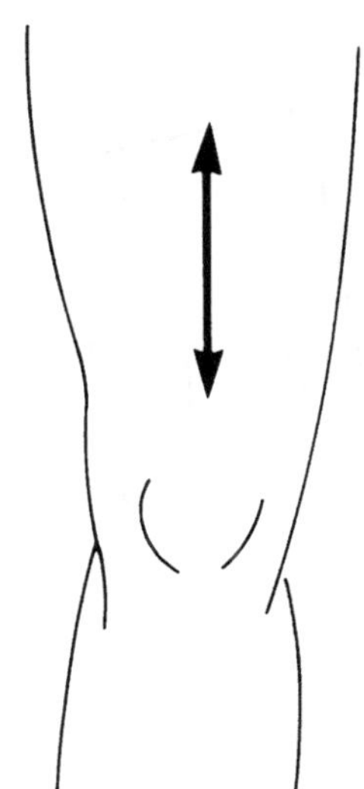

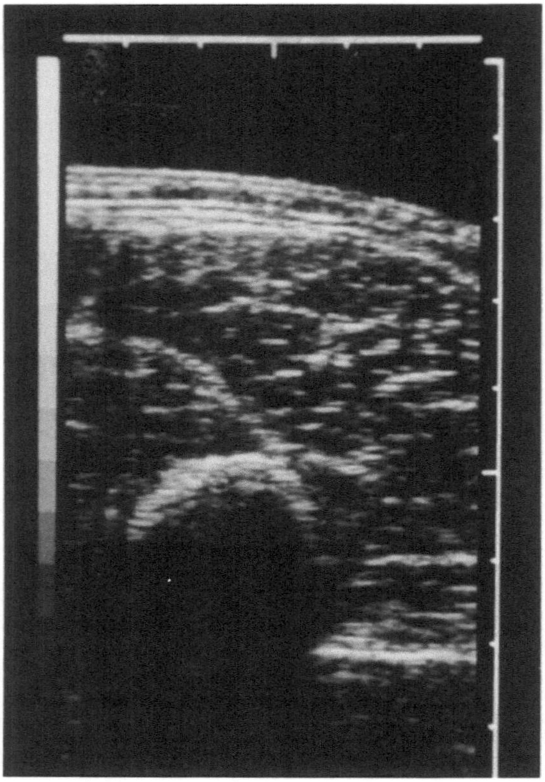

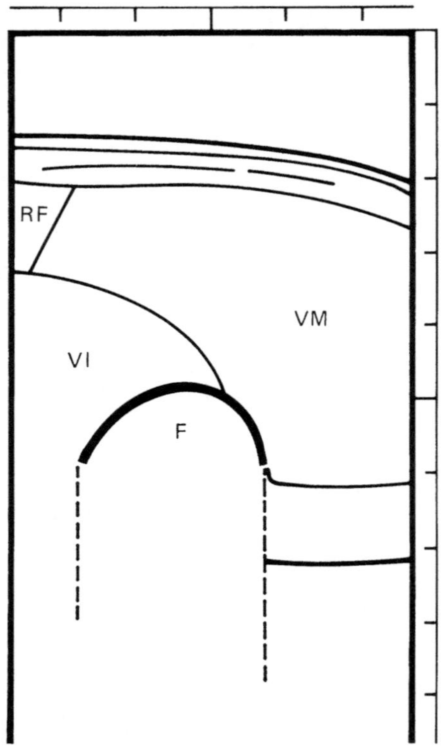
RF
VM
VI
F

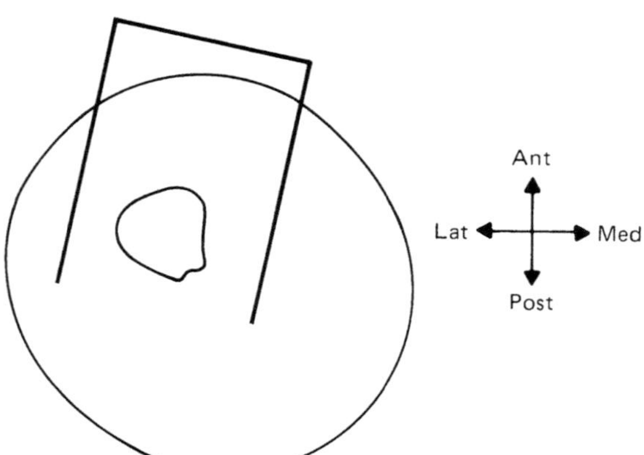
Ant
Lat
Med
Post

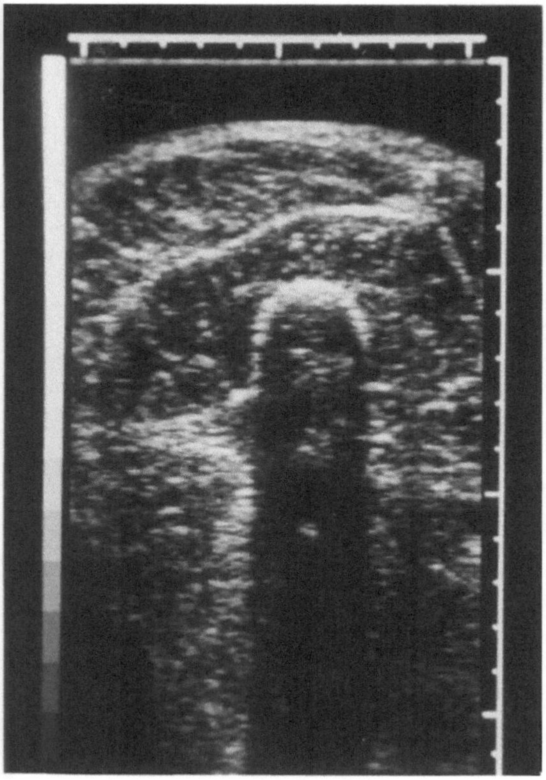

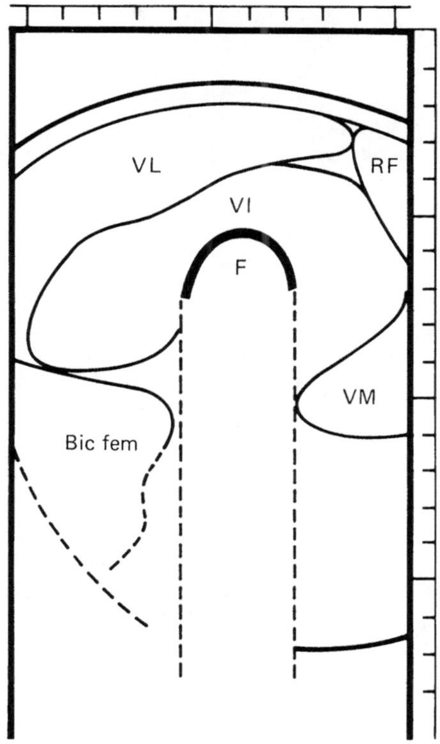
VL
RF
VI
F
VM
Bic fem

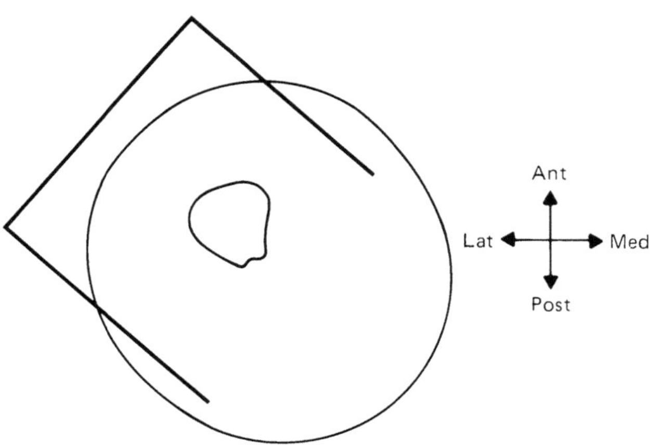
Ant
Lat
Med
Post

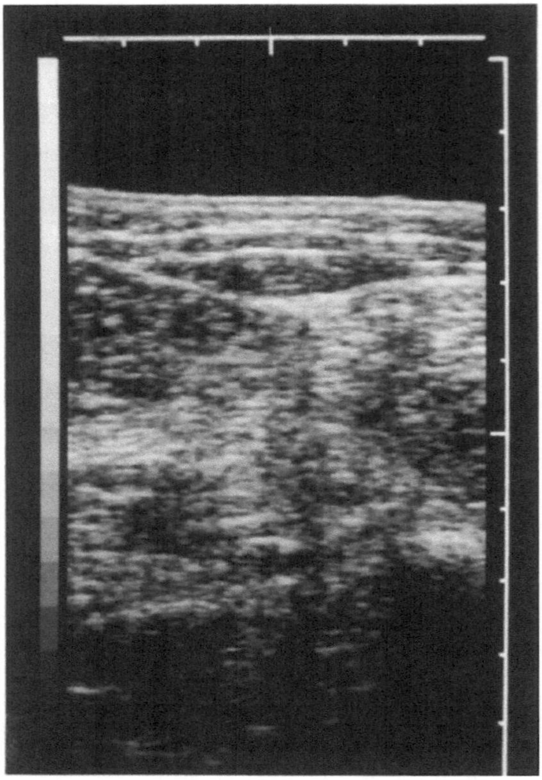

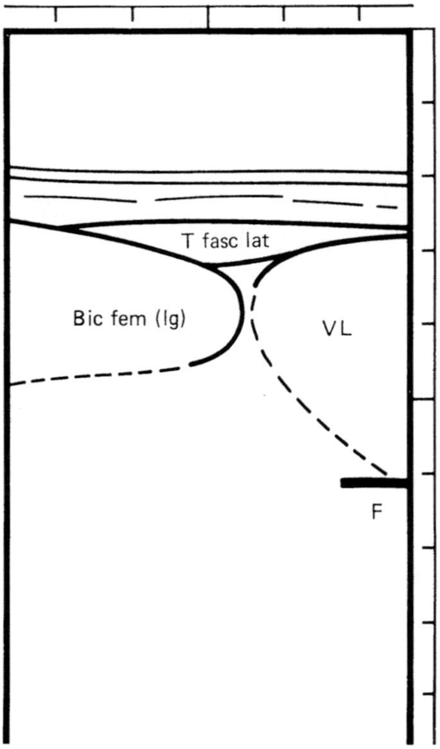
T fasc lat
Bic fem (lg)
VL
F

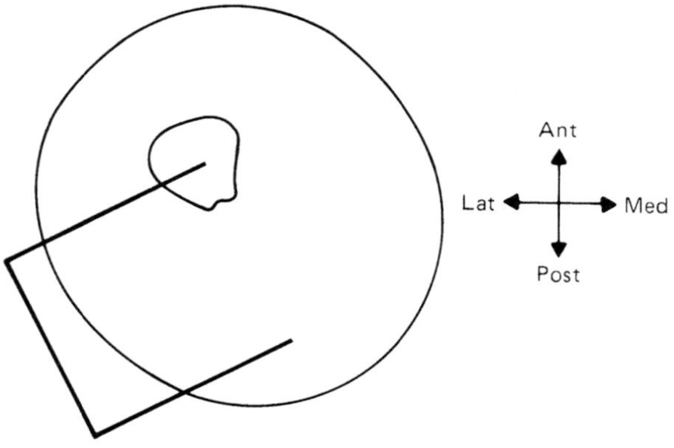
Ant
Lat
Med
Post

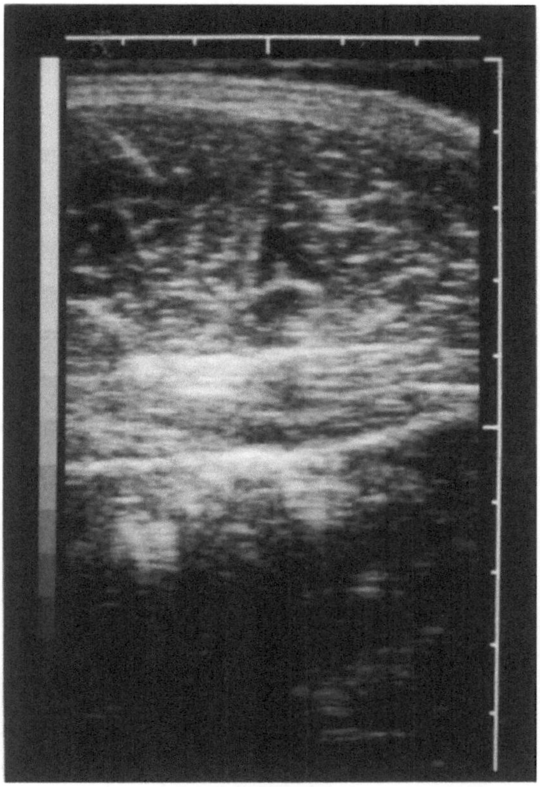

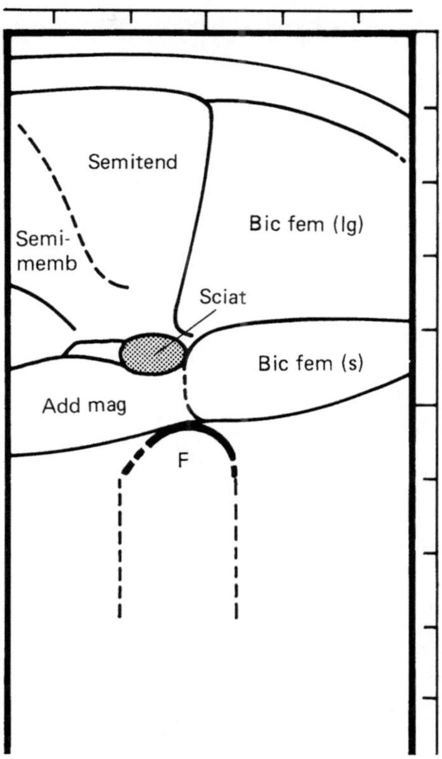

(Reprinted with permission from Fornage BD: Peripheral nerves of the extremities: Imaging with US. Radiology, 1988;167:179–182.)

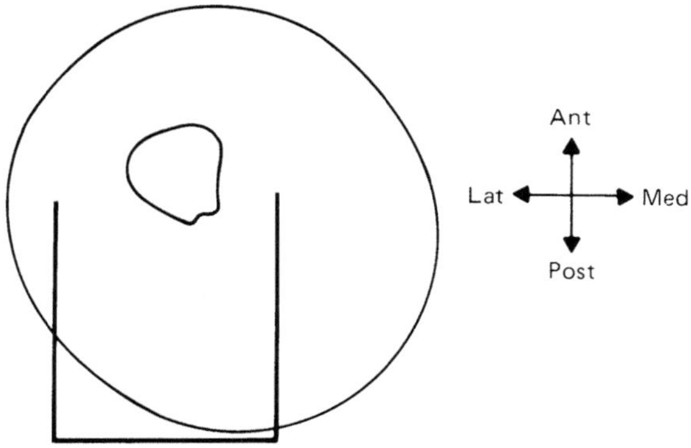

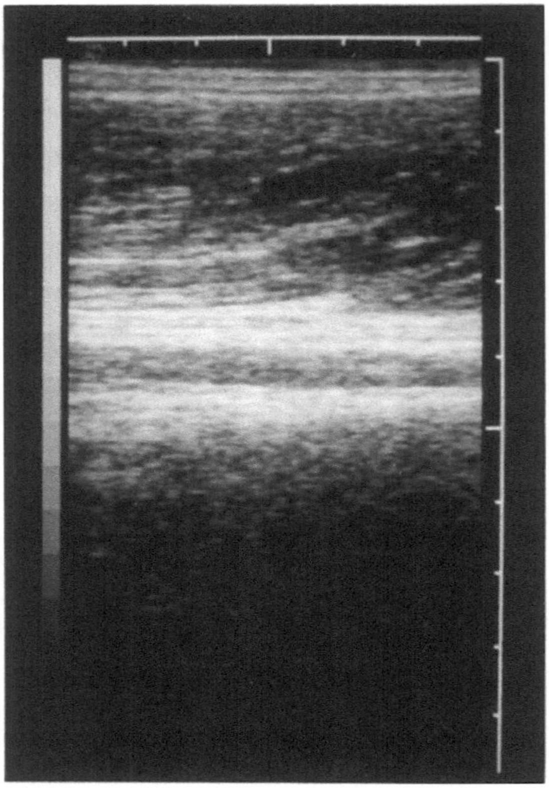

(Reprinted with permission from Fornage BD: Peripheral nerves of the extremities: Imaging with US. Radiology, 1988;167:179–182.)

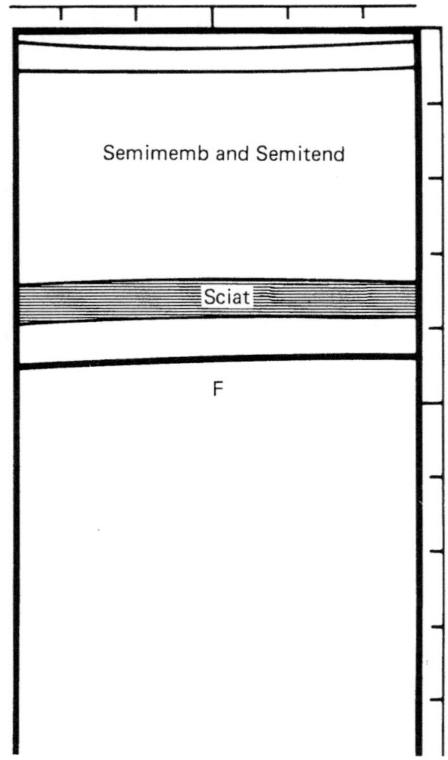

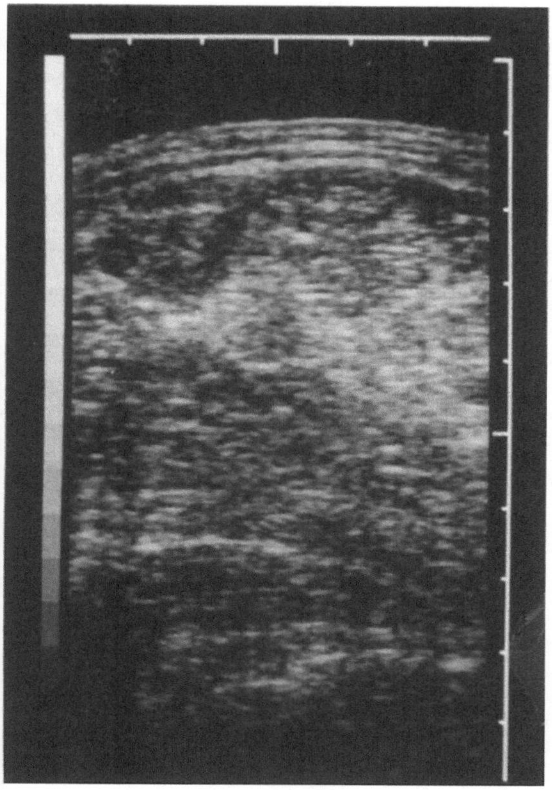

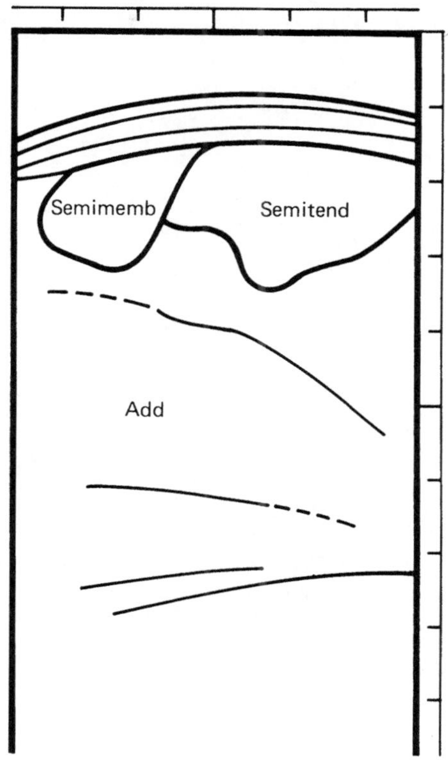
Semimemb
Semitend
Add

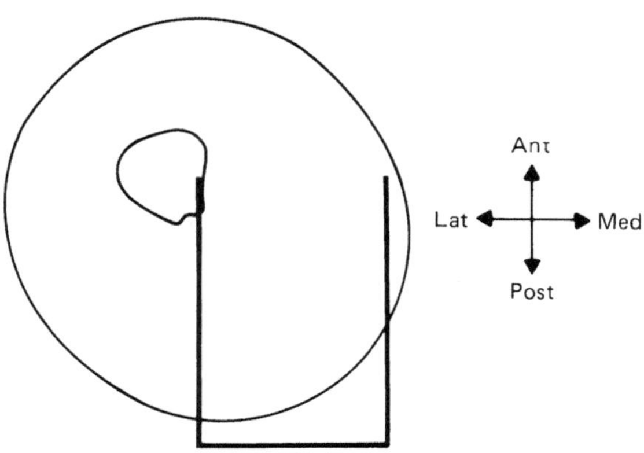
Ant
Lat
Med
Post

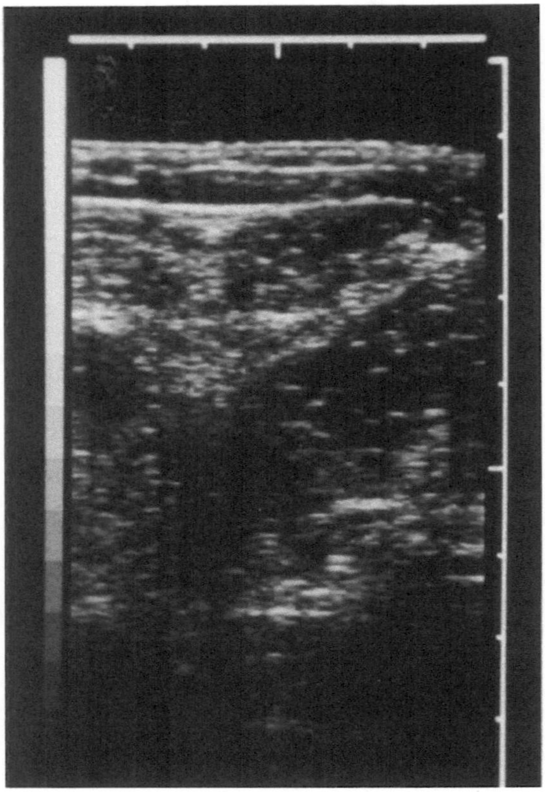

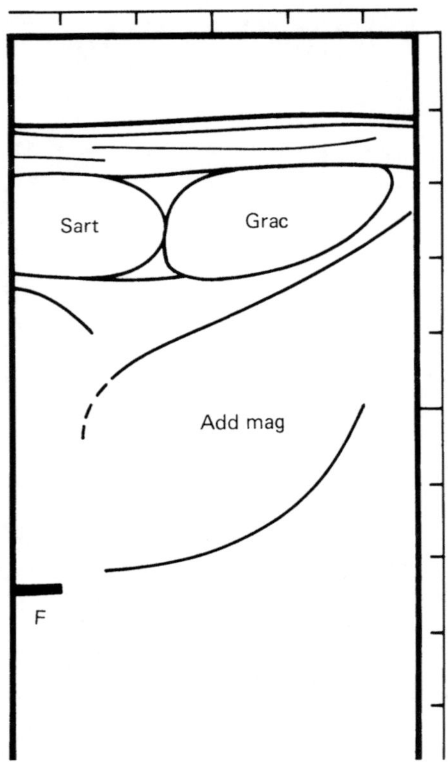
Sart
Grac
Add mag
F

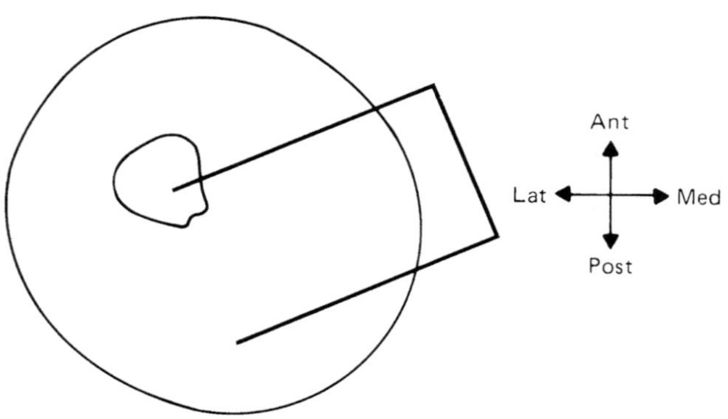
Ant
Lat
Med
Post

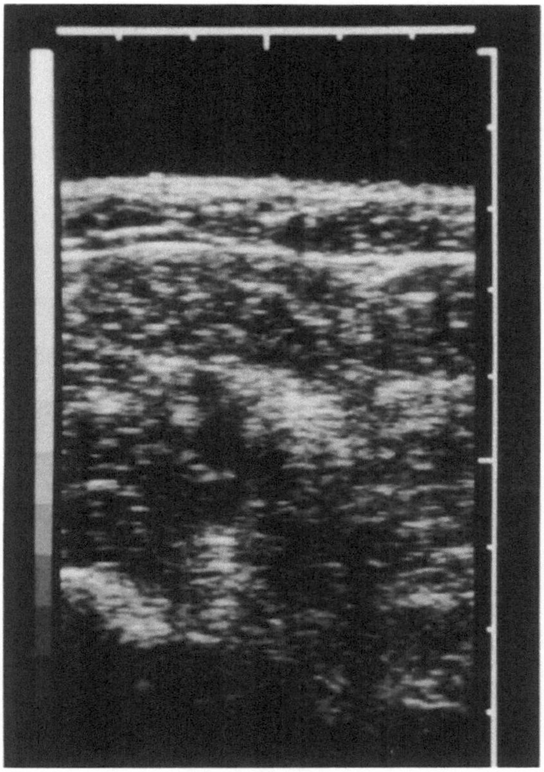

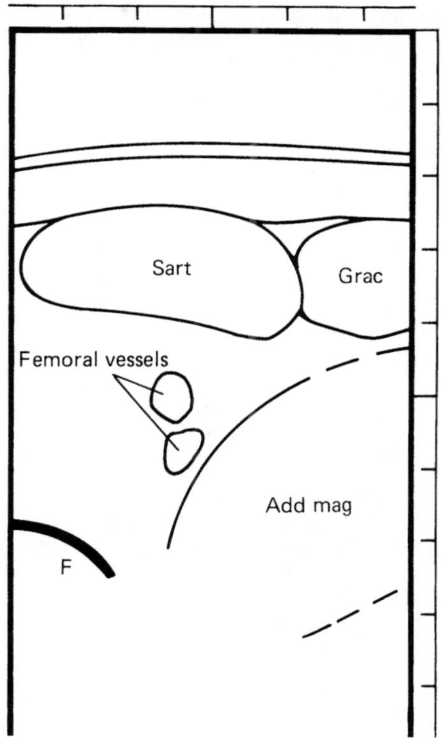
Sart
Grac
Femoral vessels
Add mag
F

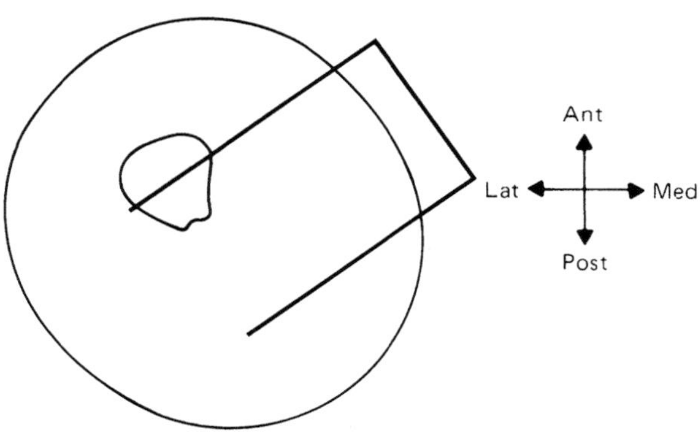
Ant
Lat
Med
Post

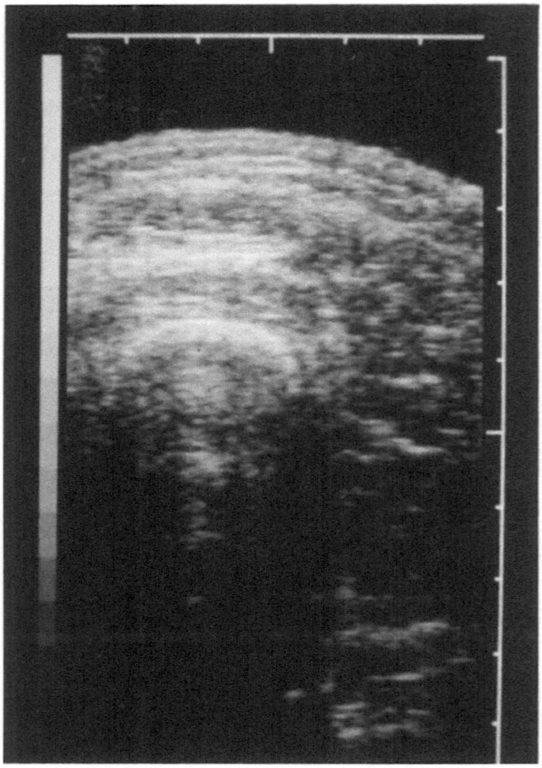

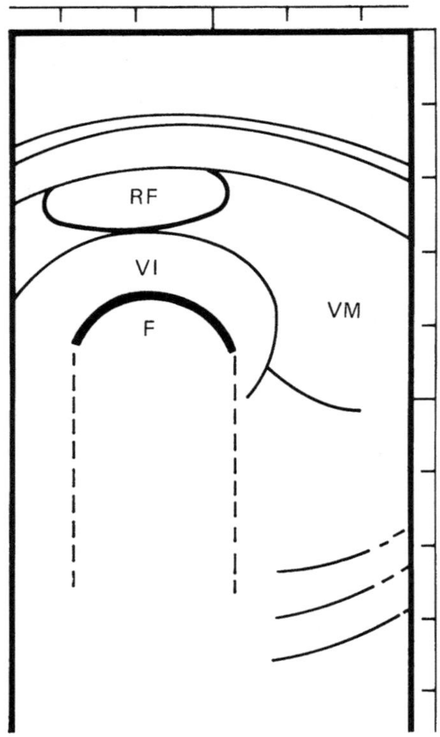
RF
VI
VM
F

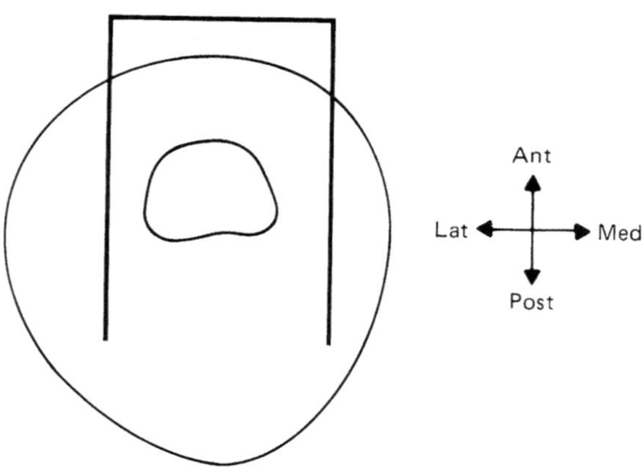
Ant
Lat
Med
Post

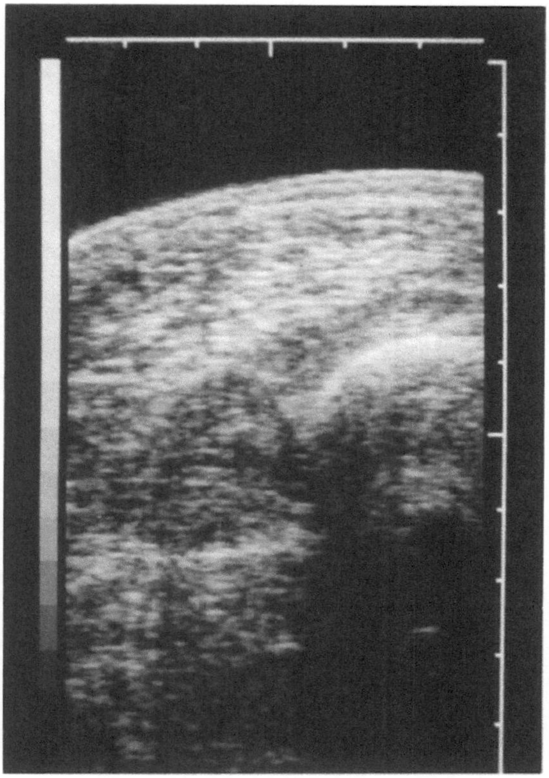

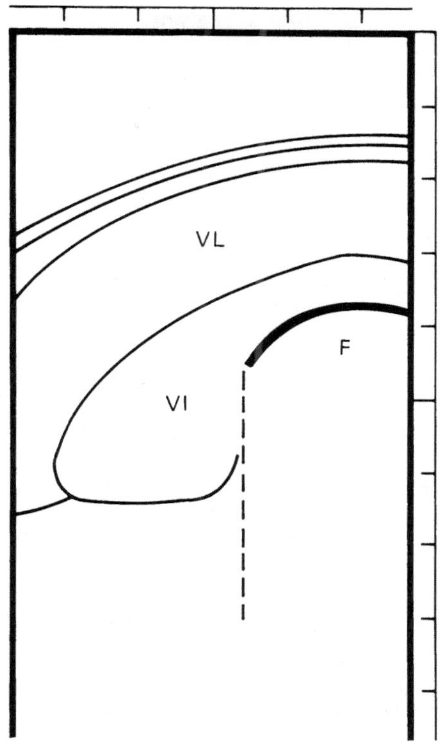
VL
F
VI

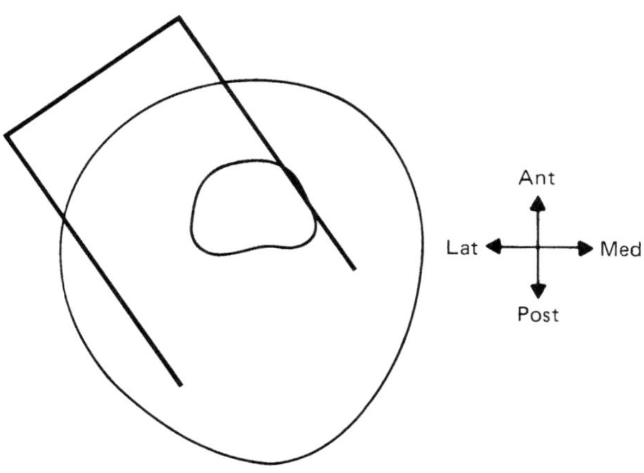
Ant
Lat
Med
Post

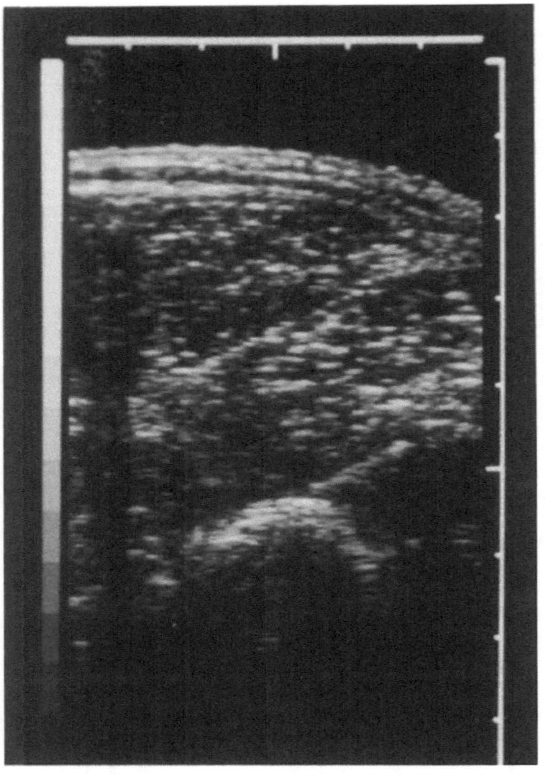

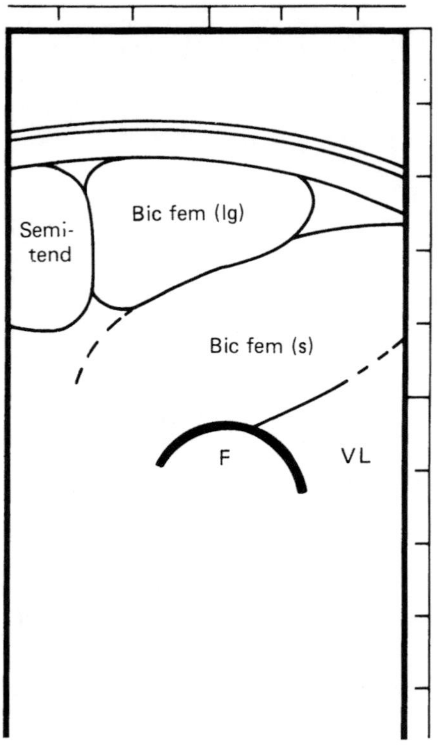
Semi-
tend
Bic fem (lg)
Bic fem (s)
F
VL

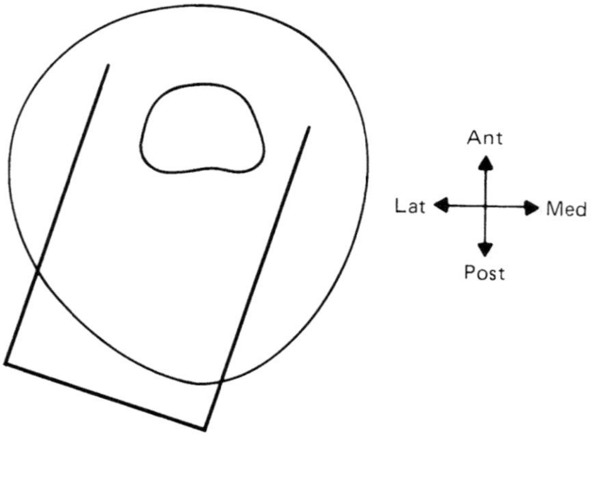
Ant
Lat
Med
Post

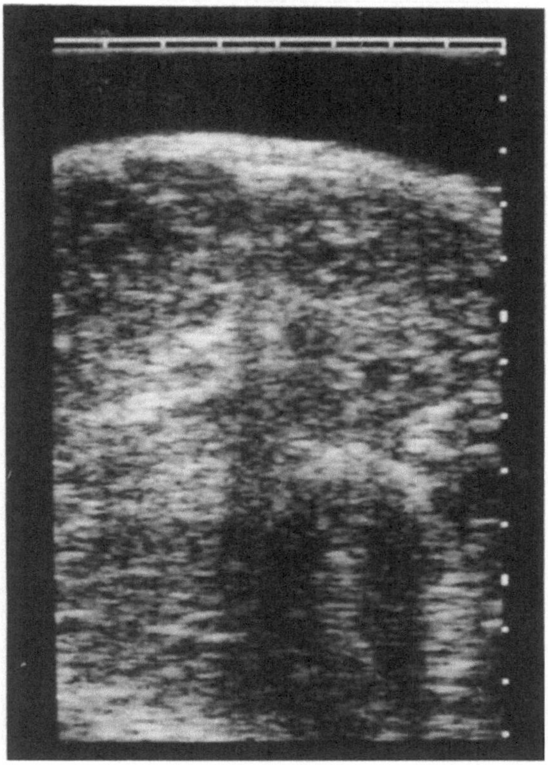

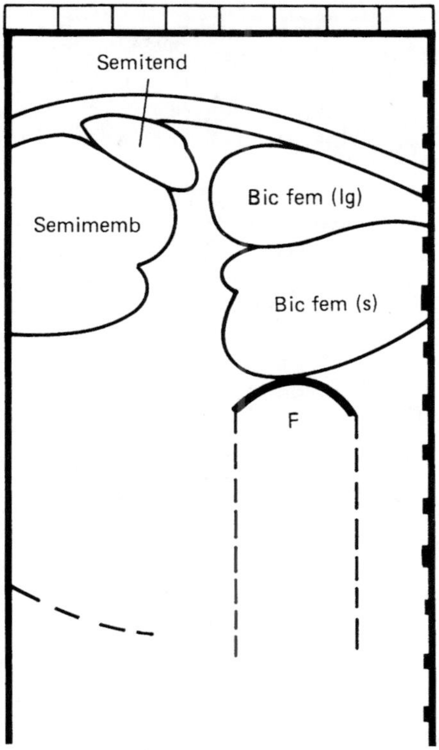
Semitend
Bic fem (lg)
Semimemb
Bic fem (s)
F

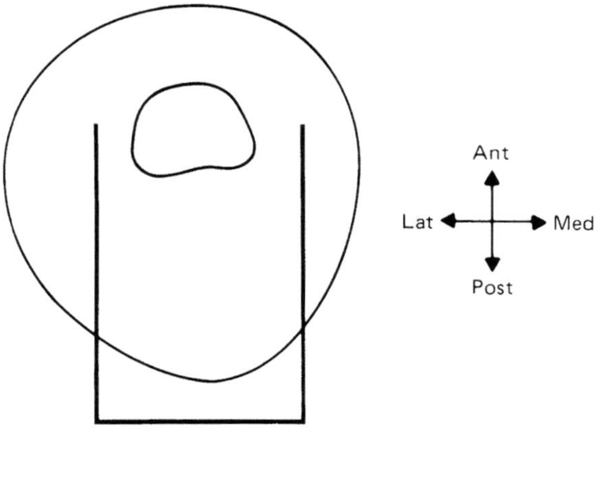
Ant
Lat
Med
Post

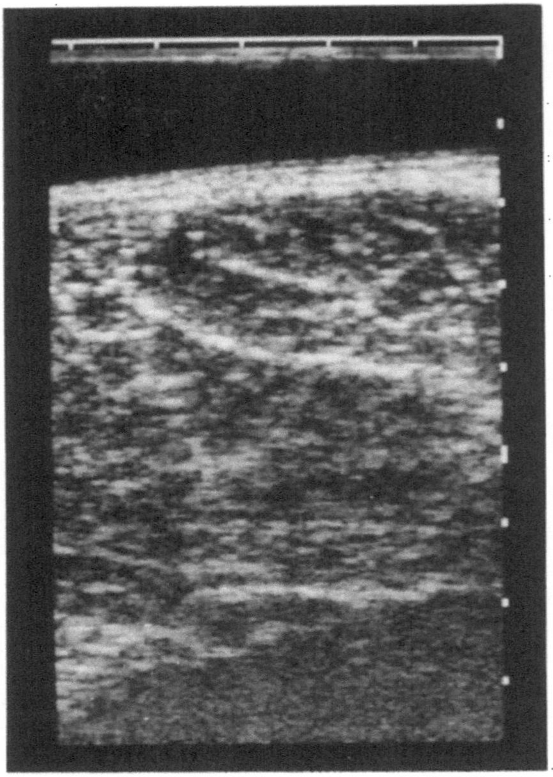

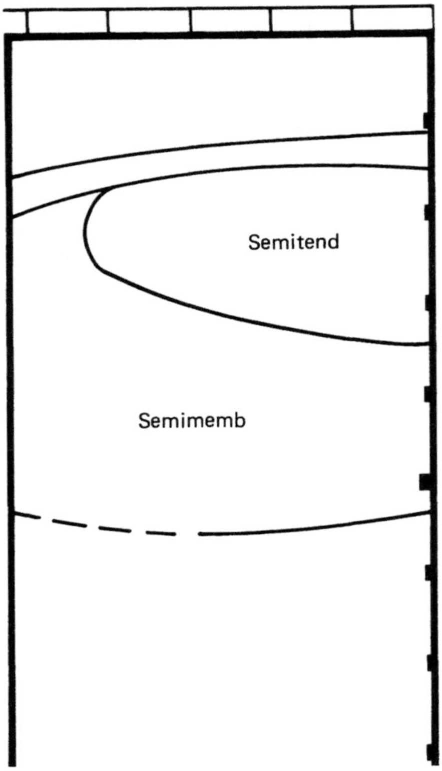
Semitend
Semimemb

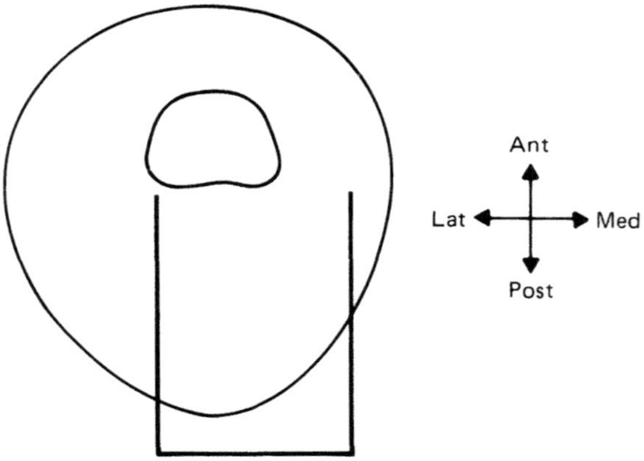
Ant
Lat
Med
Post

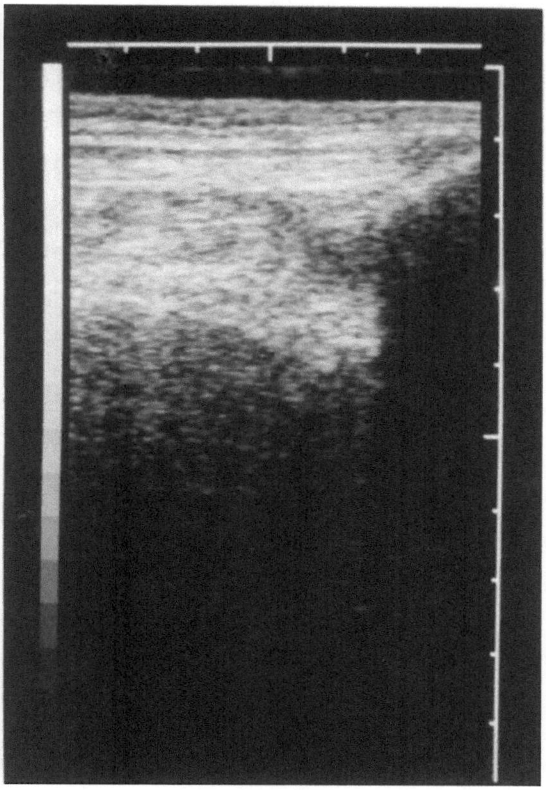

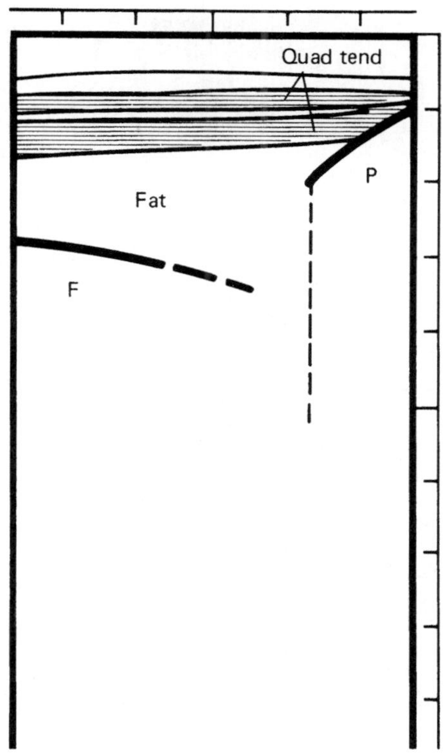
Quad tend
P
Fat
F

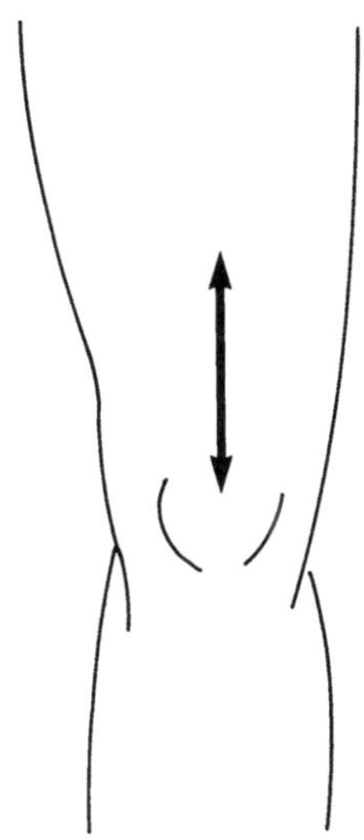

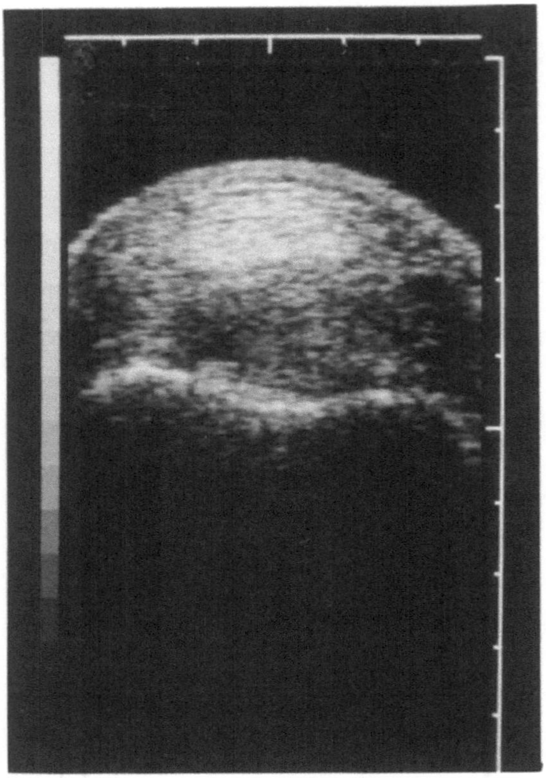

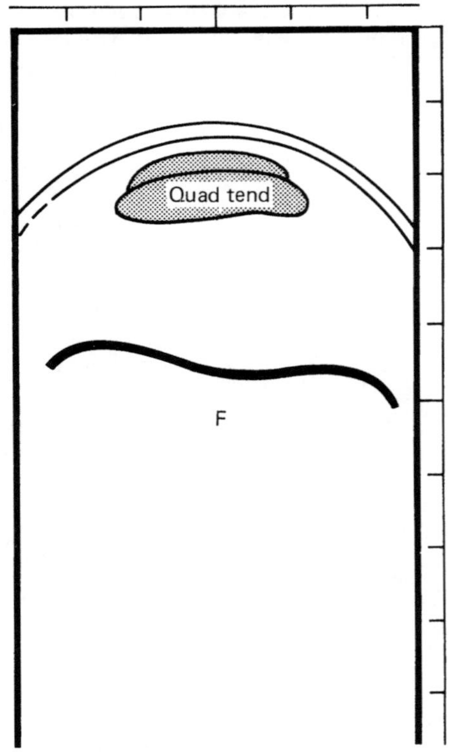

(Reprinted with permission from Fornage BD, Rifkin MD: Ultrasound examination of tendons. Radiol. Clin. North Am., 1988;26(1):87–107.)

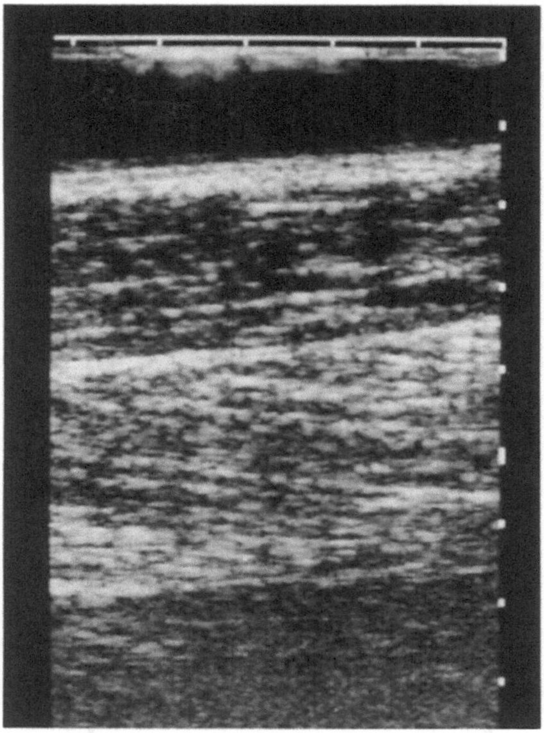

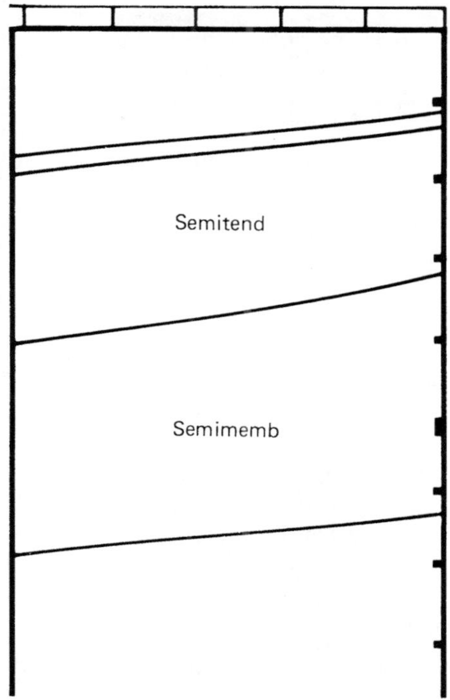
Semitend
Semimemb

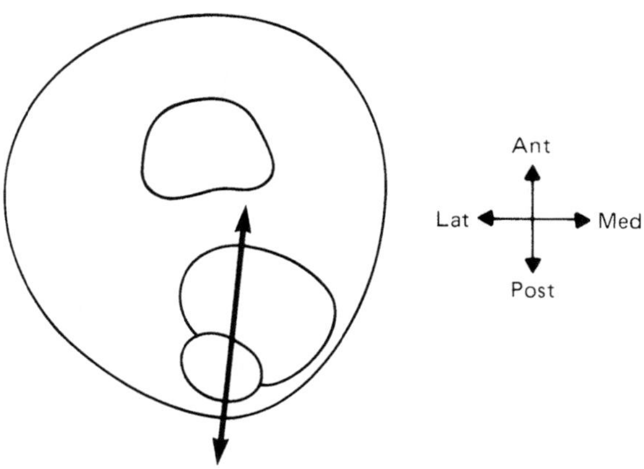
Ant
Lat
Med
Post

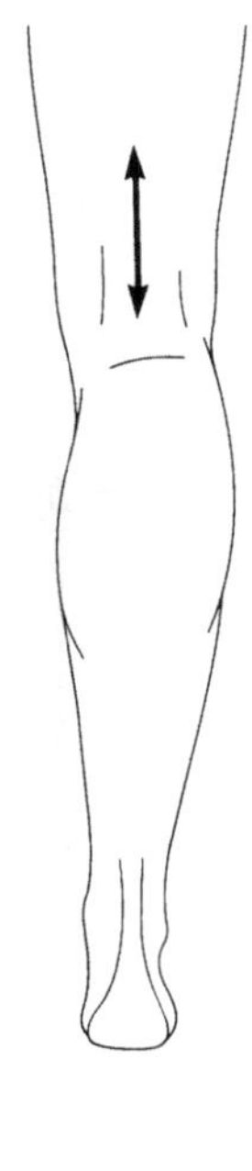

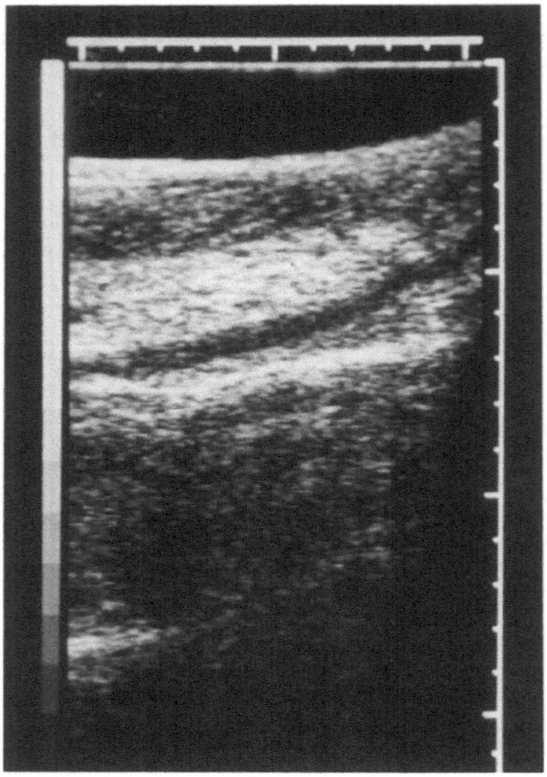

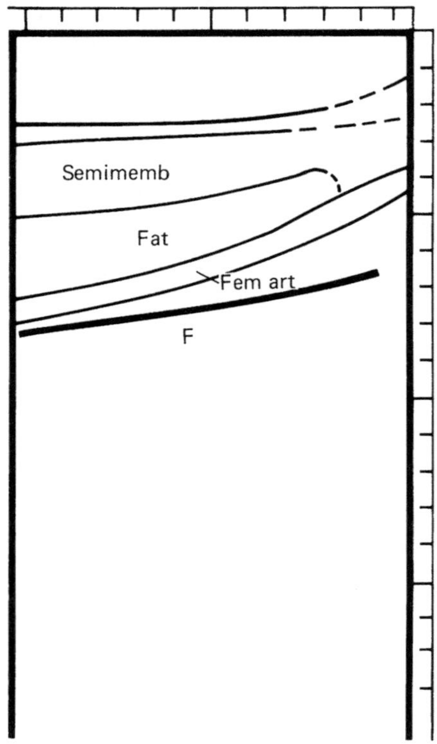
Semimemb
Fat
Fem art
F

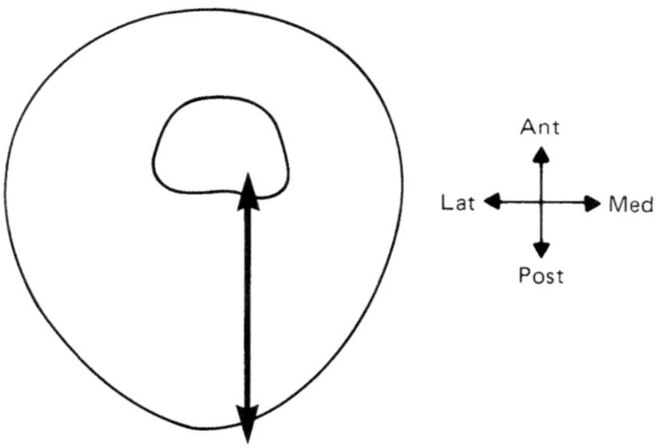
Ant
Lat
Med
Post

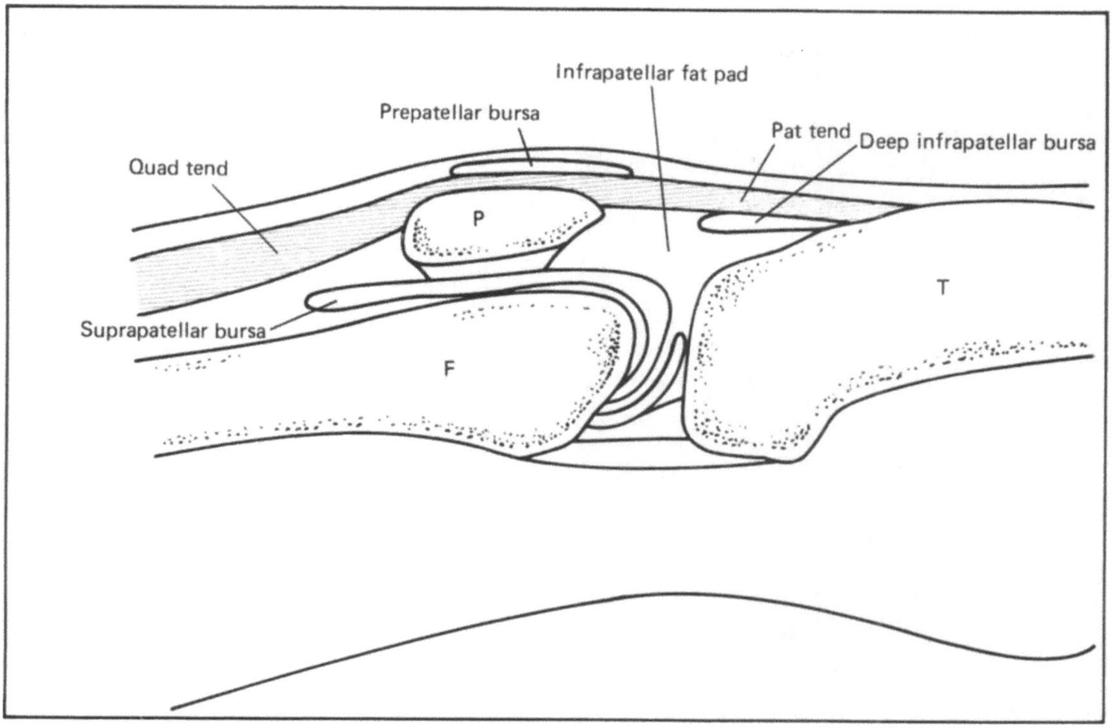
Infrapatellar fat pad
Prepatellar bursa
Pat tend
Deep infrapatellar bursa
Quad tend
P
T
Suprapatellar bursa
F

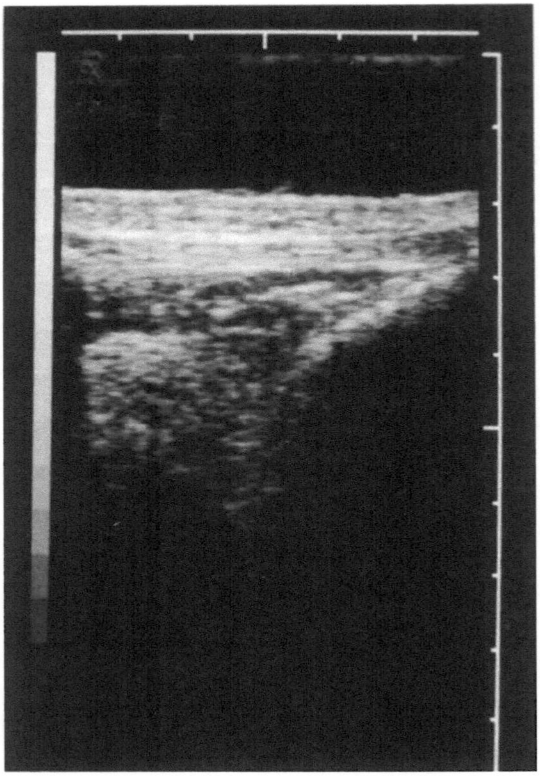

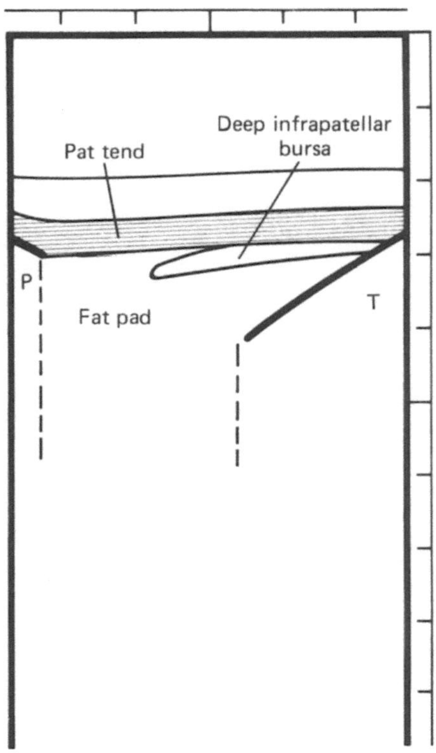
Deep infrapatellar
bursa
Pat tend
P
Fat pad
T

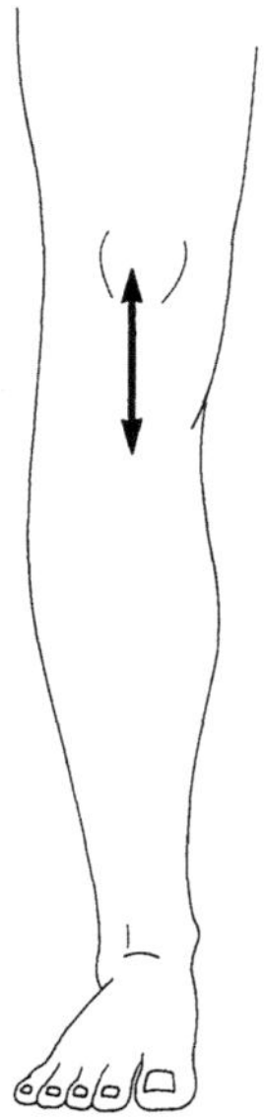

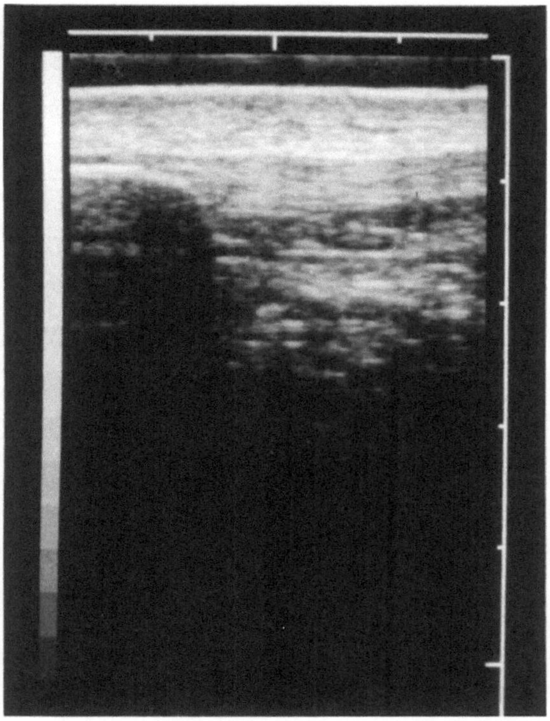

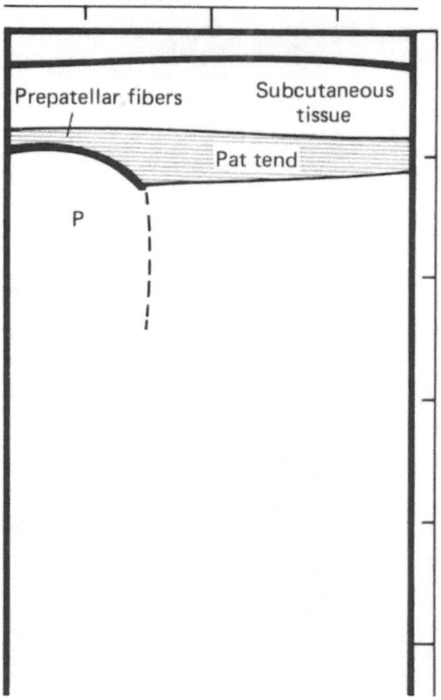
Prepatellar fibers
Subcutaneous tissue
Pat tend
P

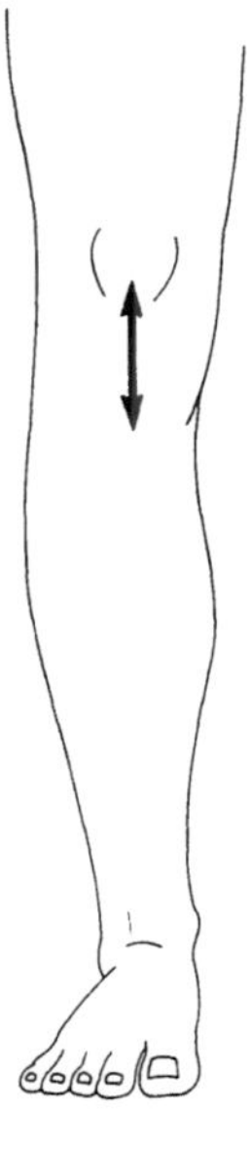

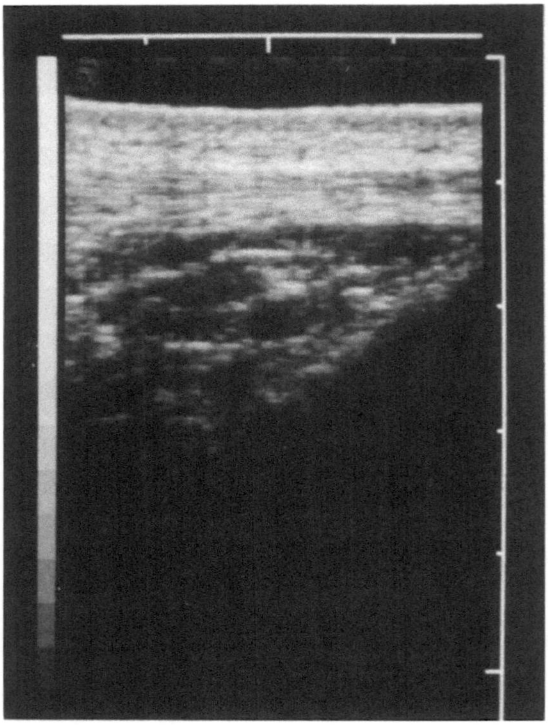

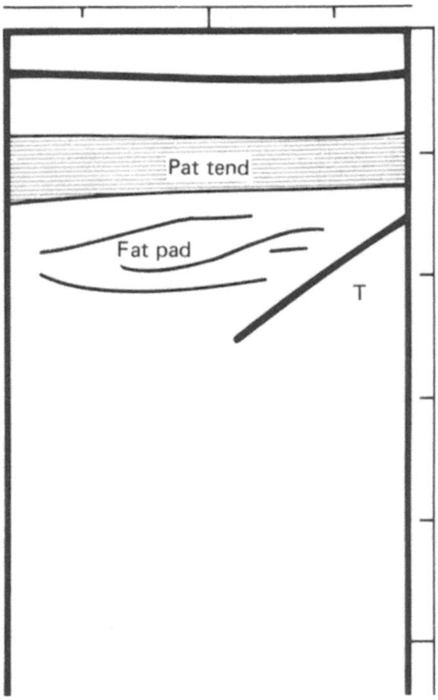
Pat tend
Fat pad
T

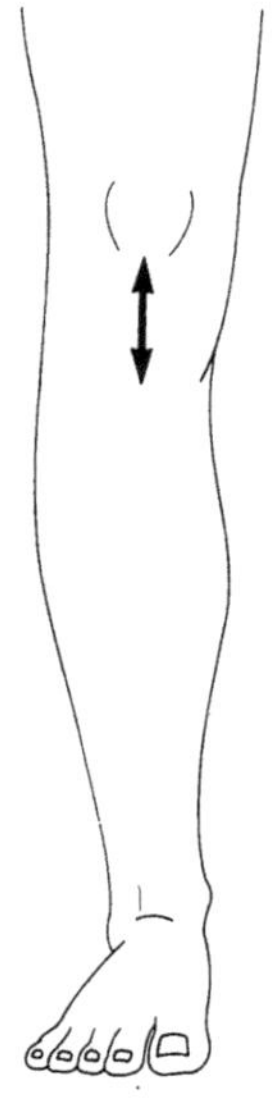

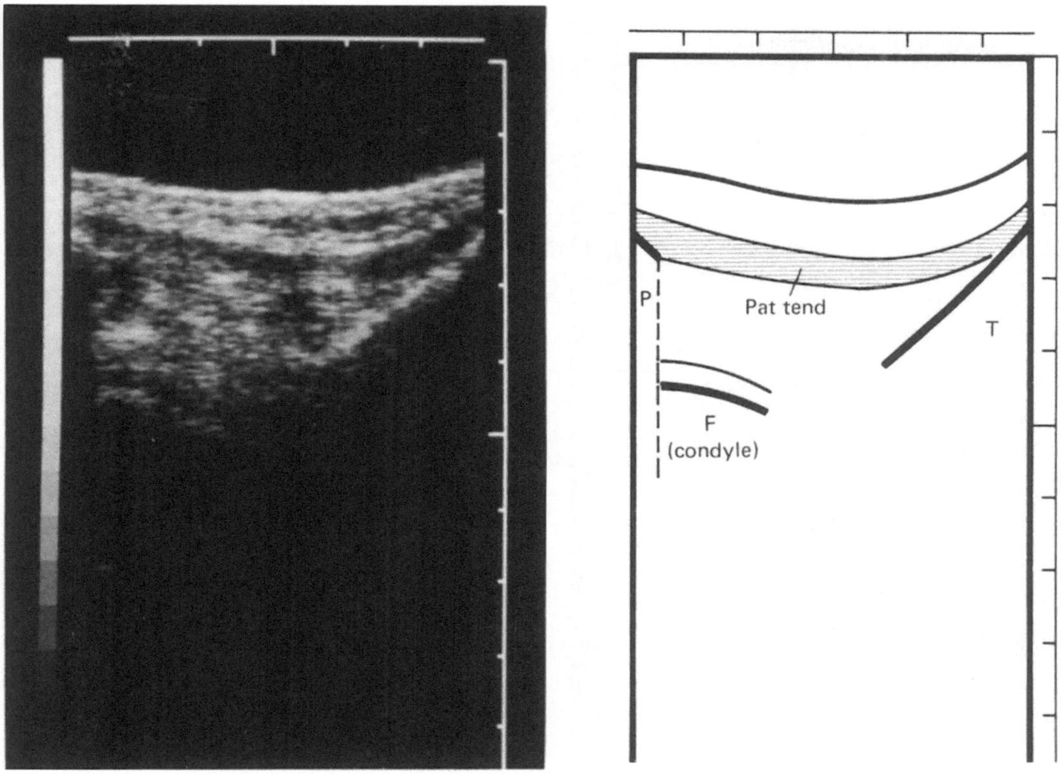

a. At rest

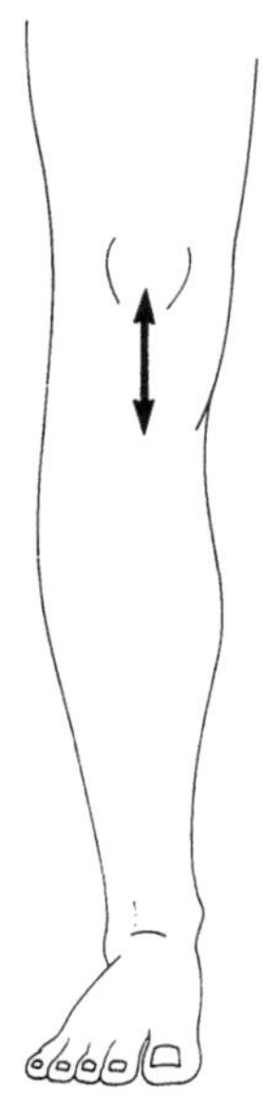

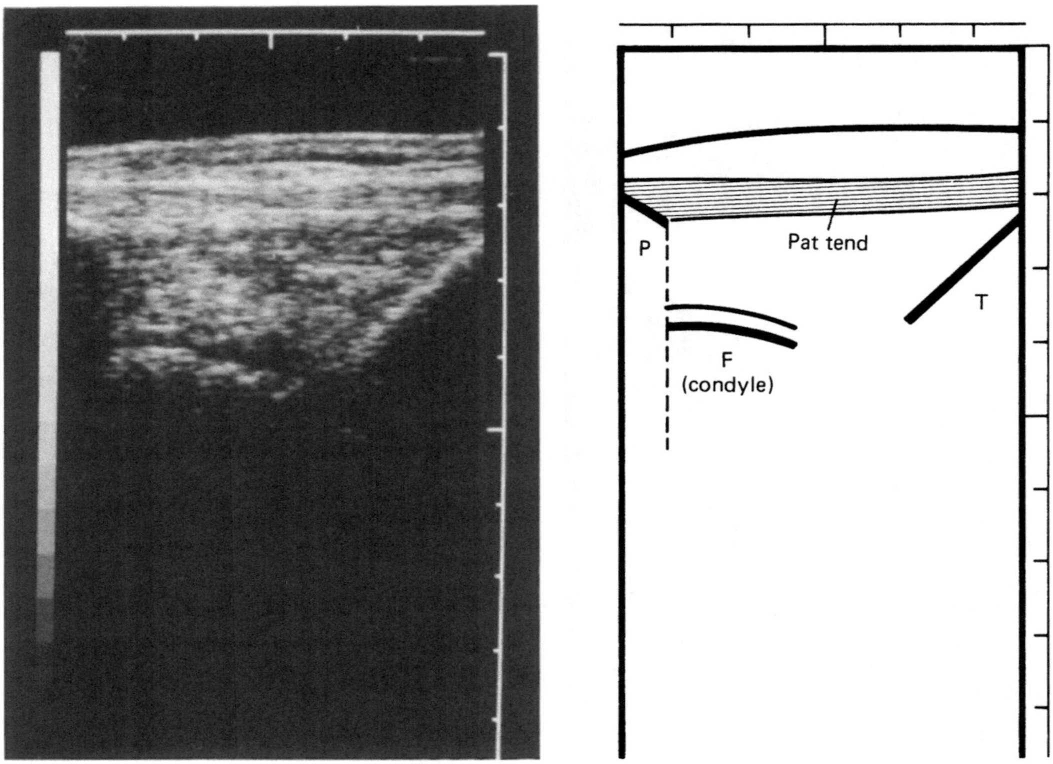

b. During contraction of quadriceps

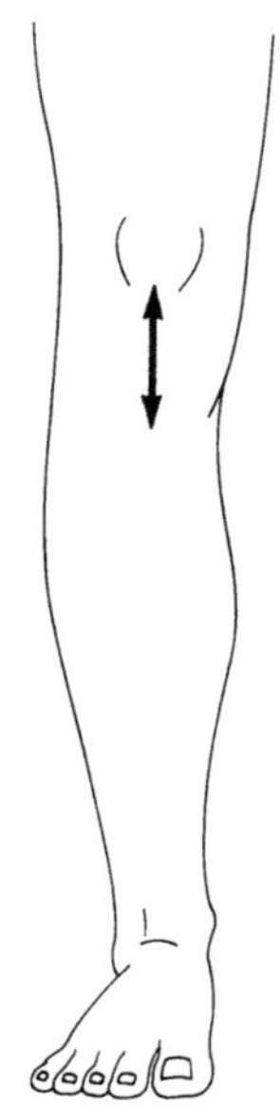

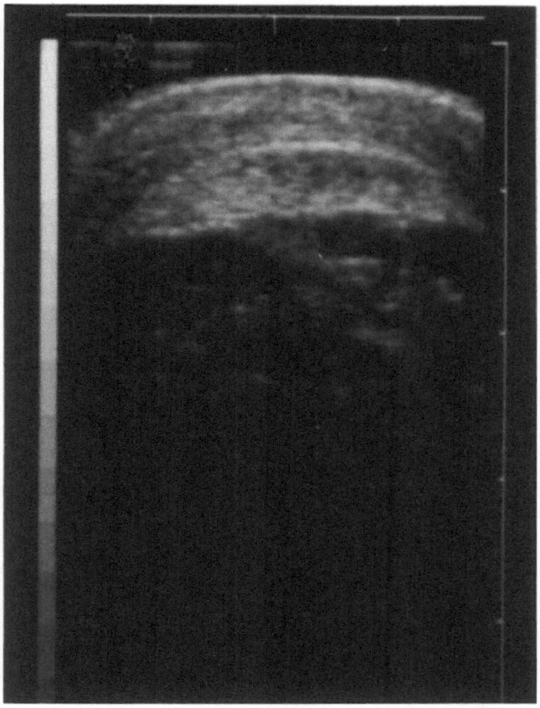

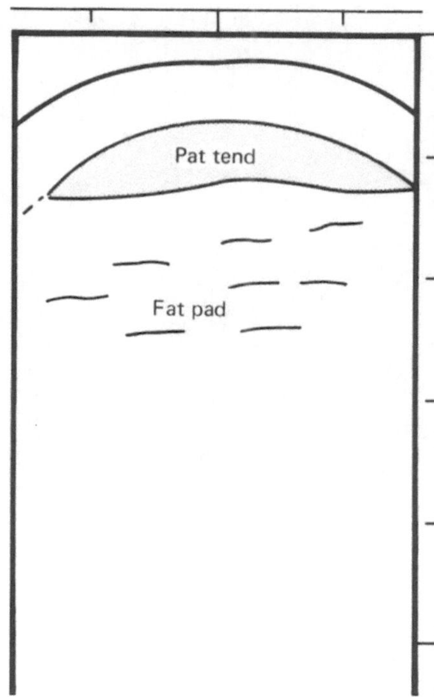

(Reprinted with permission from Fornage BD, Rifkin MD: Ultrasound examination of tendons. Radiol. Clin. North Am., 1988;26(1):87–107.)

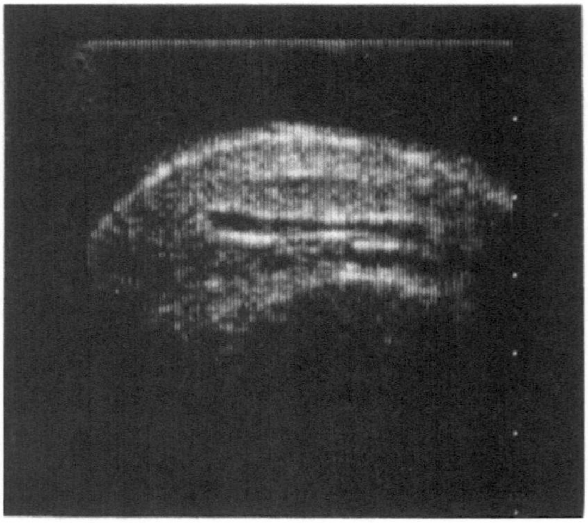

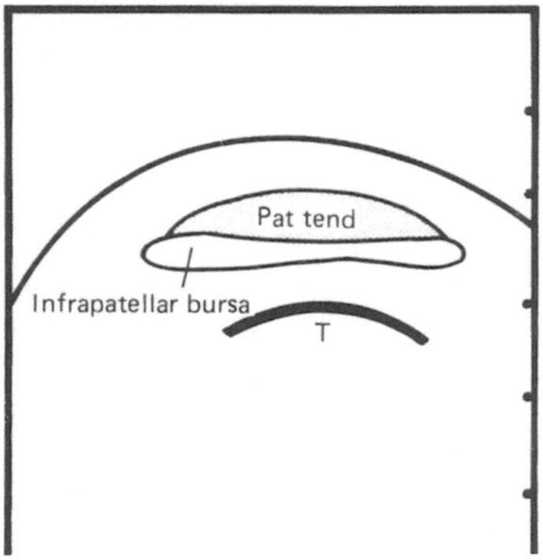
Pat tend
Infrapatellar bursa
T

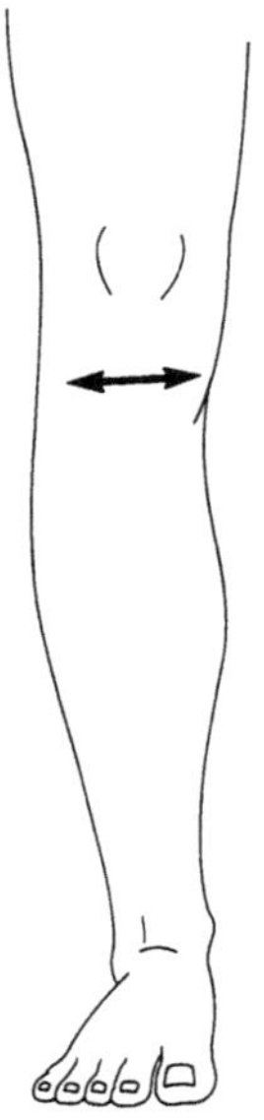

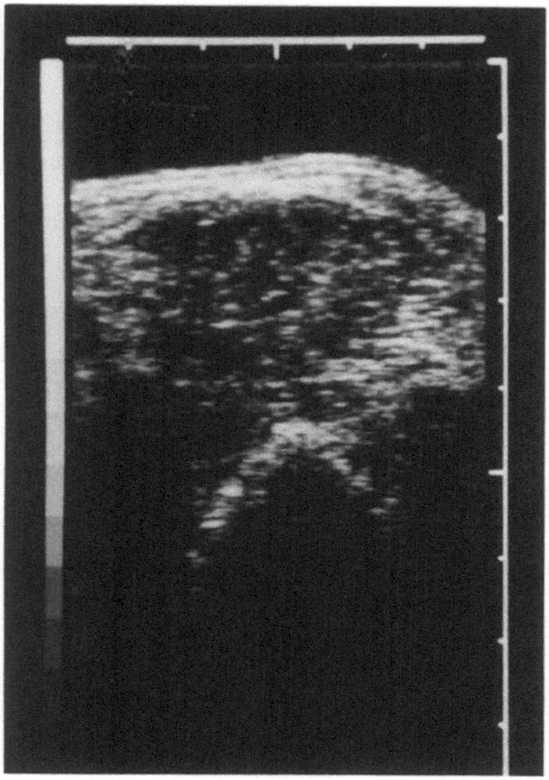

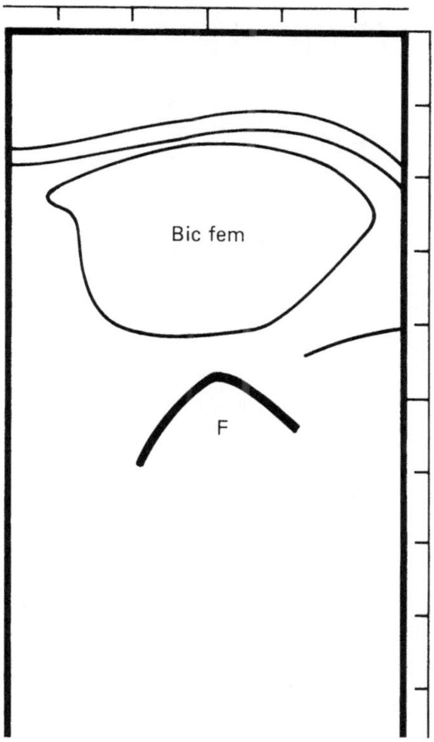
Bic fem
F

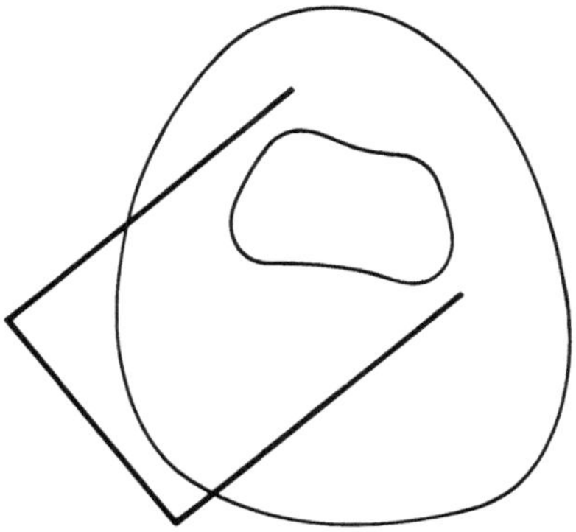
Ant
Lat
Med
Post

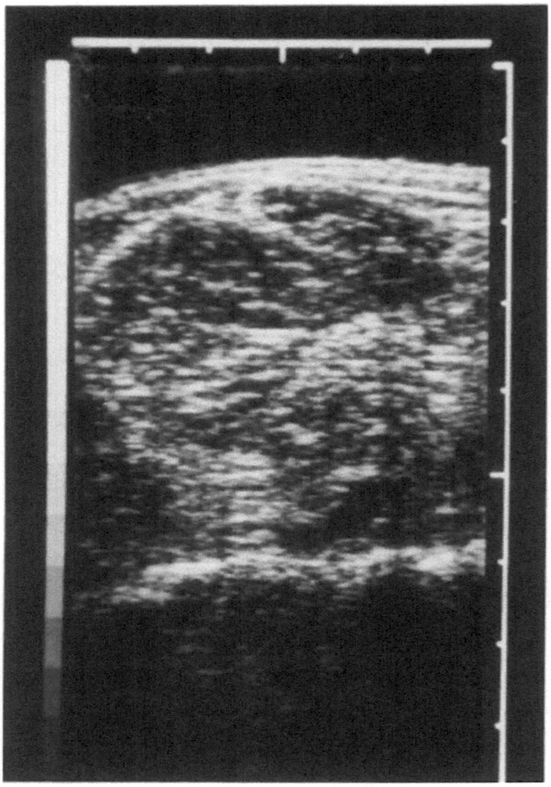

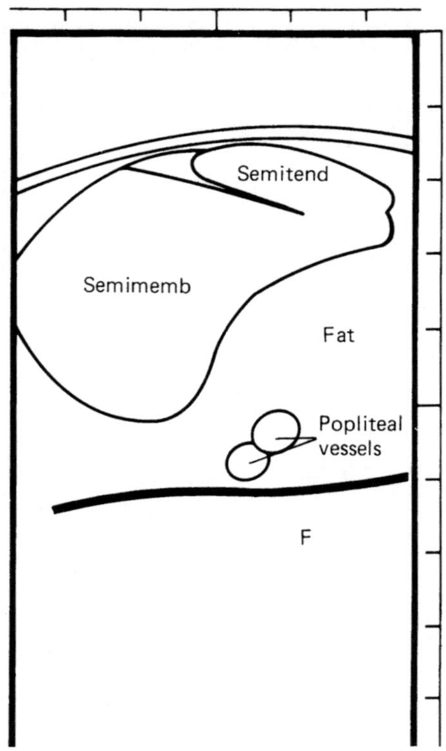
Semitend
Semimemb
Fat
Popliteal vessels
F

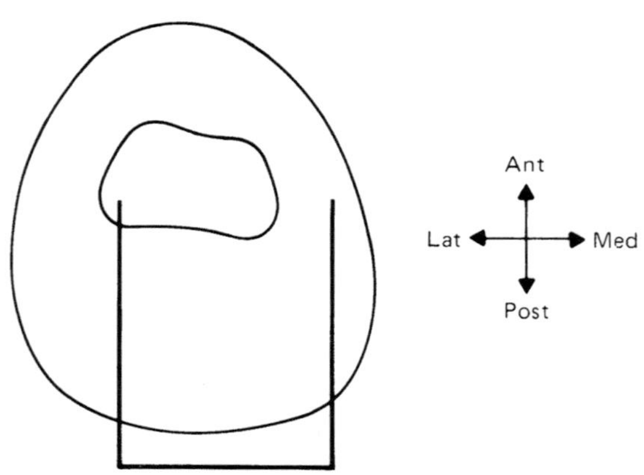
Ant
Lat
Med
Post

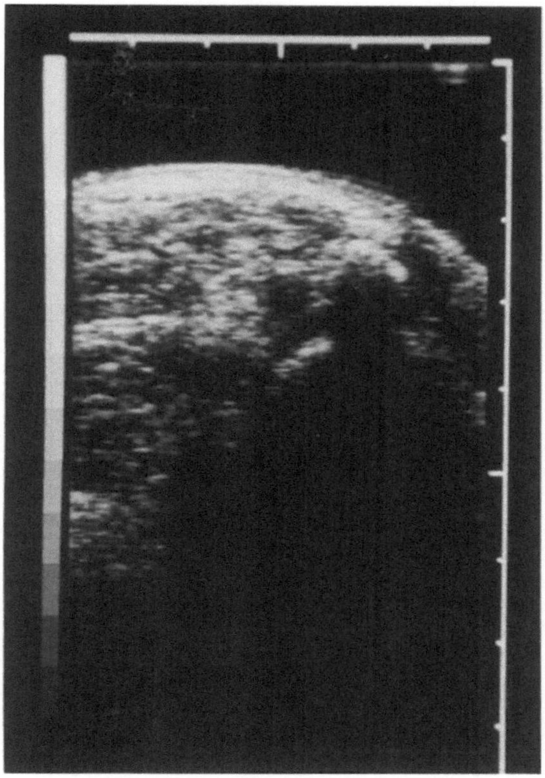

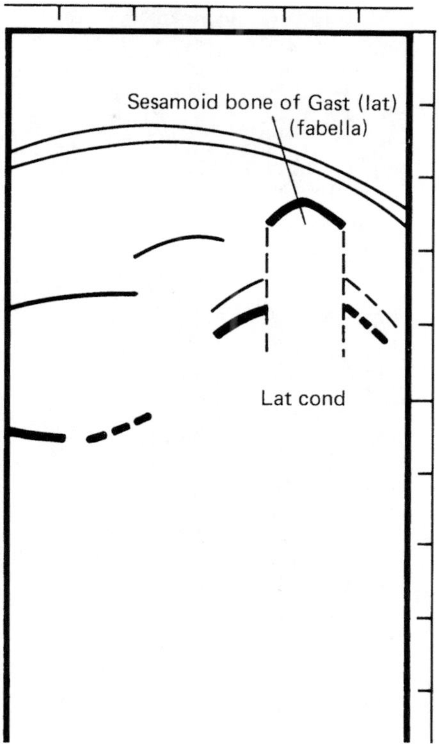
Sesamoid bone of Gast (lat)
(fabella)
Lat cond

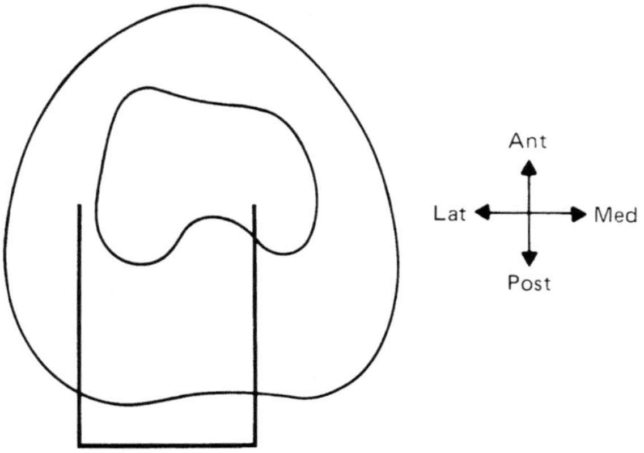
Ant
Lat
Med
Post

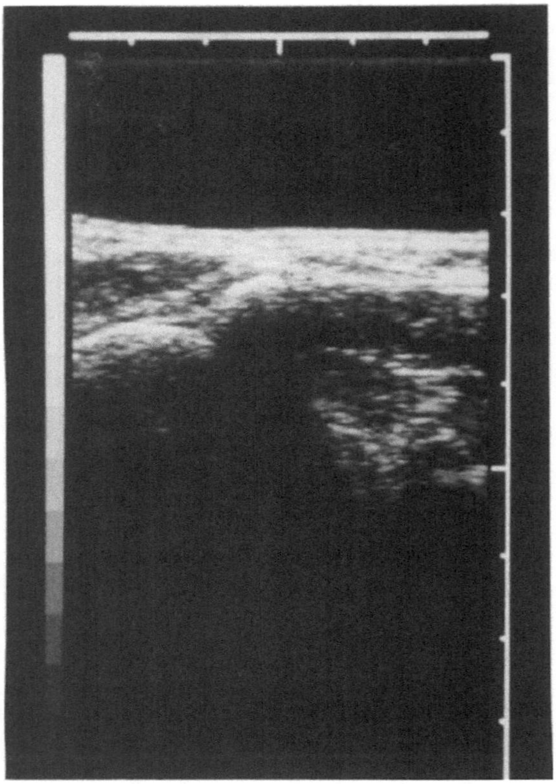

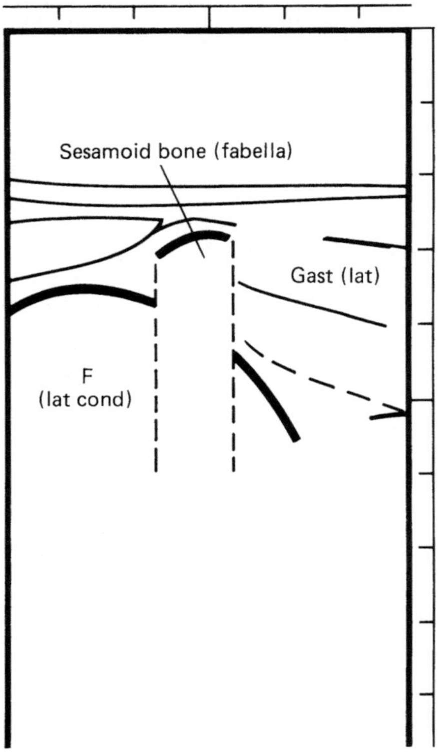
Sesamoid bone (fabella)
Gast (lat)
F
(lat cond)

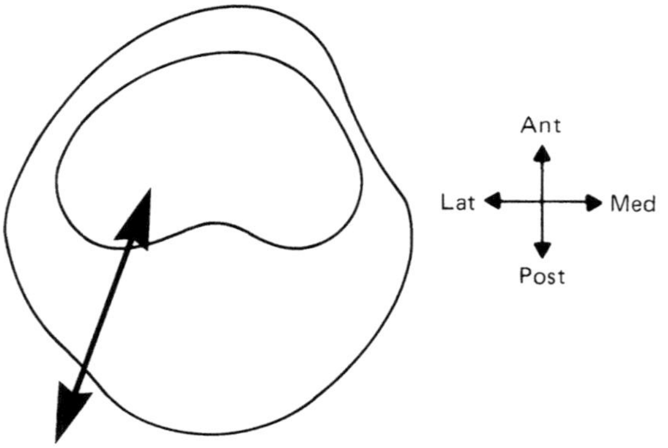
Ant
Lat
Med
Post

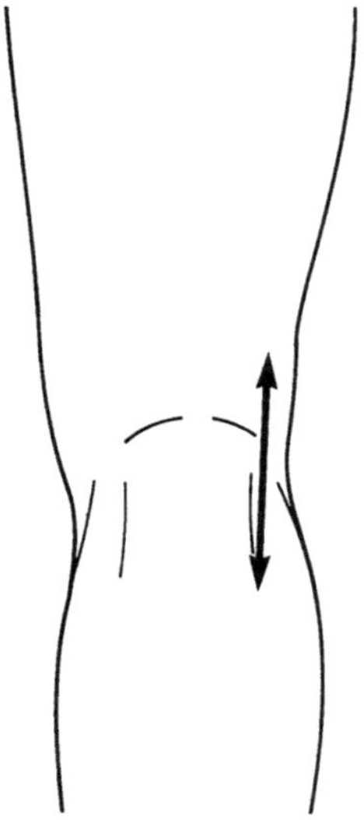

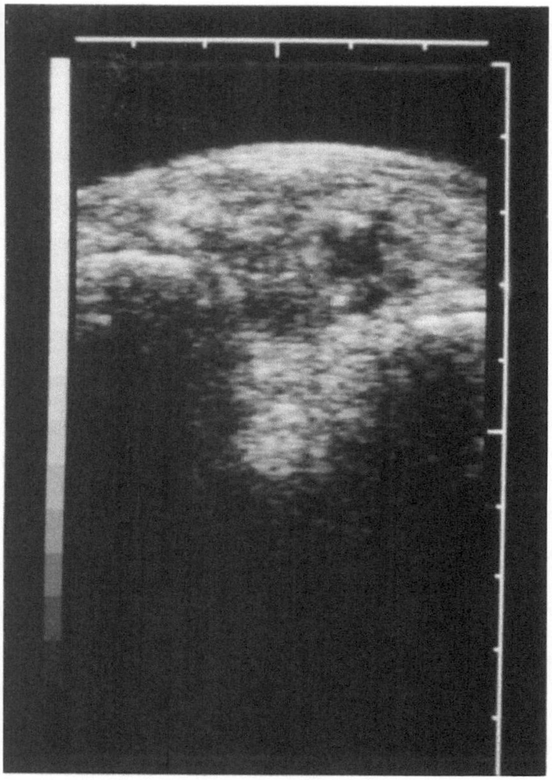

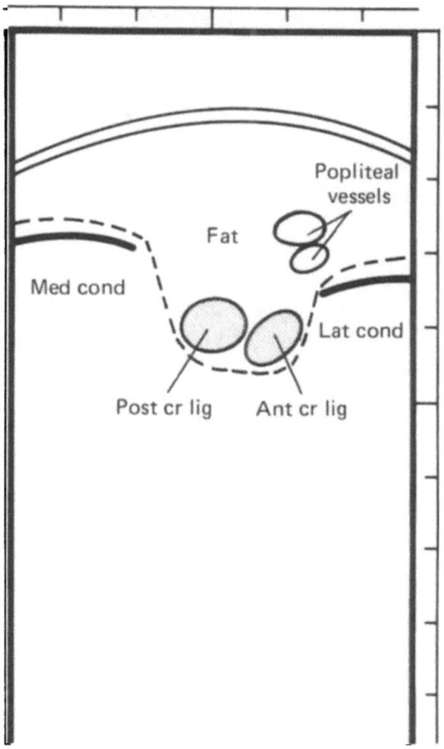
Popliteal vessels
Fat
Med cond
Lat cond
Post cr lig
Ant cr lig

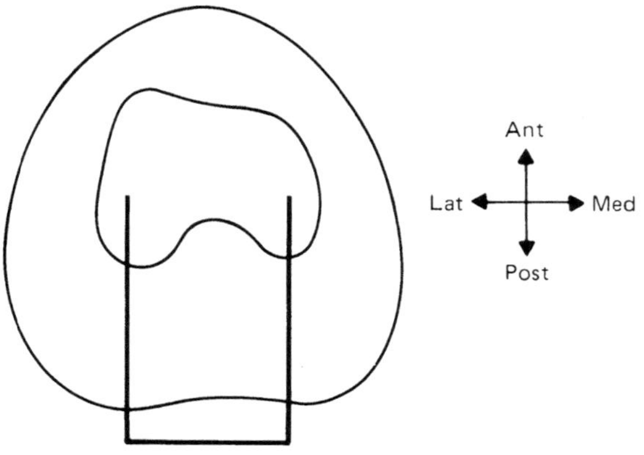
Ant
Lat
Med
Post

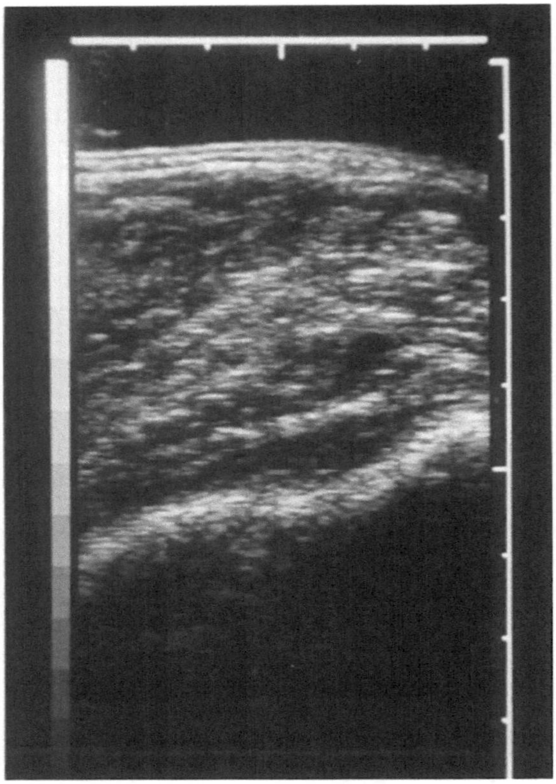

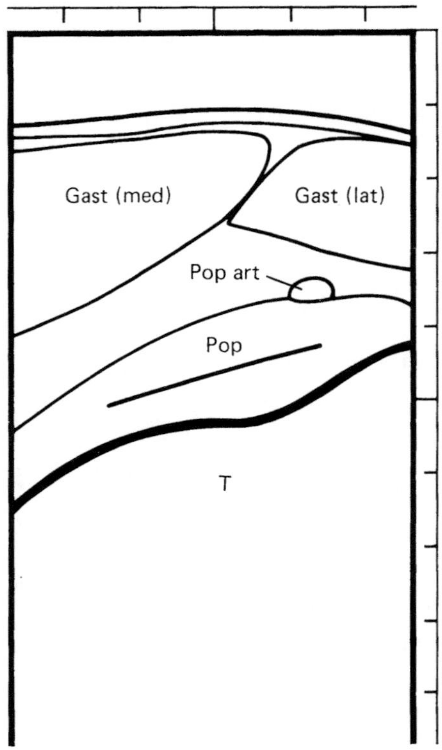
Gast (med)
Gast (lat)
Pop art
Pop
T

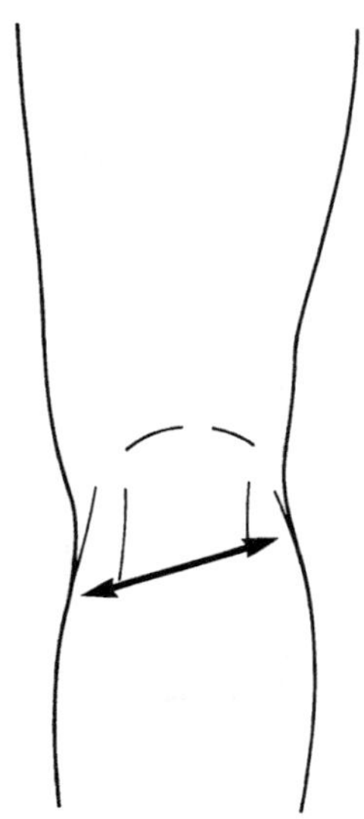

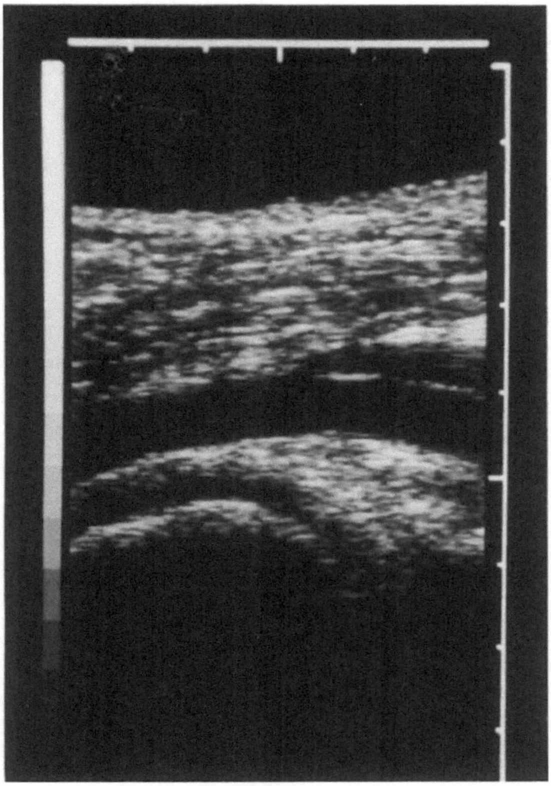

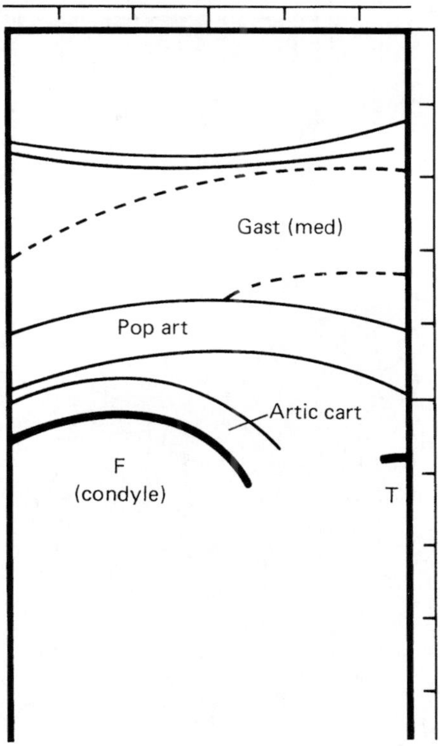
Gast (med)
Pop art
Artic cart
F
(condyle)
T

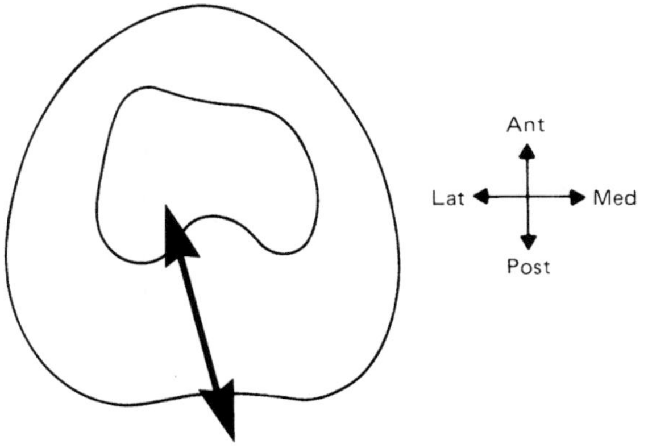
Ant
Lat
Med
Post

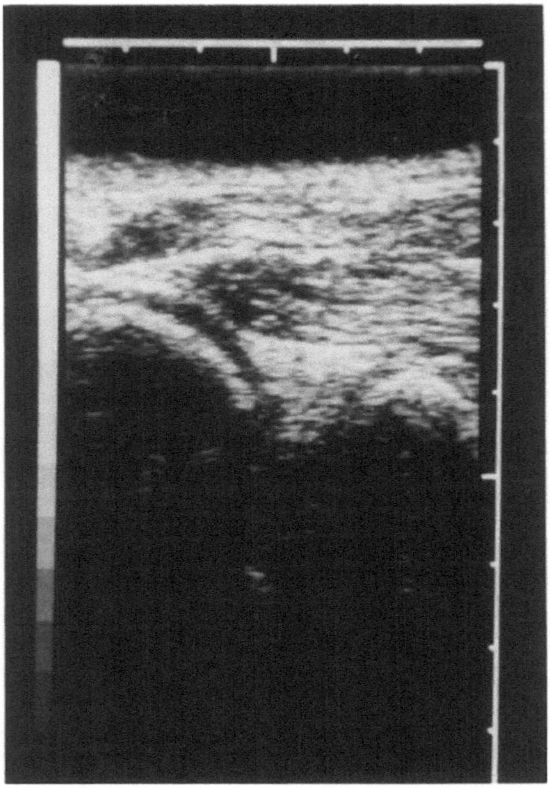

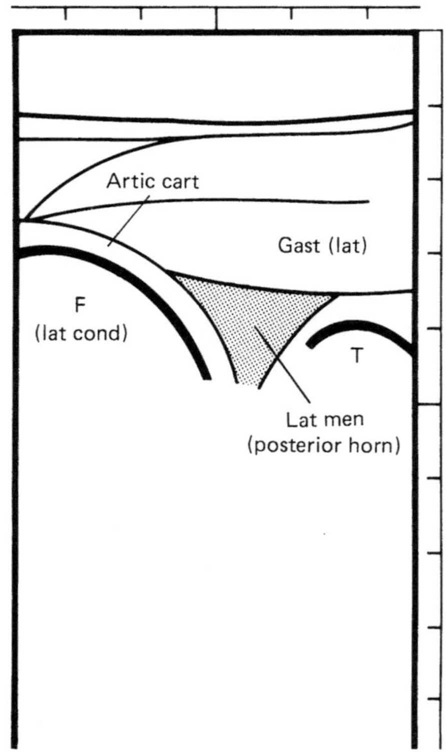
Artic cart
Gast (lat)
F
(lat cond)
T
Lat men
(posterior horn)

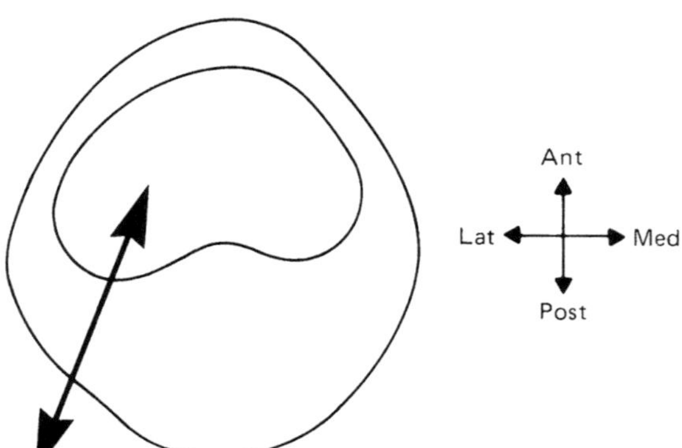
Ant
Lat
Med
Post

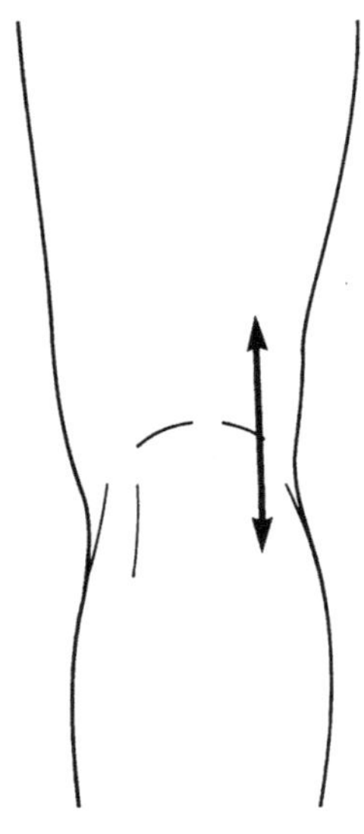

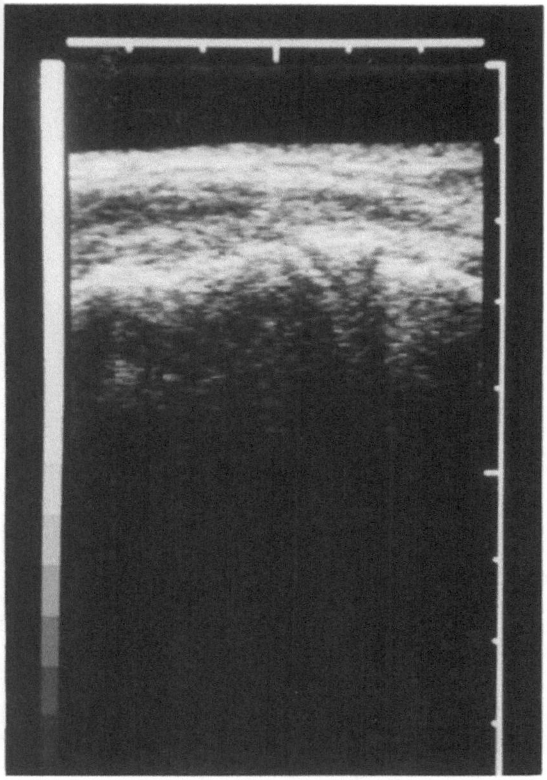

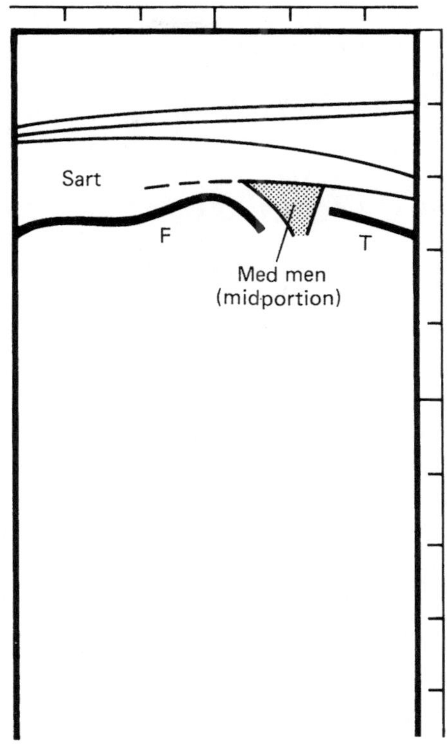
Sart
F
T
Med men
(midportion)

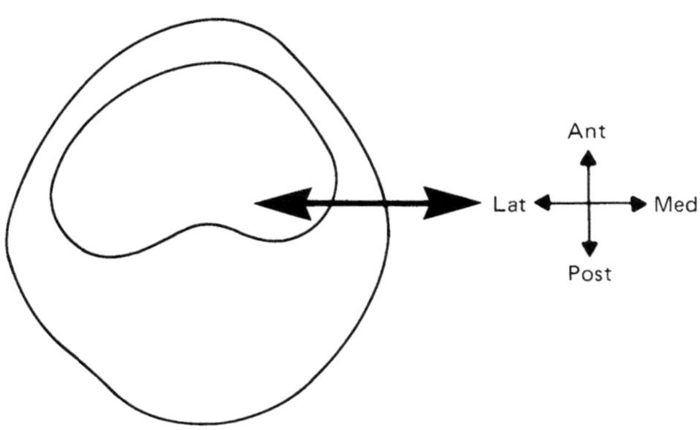
Ant
Lat
Med
Post

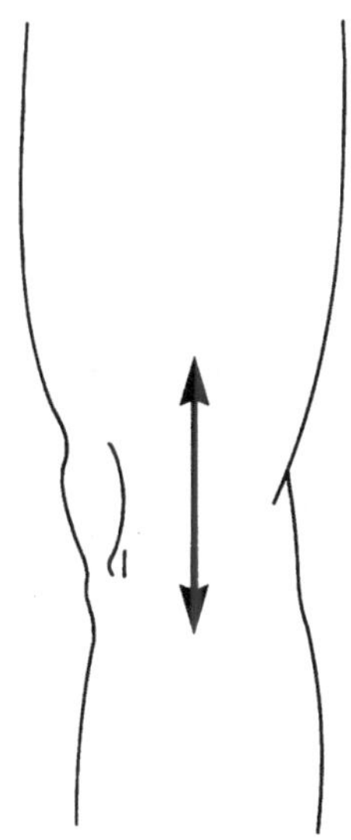

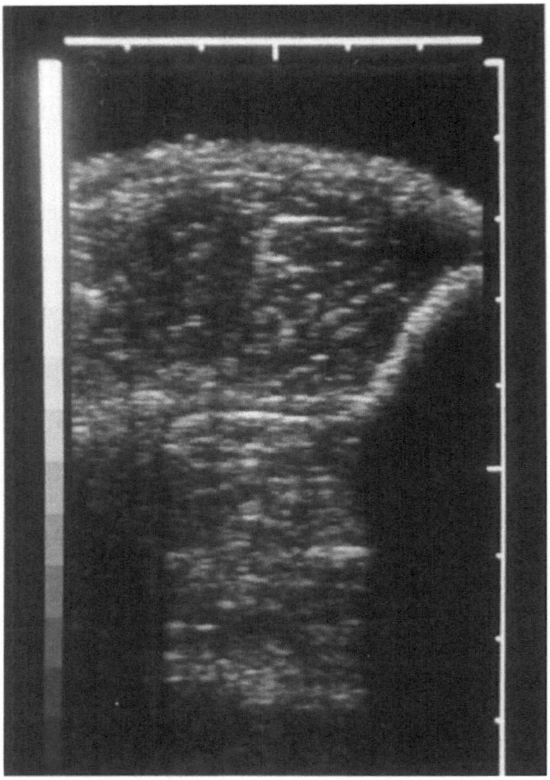

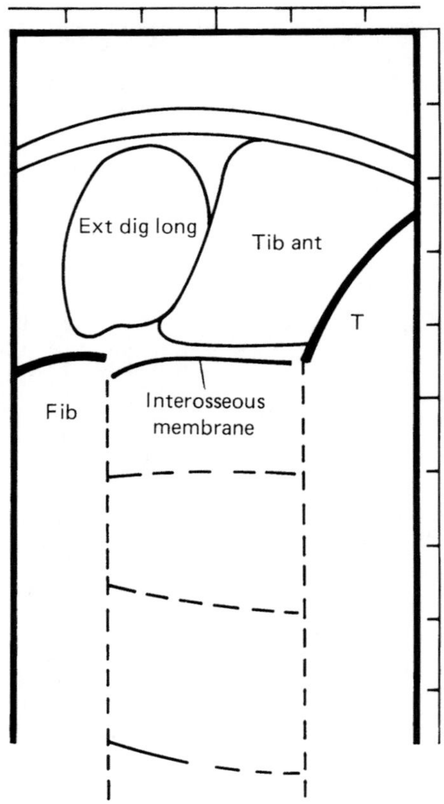
Ext dig long
Tib ant
T
Fib
Interosseous
membrane

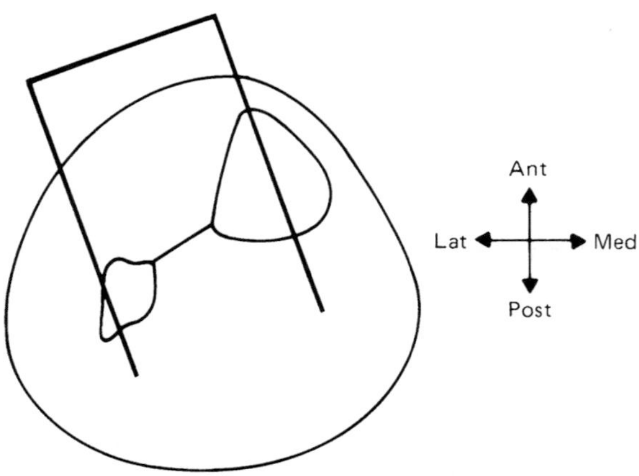
Ant
Lat
Med
Post

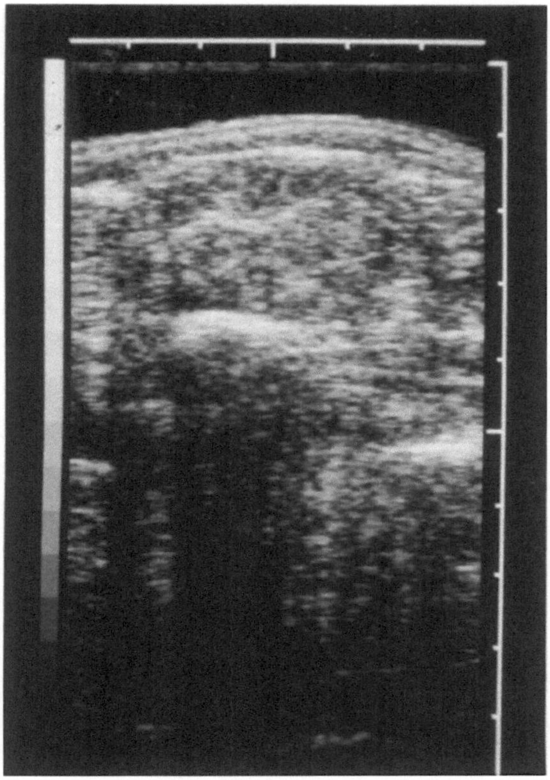

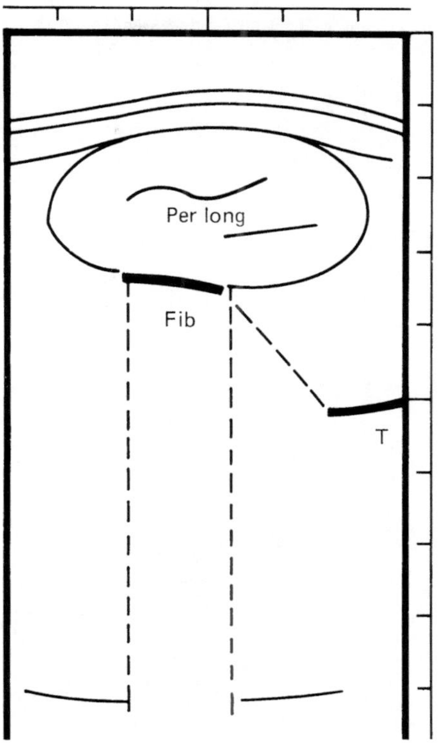
Per long
Fib
T

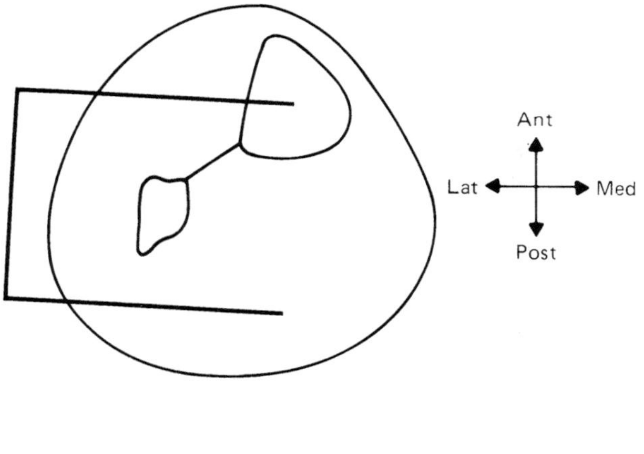
Ant
Lat
Med
Post

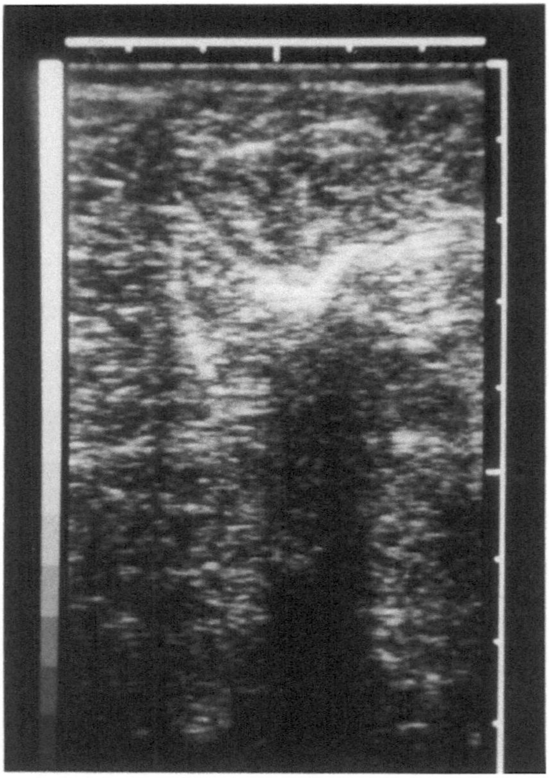

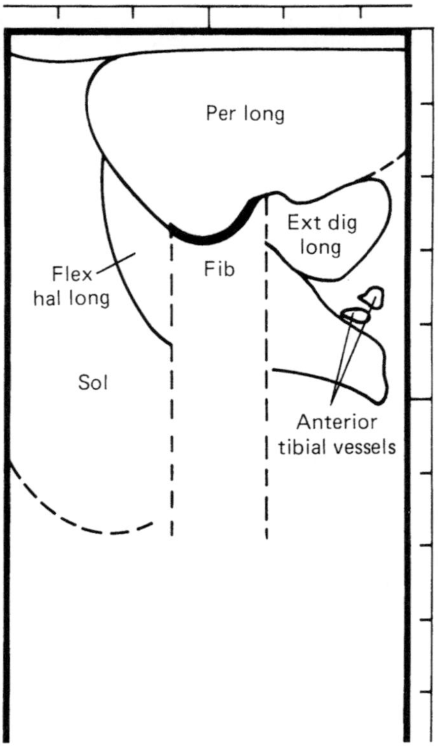
Per long
Ext dig long
Flex hal long
Fib
Sol
Anterior tibial vessels

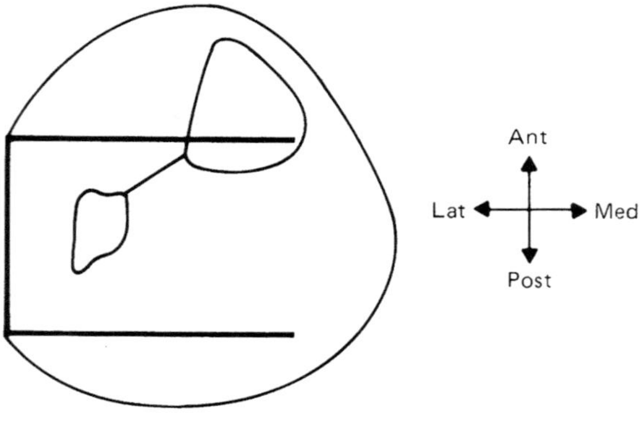
Ant
Lat
Med
Post

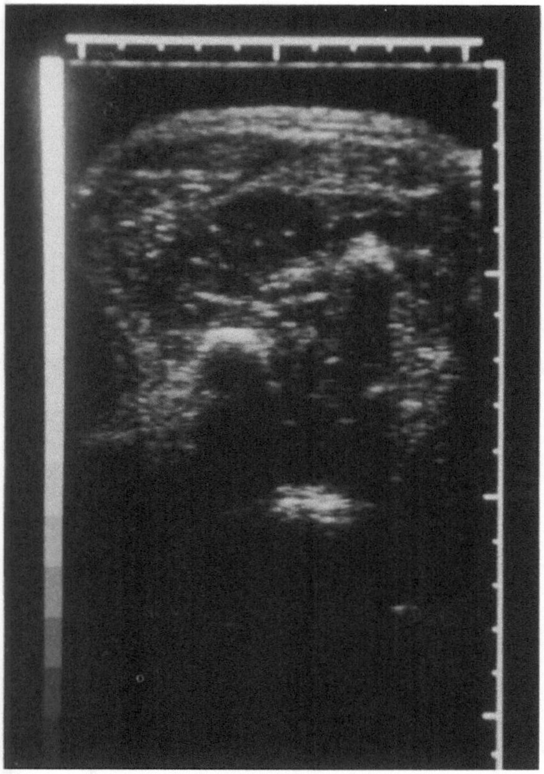

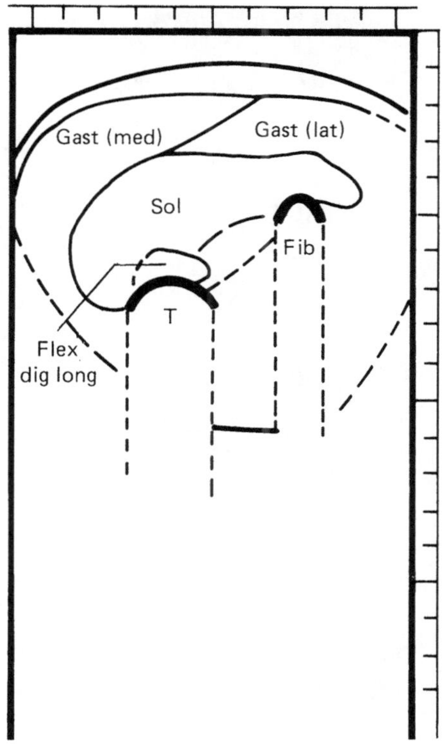
Gast (med)
Gast (lat)
Sol
Fib
T
Flex dig long

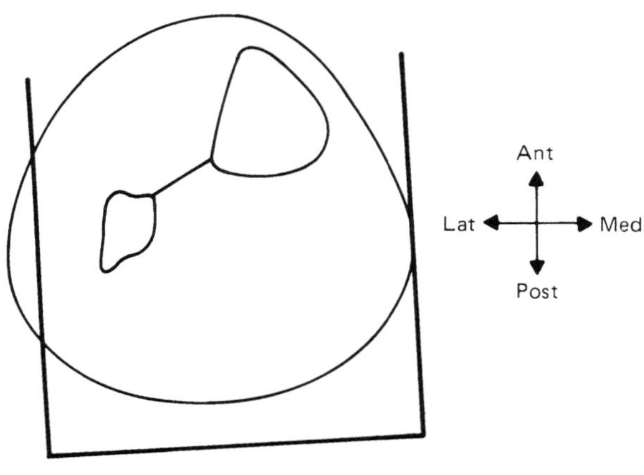
Ant
Lat
Med
Post

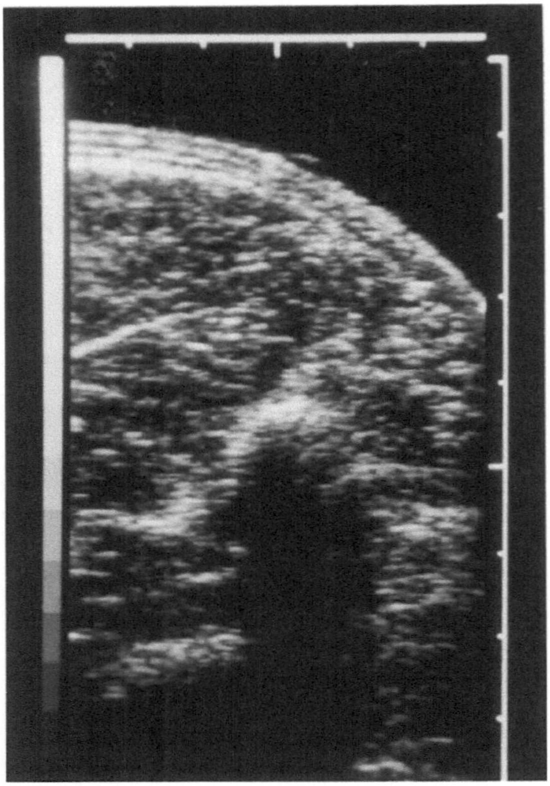

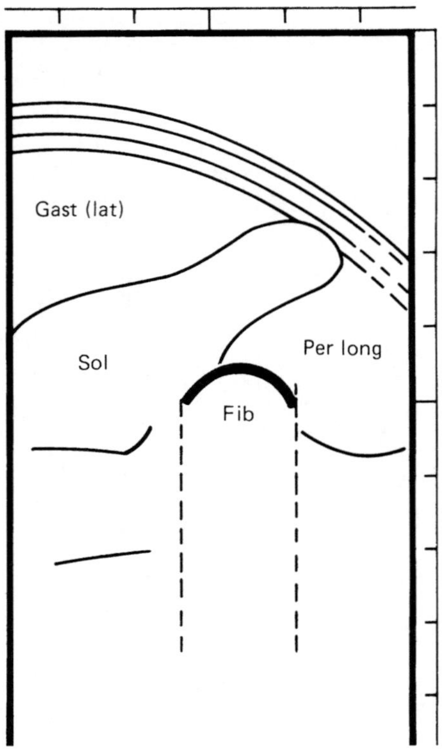
Gast (lat)
Sol
Per long
Fib

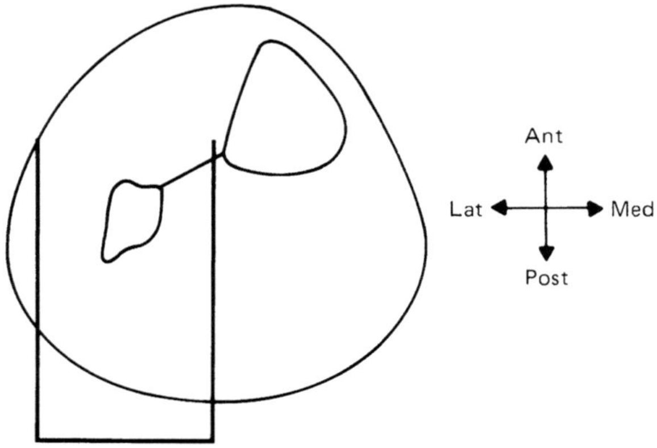
Ant
Lat
Med
Post

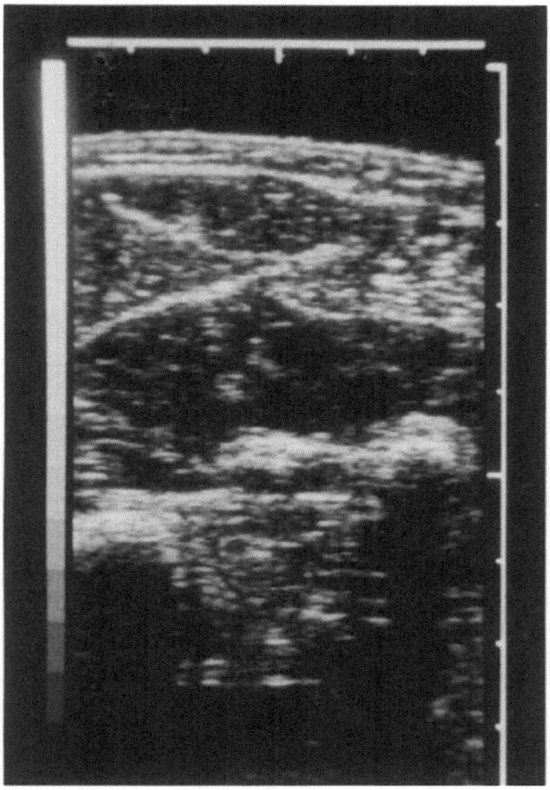

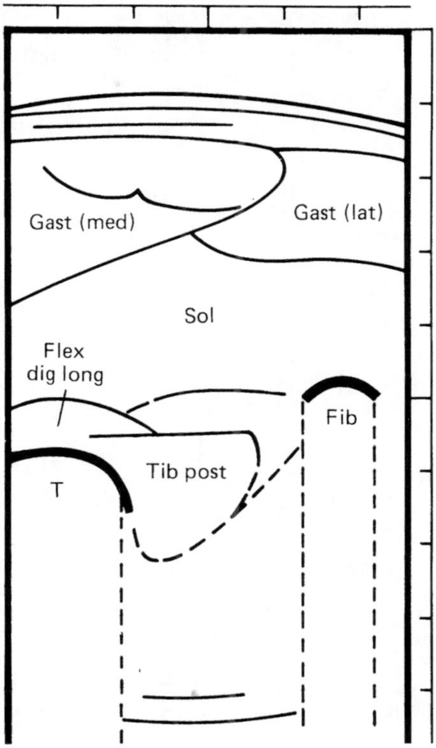
Gast (med)
Gast (lat)
Sol
Flex
dig long
Fib
Tib post
T

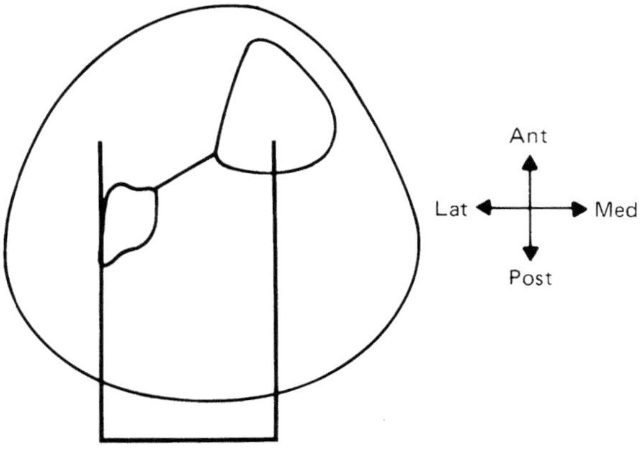
Ant
Lat
Med
Post

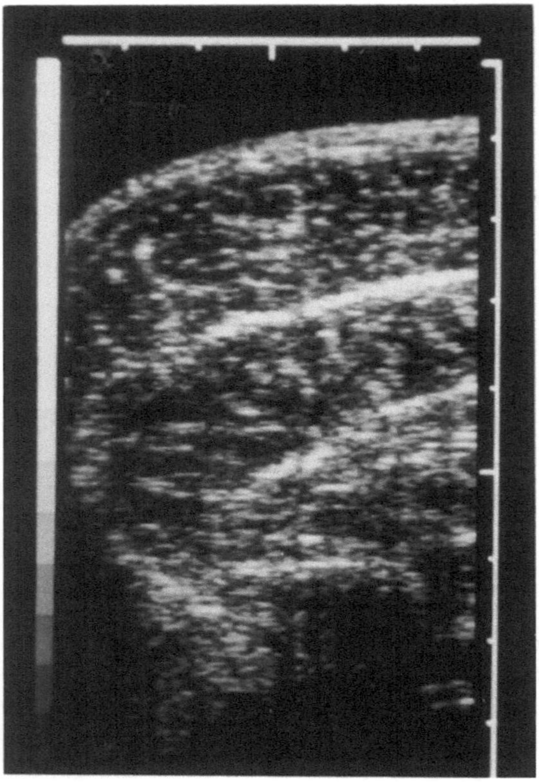

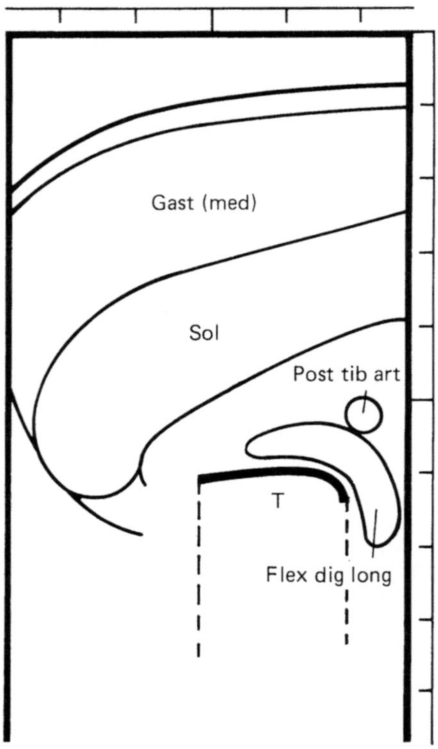
Gast (med)
Sol
Post tib art
T
Flex dig long

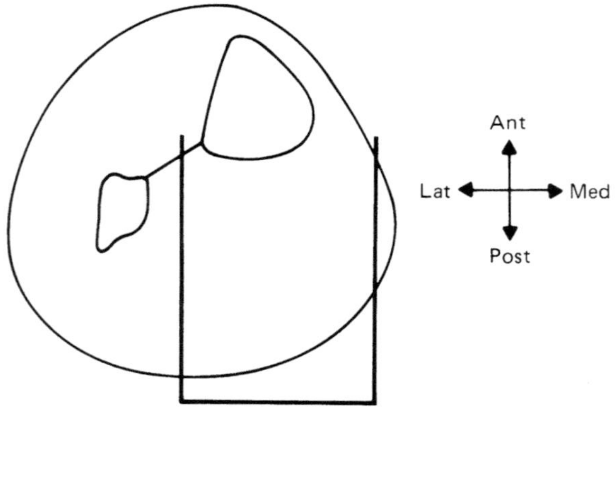
Ant
Lat
Med
Post

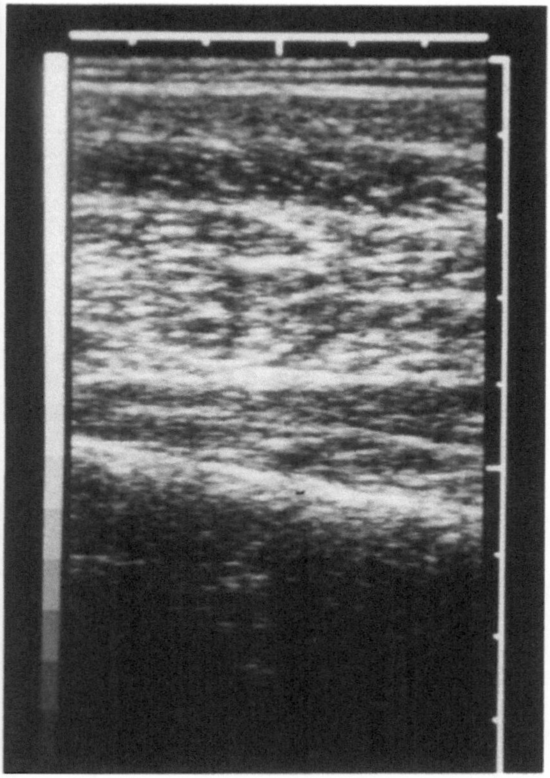

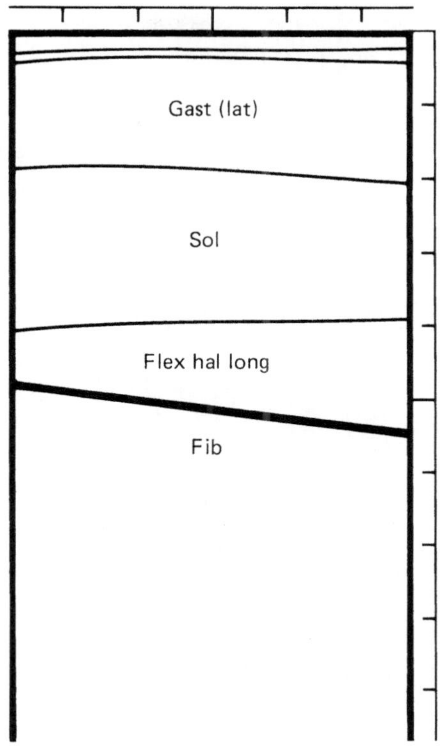
Gast (lat)
Sol
Flex hal long
Fib

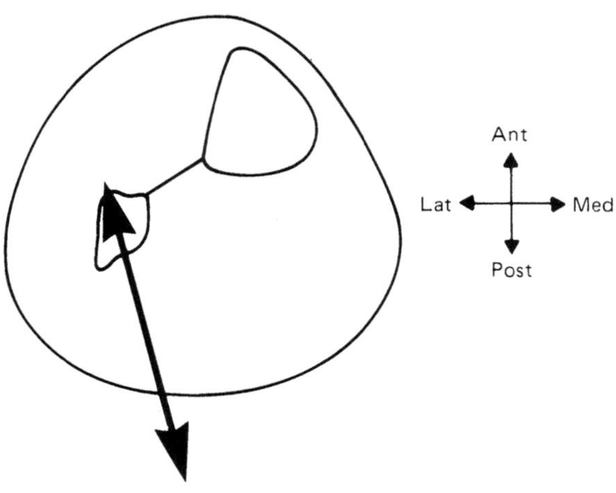
Ant
Lat
Med
Post

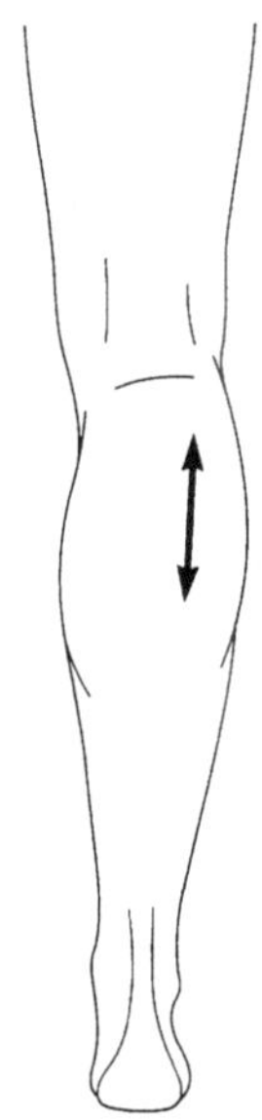

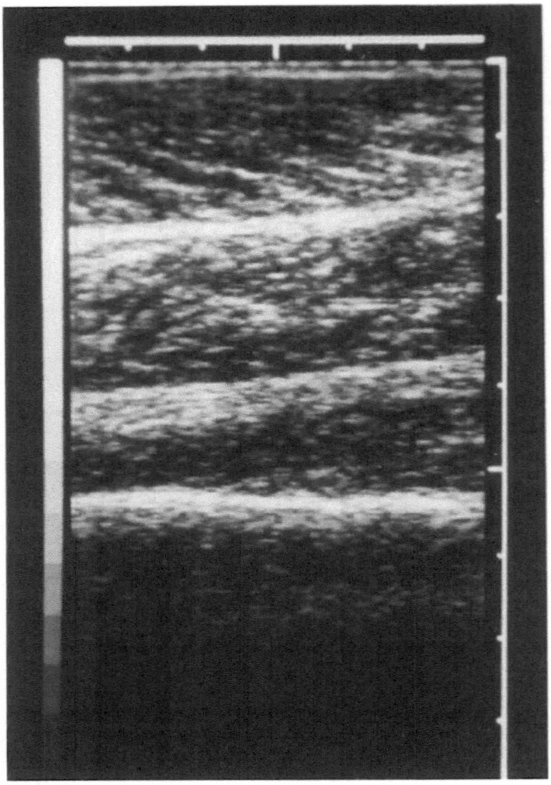

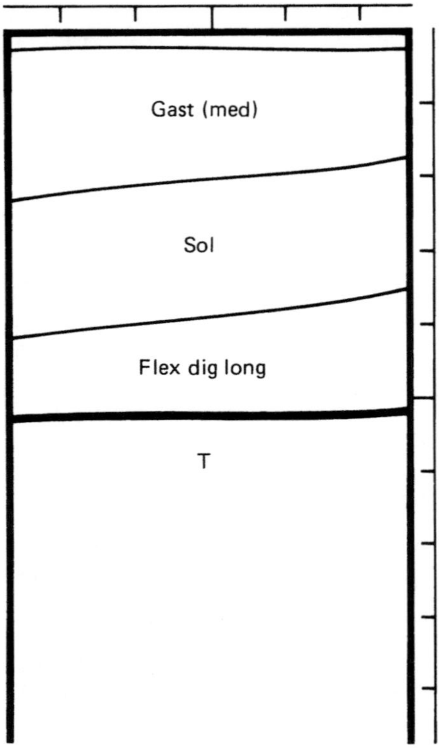
Gast (med)
Sol
Flex dig long
T

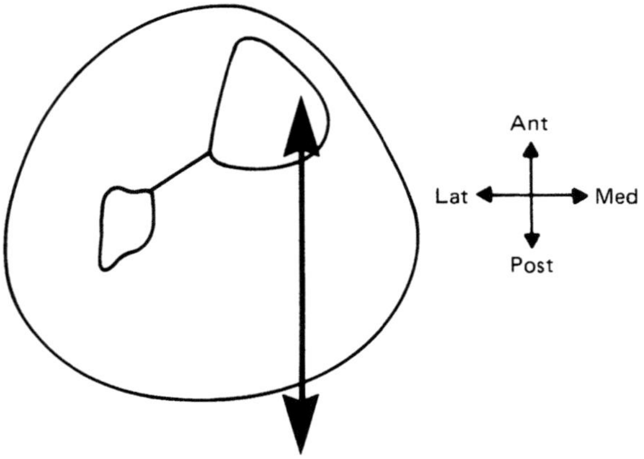
Ant
Lat
Med
Post

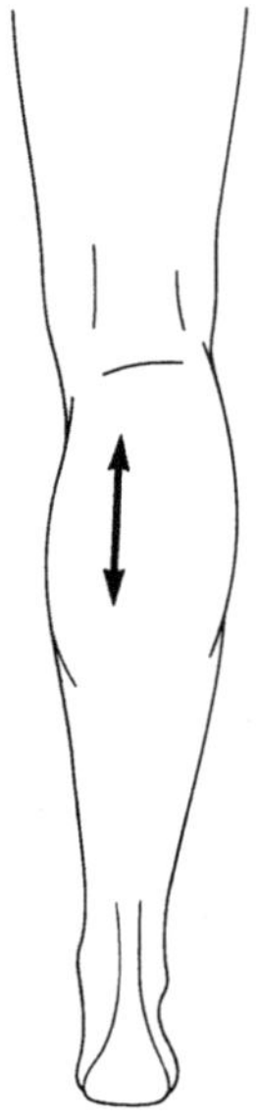

Leg, Midthird

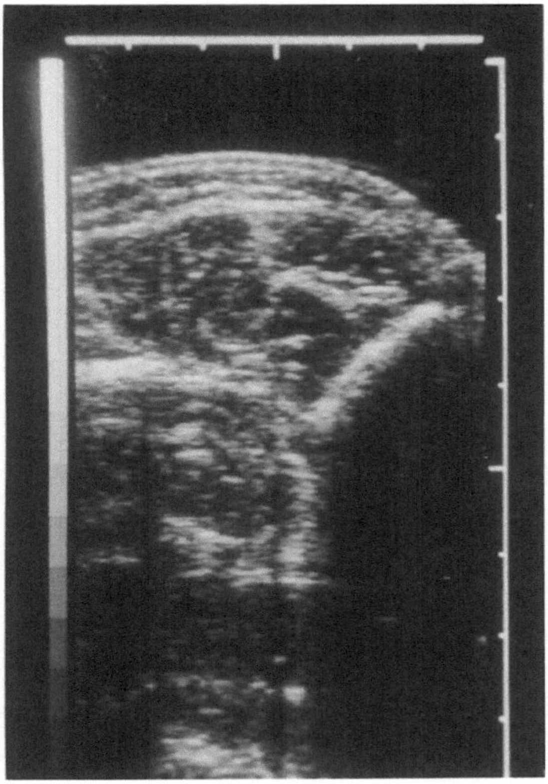

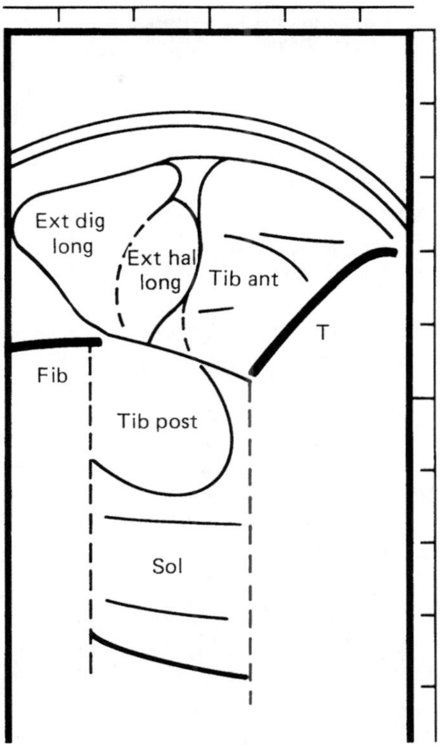

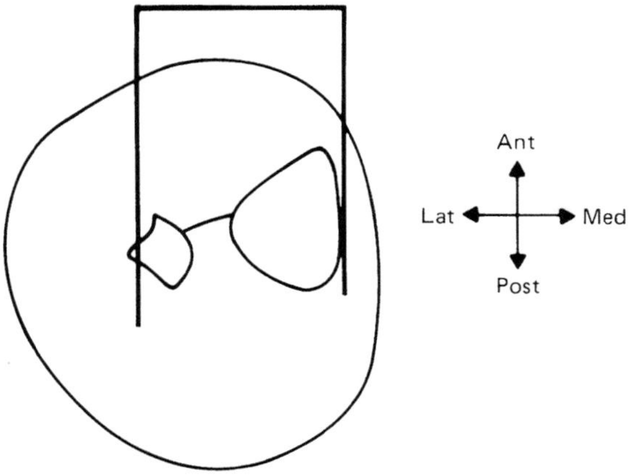

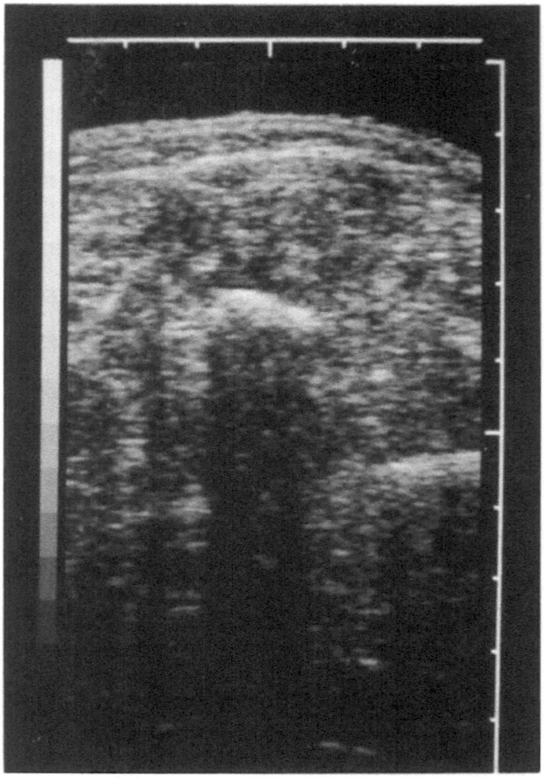

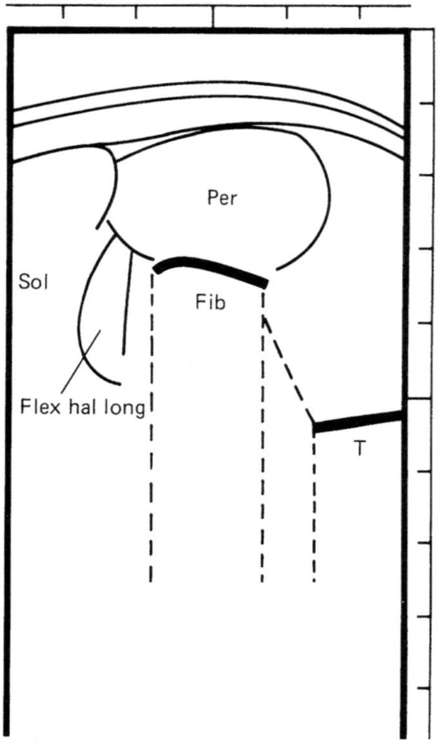
Per
Sol
Fib
Flex hal long
T

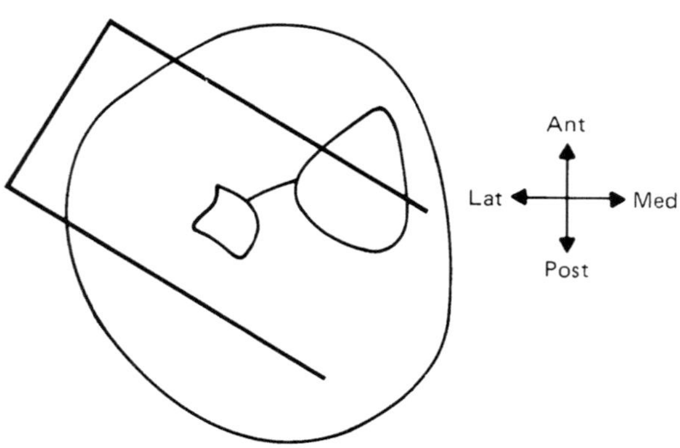
Ant
Lat
Med
Post

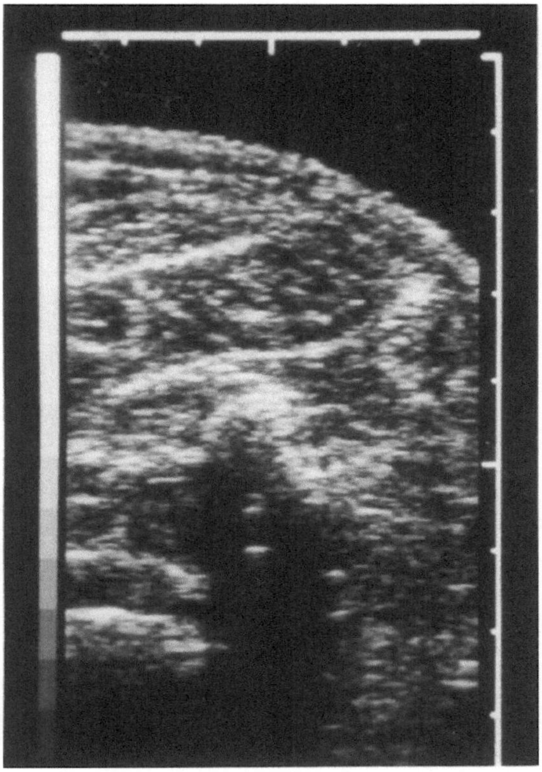

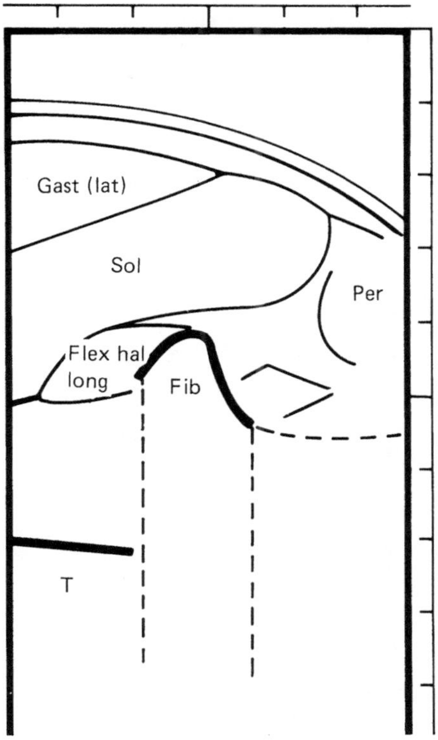
Gast (lat)
Sol
Per
Flex hal long
Fib
T

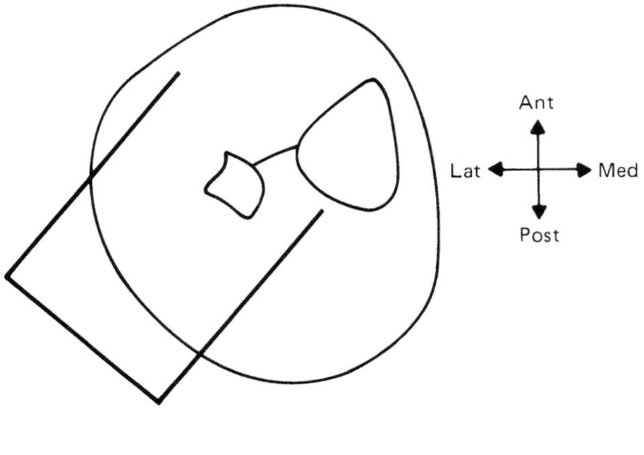
Ant
Lat
Med
Post

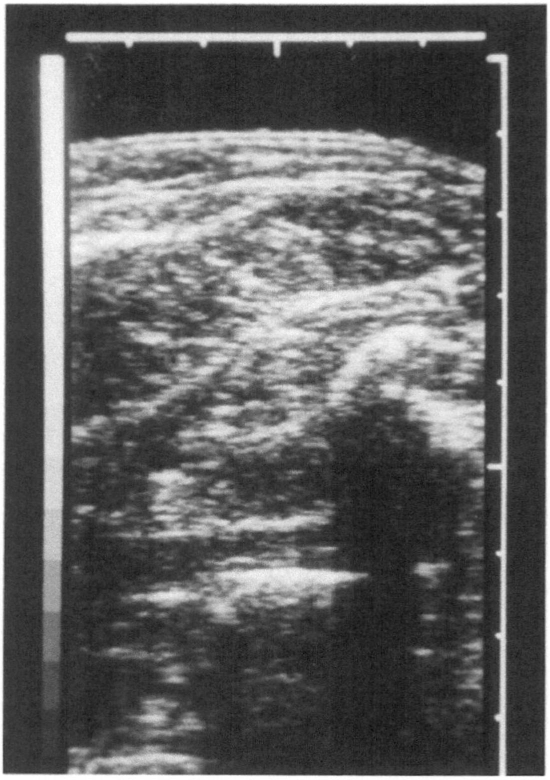

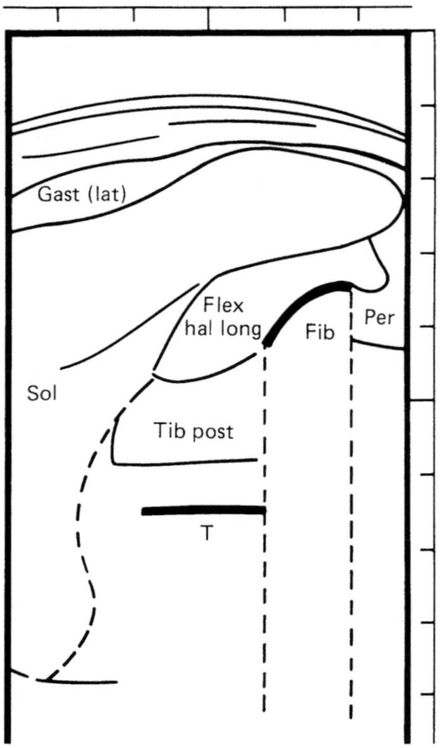
Gast (lat)
Flex
hal long
Fib
Per
Sol
Tib post
T

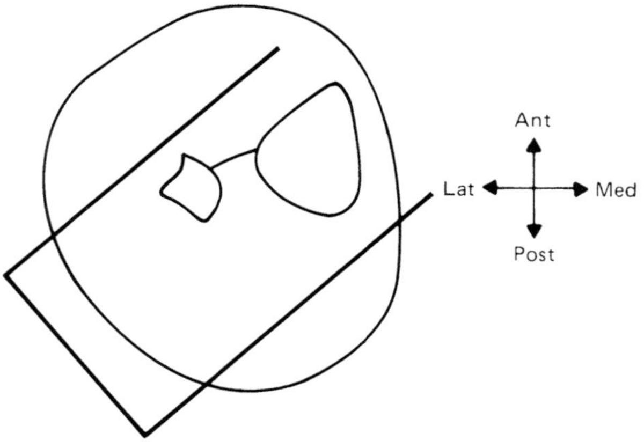
Ant
Lat
Med
Post

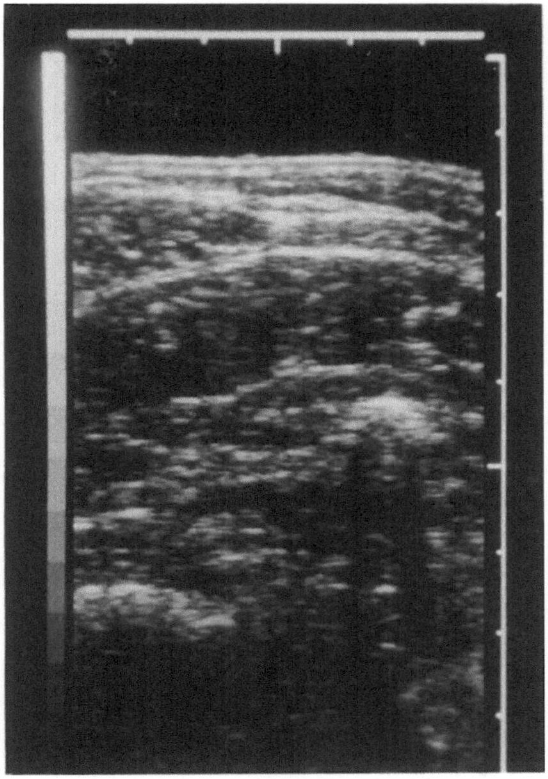

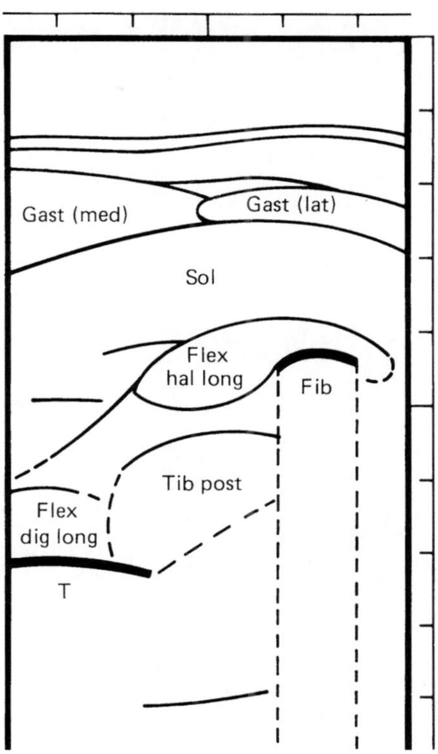
Gast (med)
Gast (lat)
Sol
Flex hal long
Fib
Tib post
Flex dig long
T

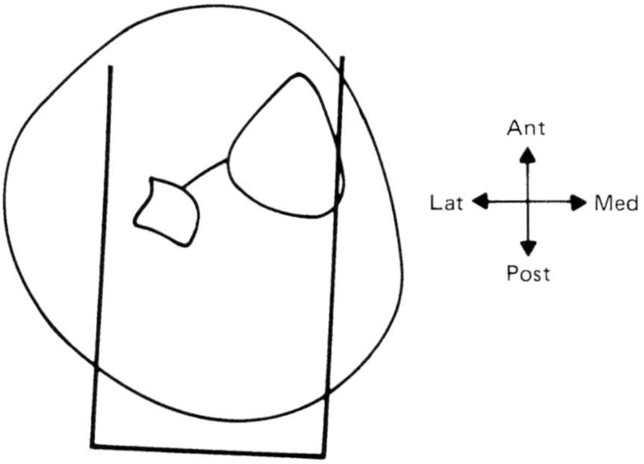
Ant
Lat
Med
Post

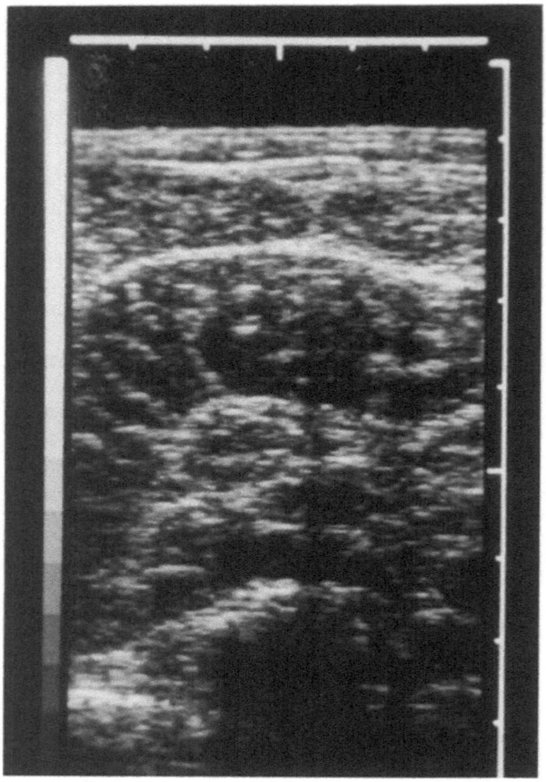

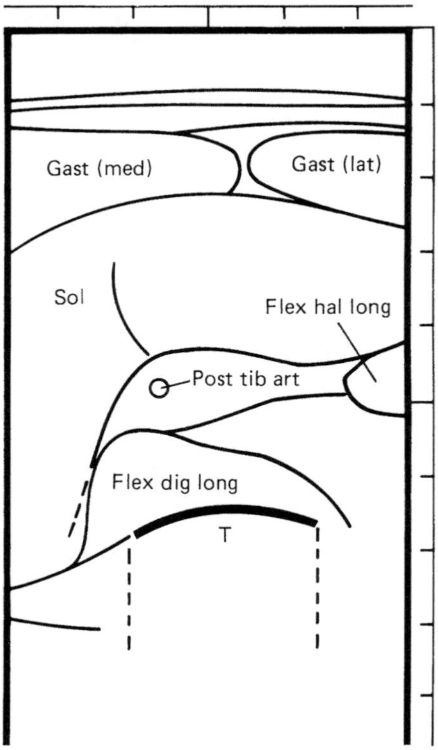
Gast (med)
Gast (lat)
Sol
Flex hal long
Post tib art
Flex dig long
T

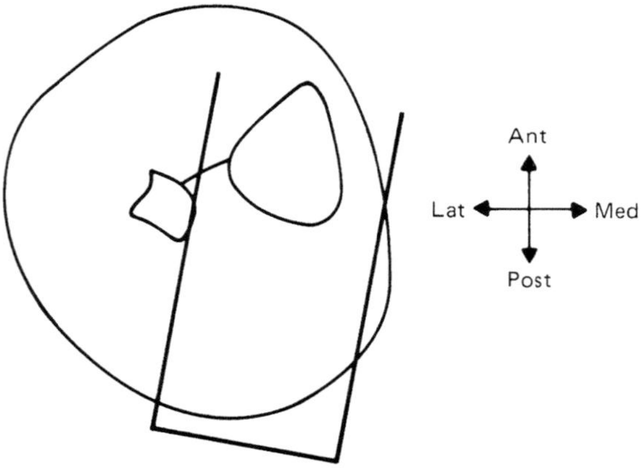
Ant
Lat
Med
Post

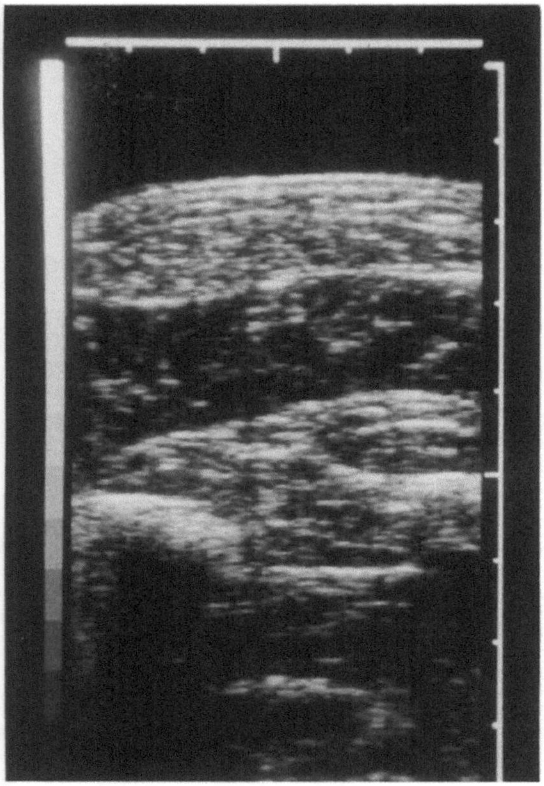

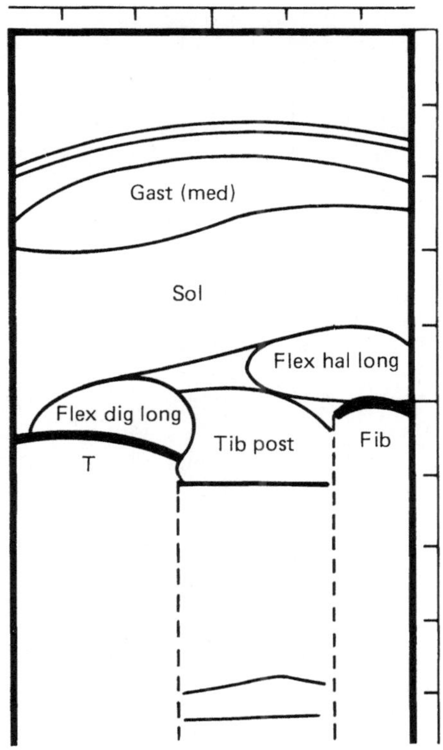
Gast (med)
Sol
Flex hal long
Flex dig long
Tib post
Fib
T

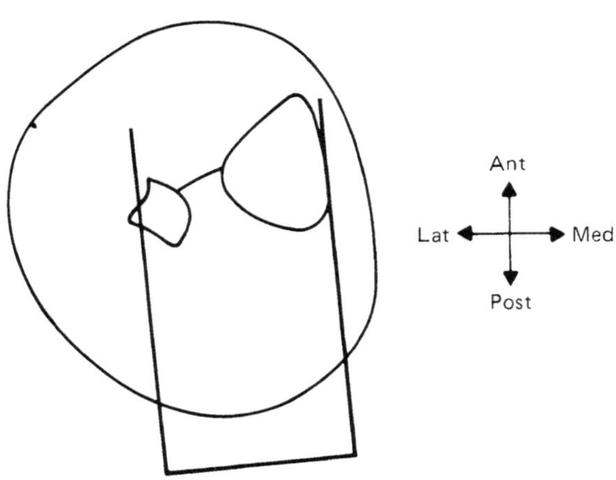
Ant
Lat
Med
Post

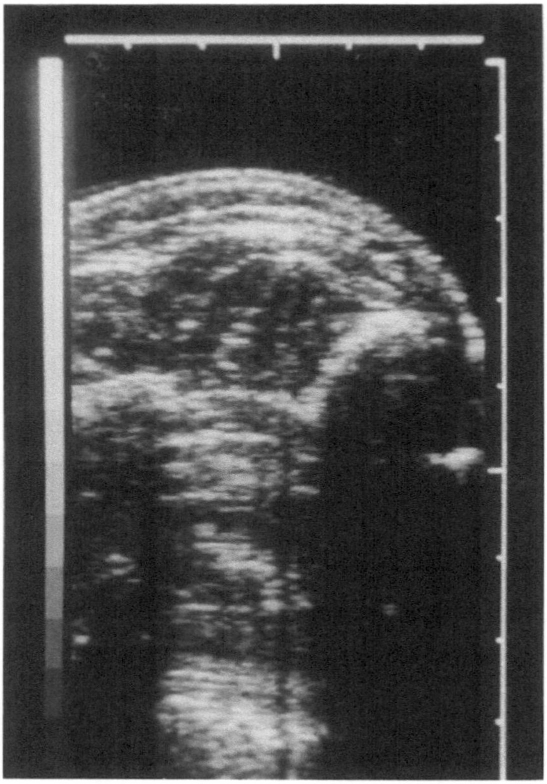

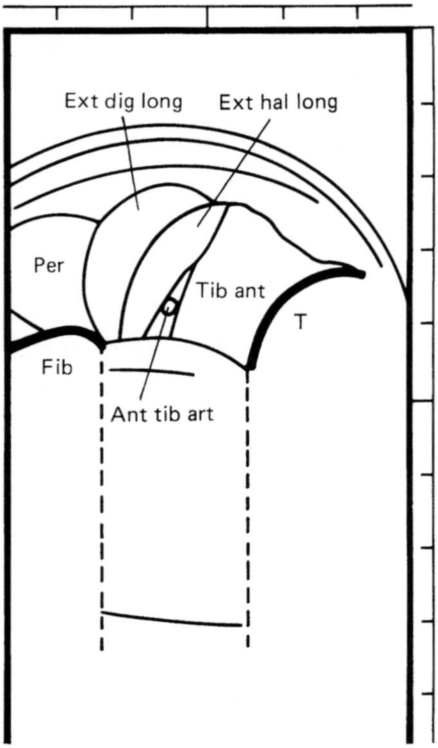
Ext dig long
Ext hal long
Per
Tib ant
T
Fib
Ant tib art

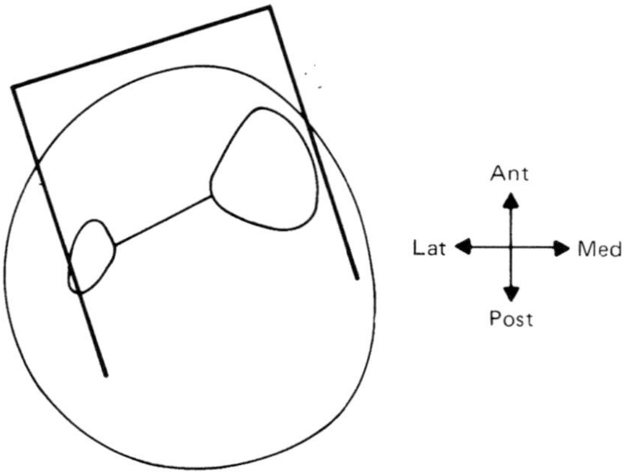
Ant
Lat
Med
Post

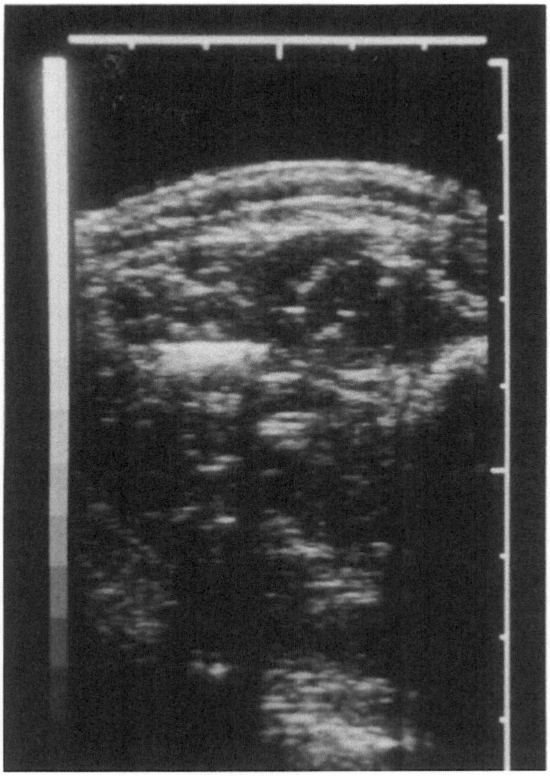

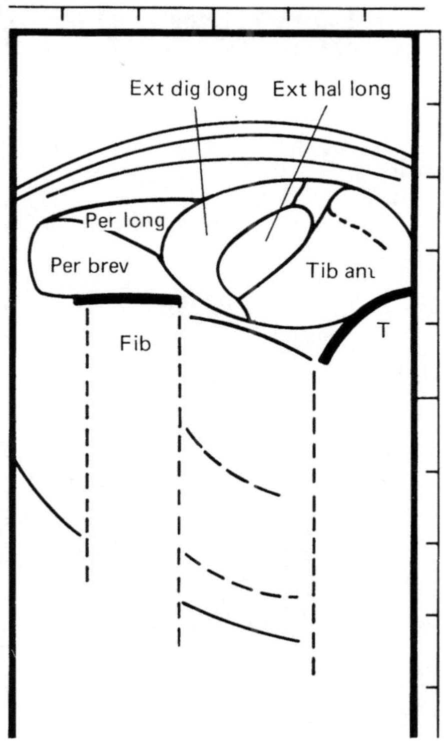
Ext dig long
Ext hal long
Per long
Per brev
Tib ant
Fib
T

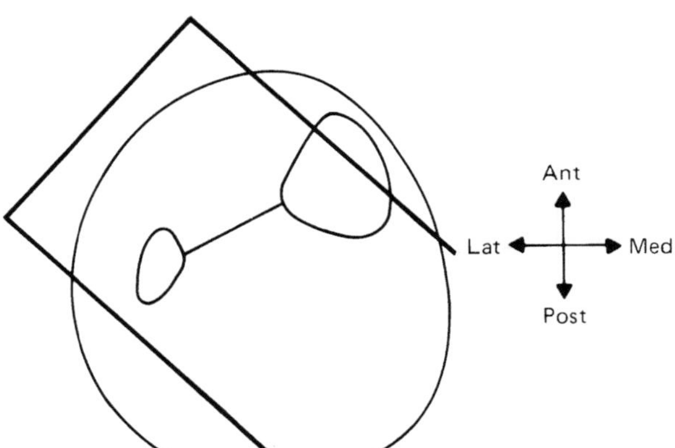
Ant
Lat
Med
Post

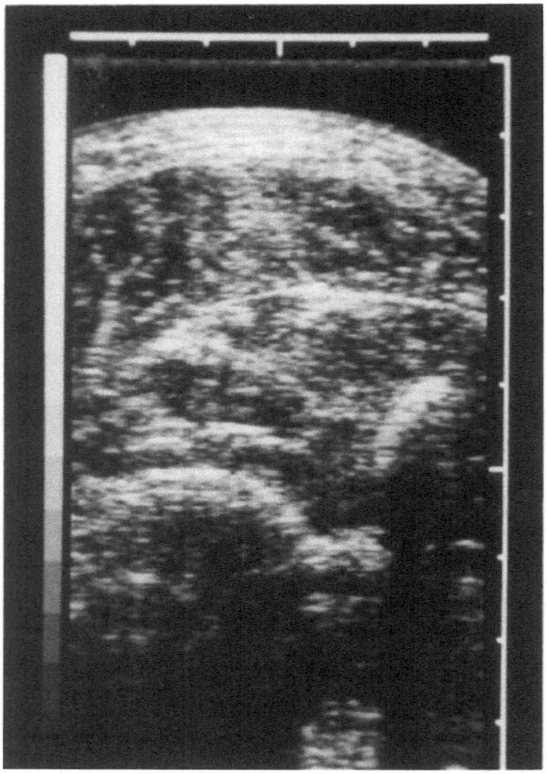

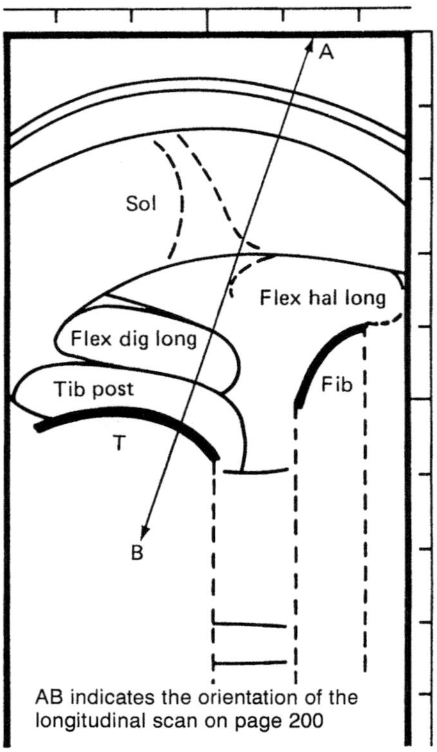
A
Sol
Flex hal long
Flex dig long
Tib post
Fib
T
B
AB indicates the orientation of the longitudinal scan on page 200

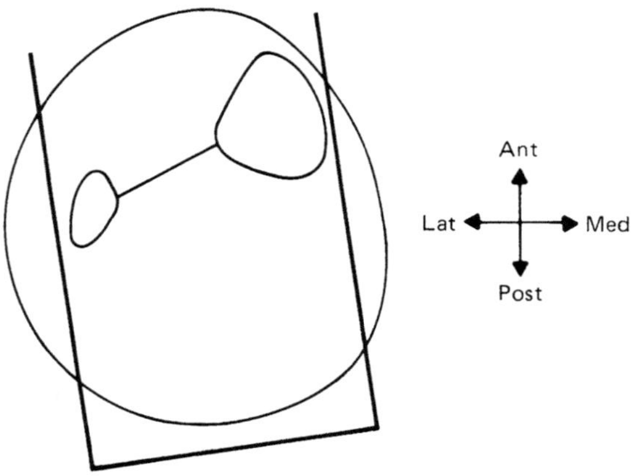
Ant
Lat
Med
Post

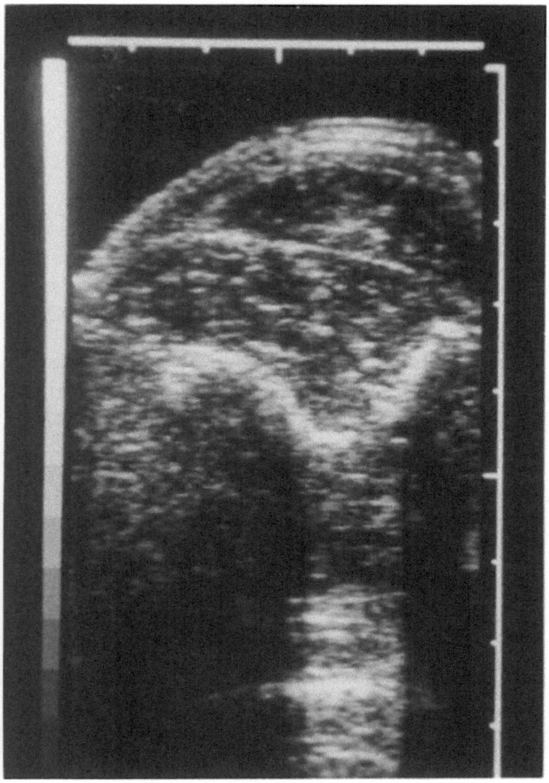

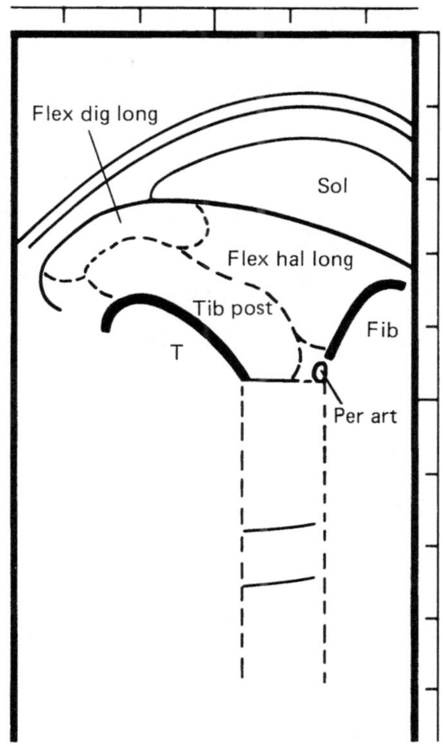
Flex dig long
Sol
Flex hal long
Tib post
T
Fib
Per art

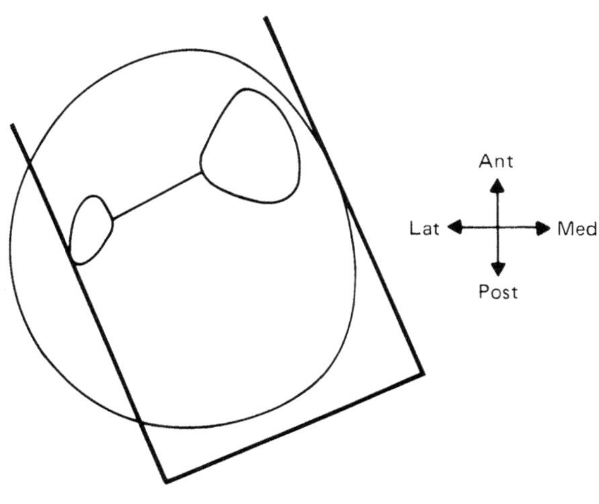
Ant
Lat
Med
Post

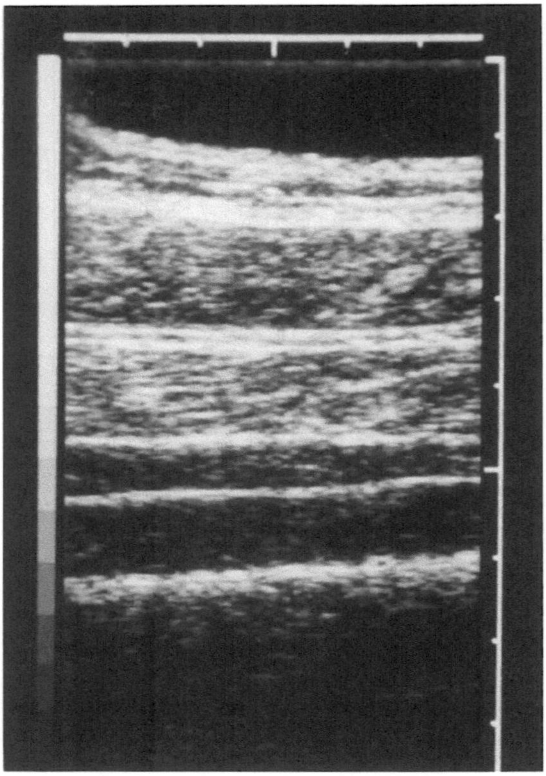

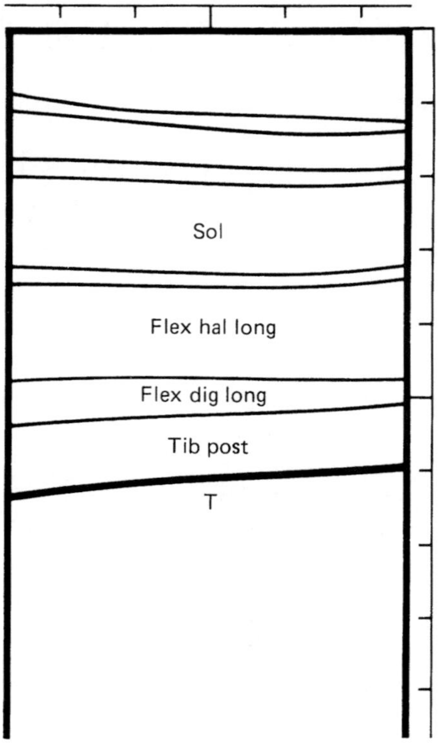
Sol
Flex hal long
Flex dig long
Tib post
T

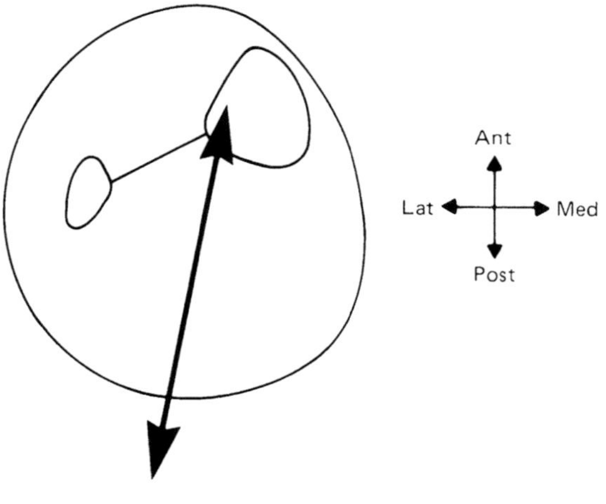
Ant
Lat
Med
Post

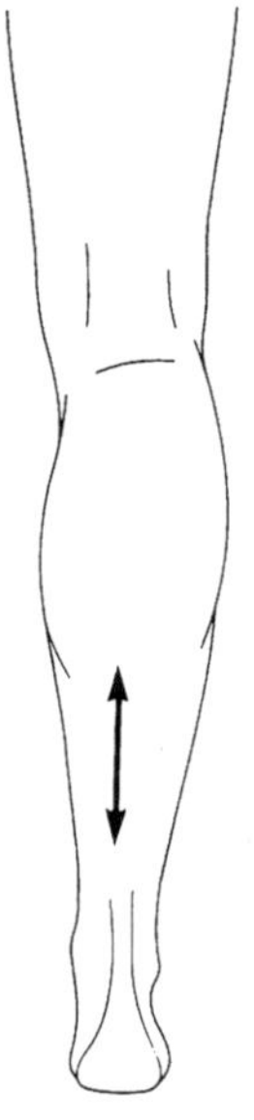

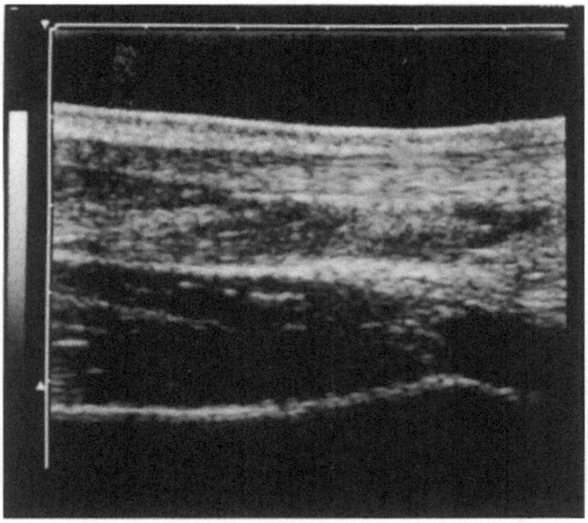

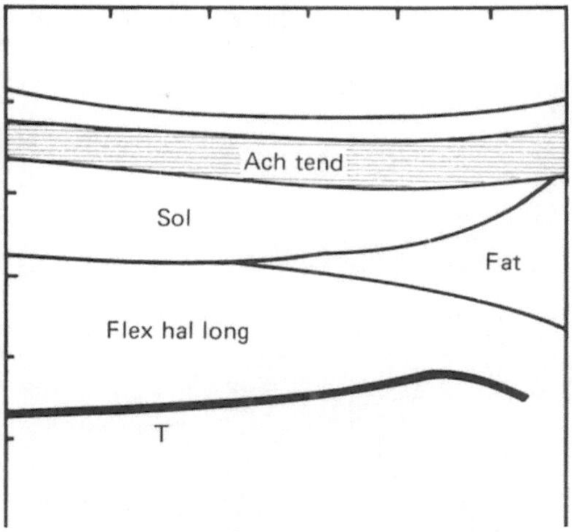
Ach tend
Sol
Fat
Flex hal long
T

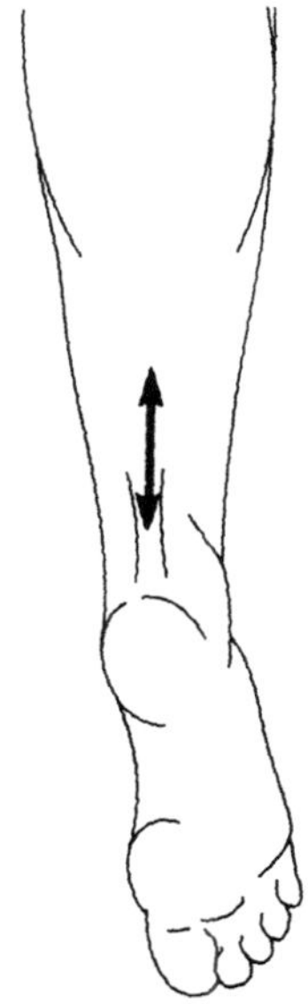

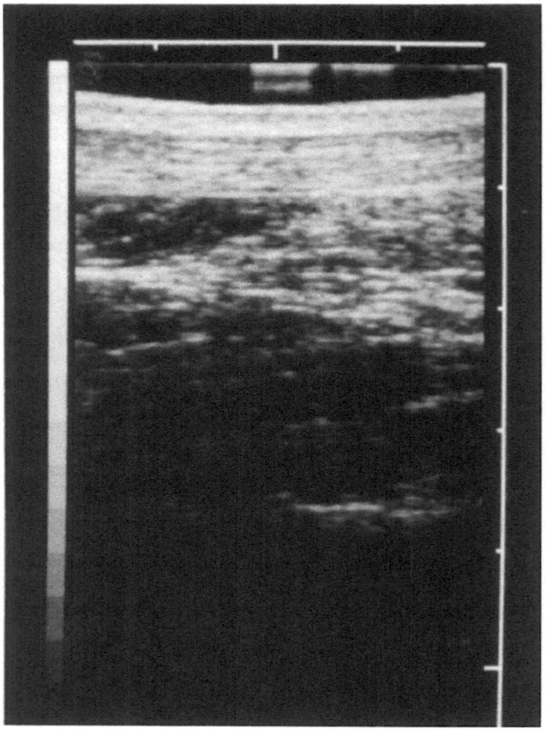

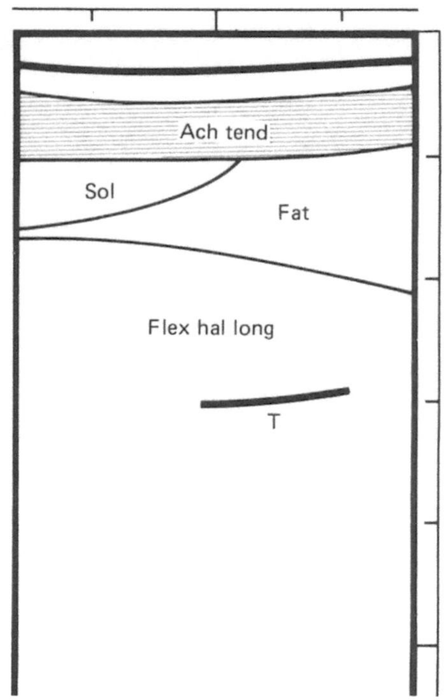
Ach tend
Sol
Fat
Flex hal long
T

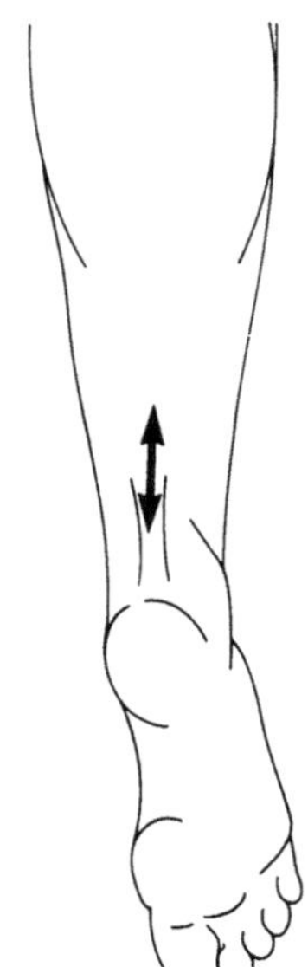

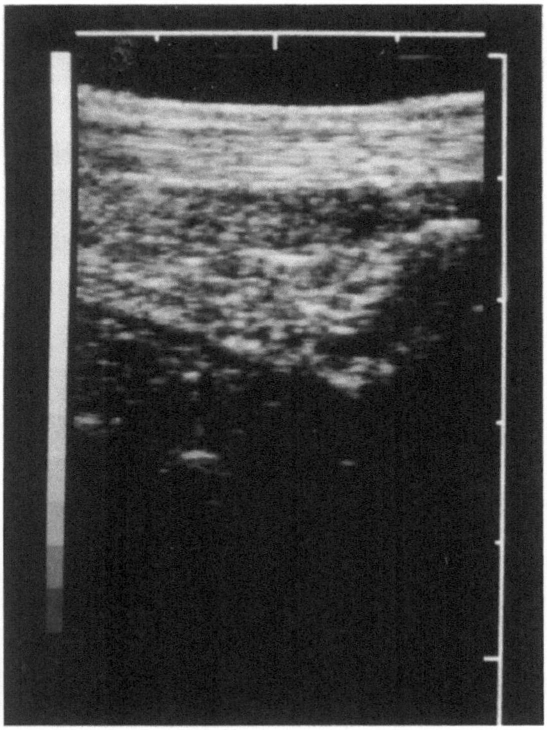

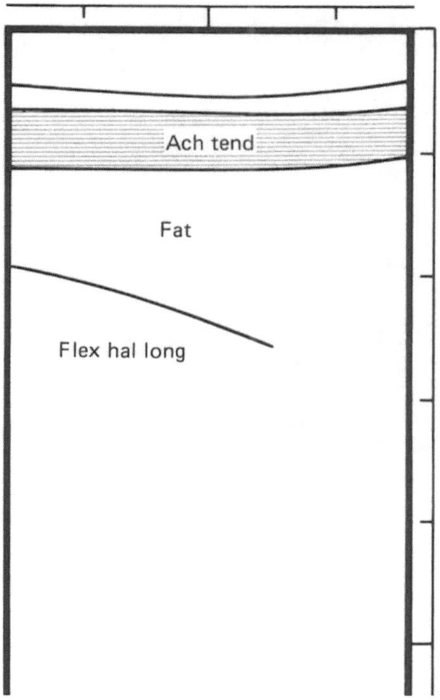
Ach tend
Fat
Flex hal long

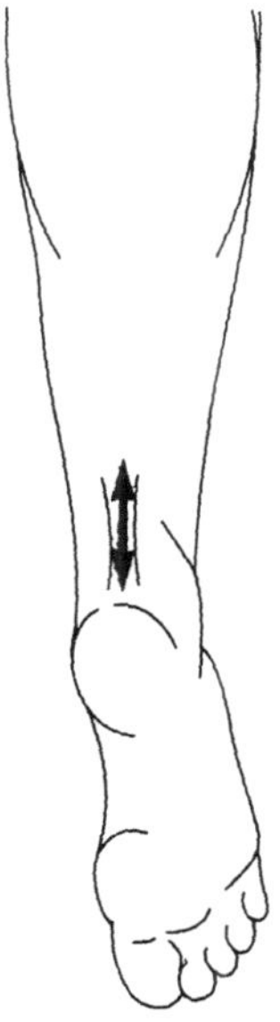

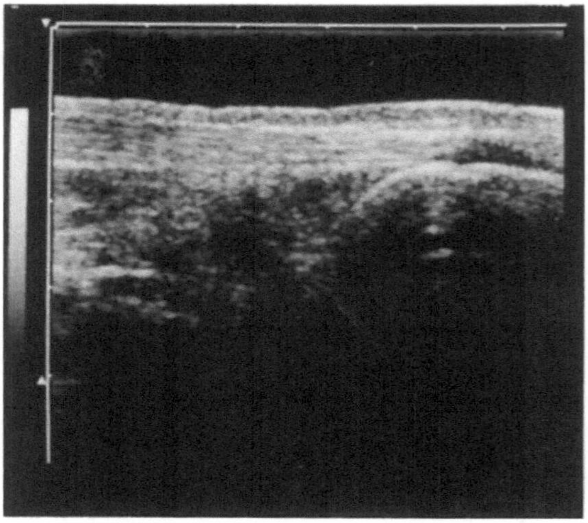

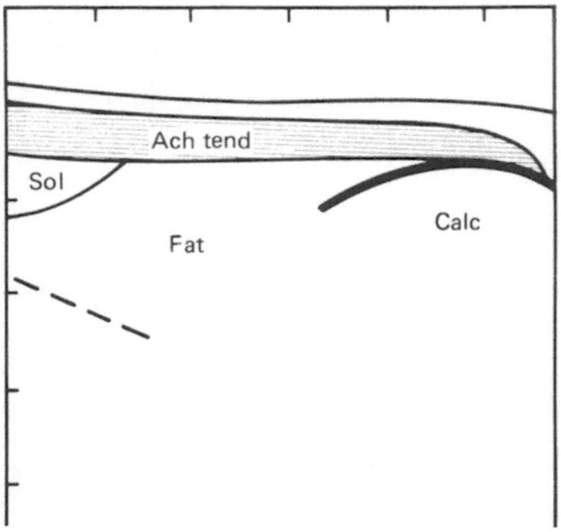
Ach tend
Sol
Fat
Calc

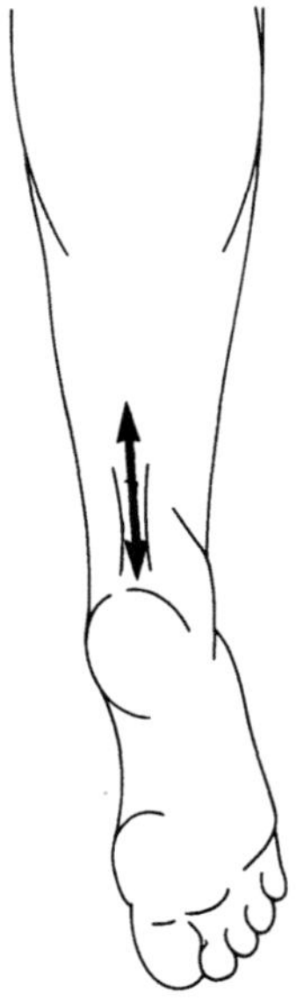

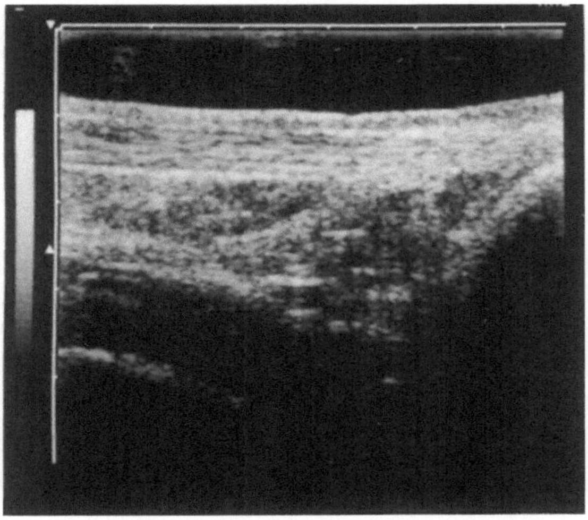

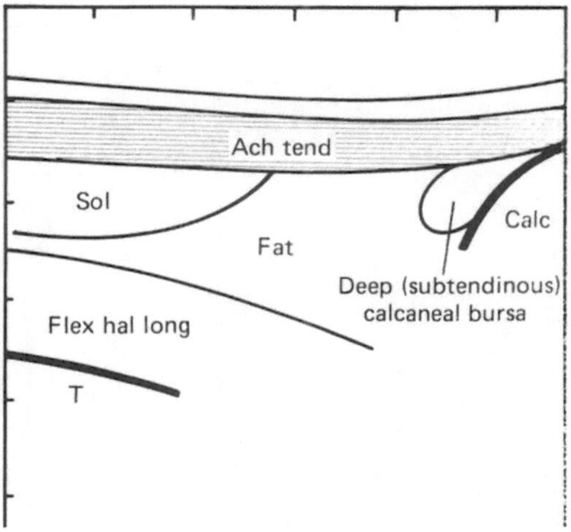
Ach tend
Sol
Fat
Calc
Deep (subtendinous) calcaneal bursa
Flex hal long
T

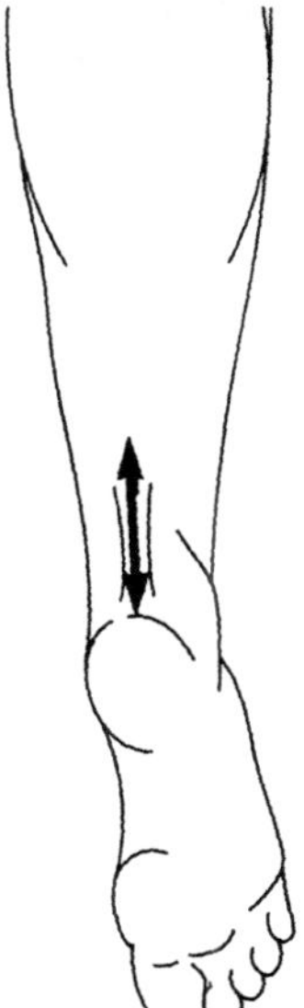

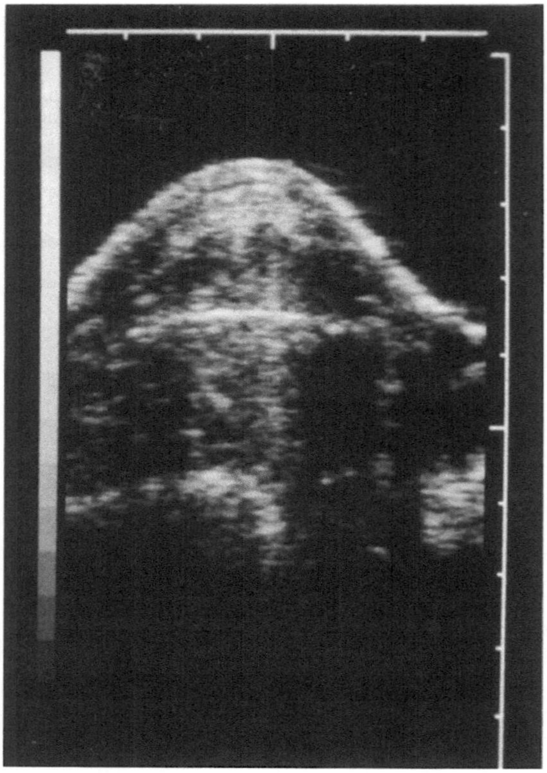

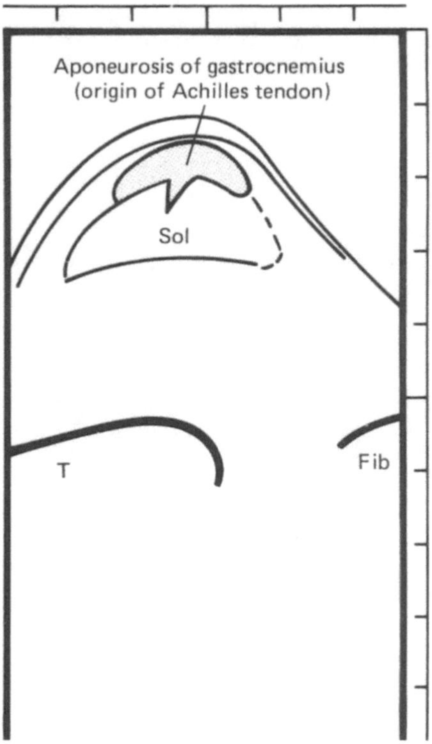
Aponeurosis of gastrocnemius
(origin of Achilles tendon)
Sol
T
Fib

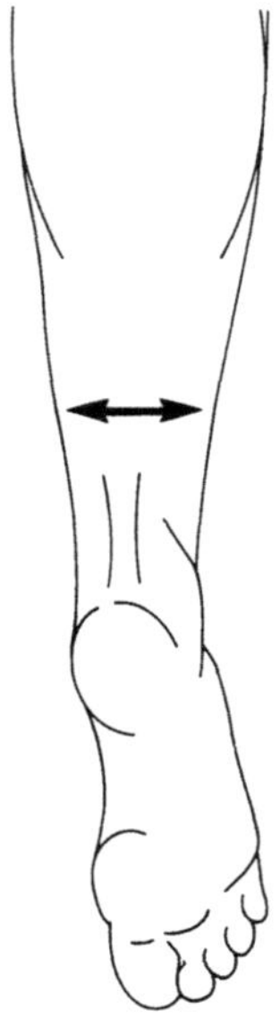

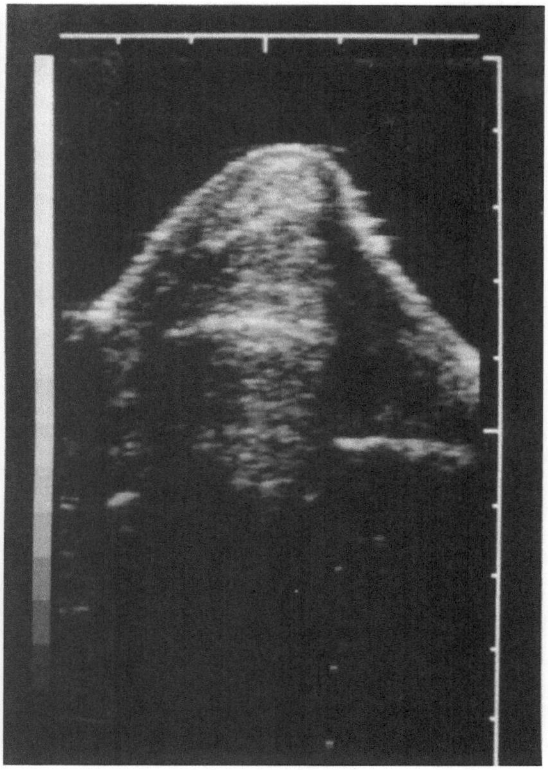

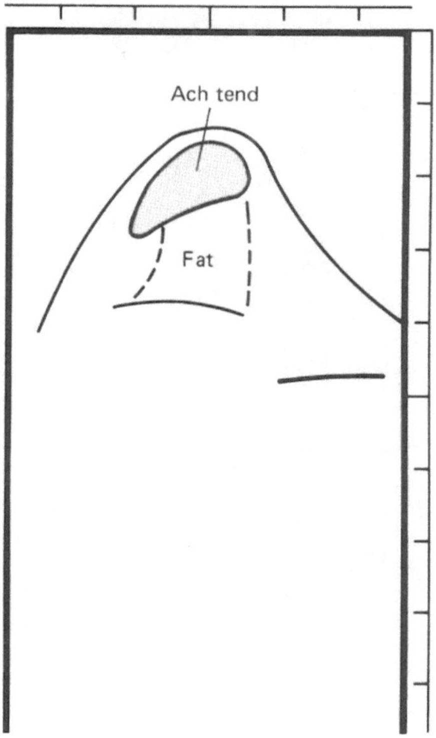
Ach tend
Fat

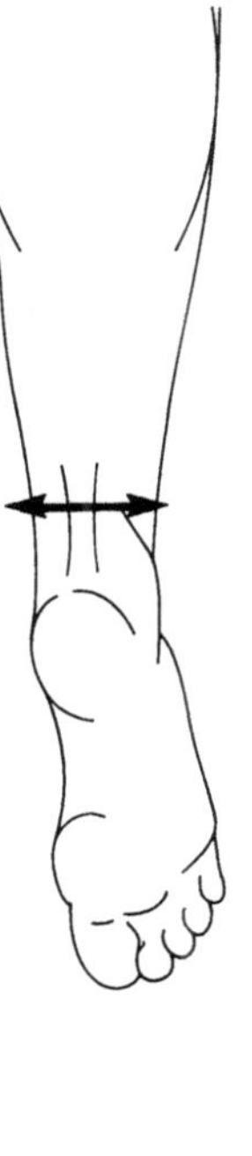

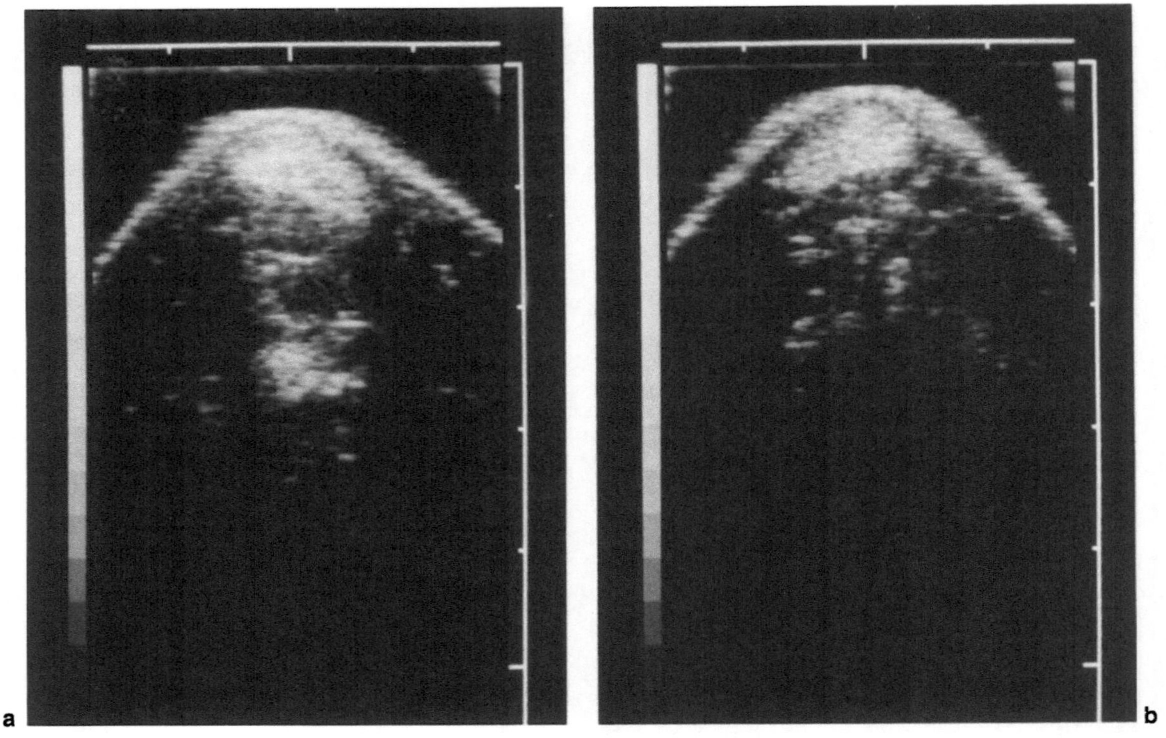

Transverse sonograms of left (a) and right (b) Achilles tendons in the same subject.

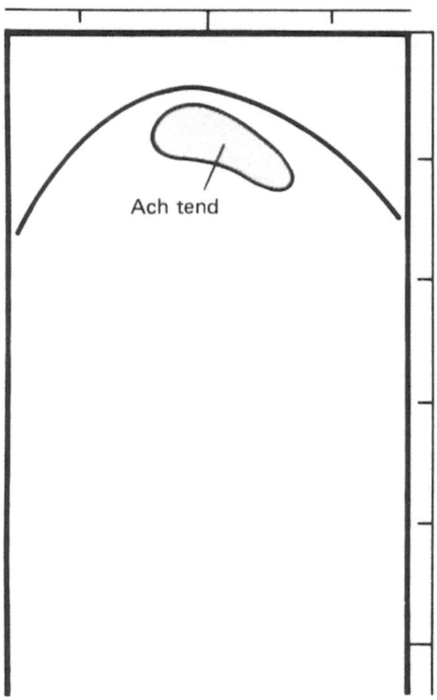
Ach tend

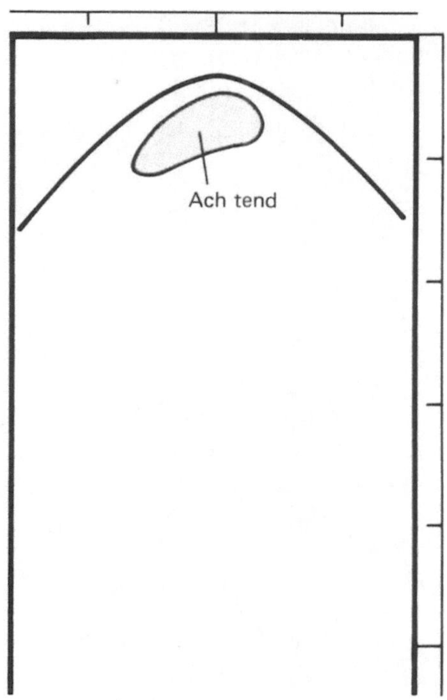
Ach tend

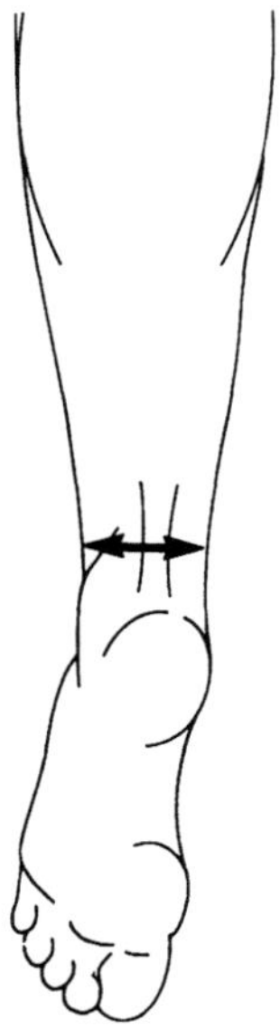

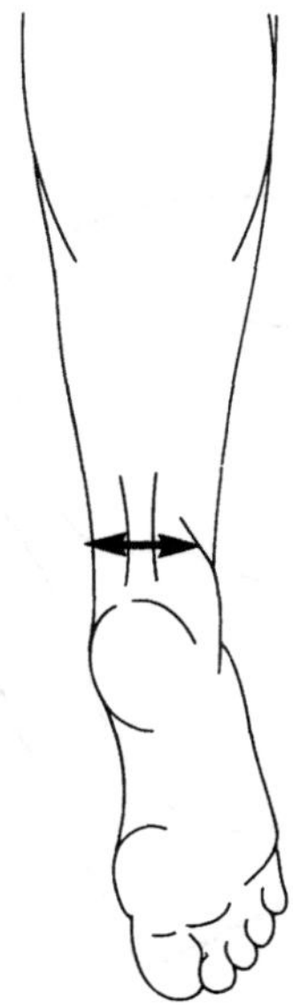

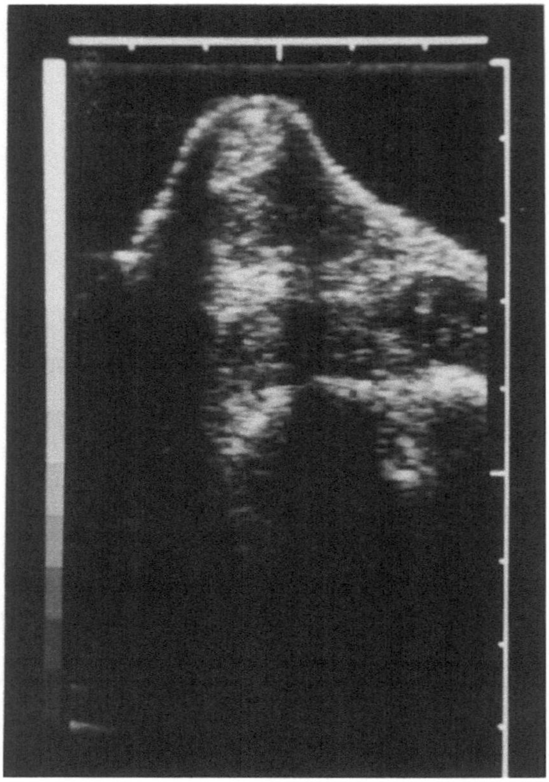

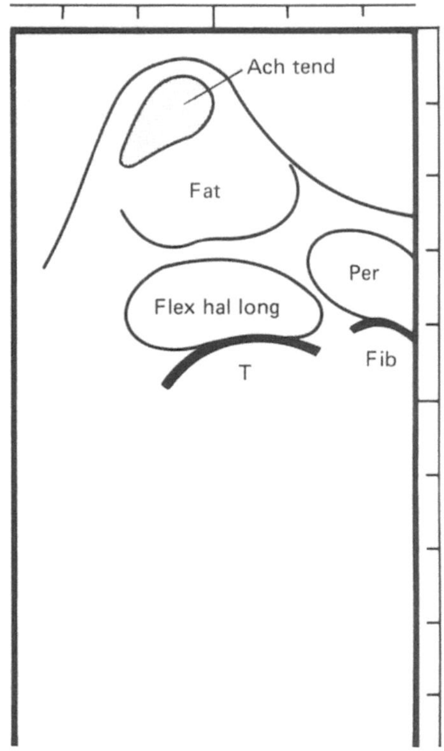
Ach tend
Fat
Per
Flex hal long
T
Fib

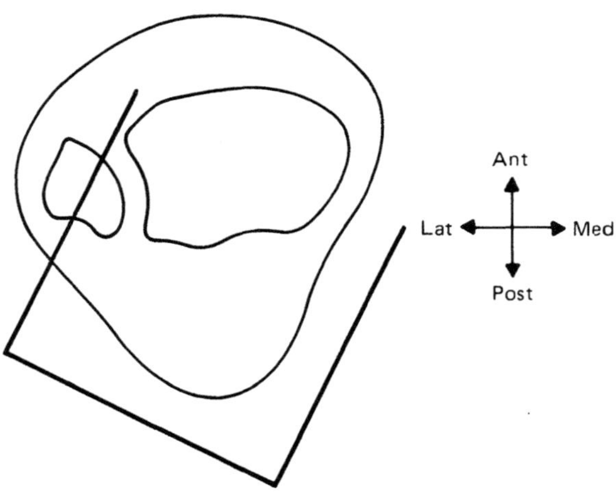
Ant
Lat
Med
Post

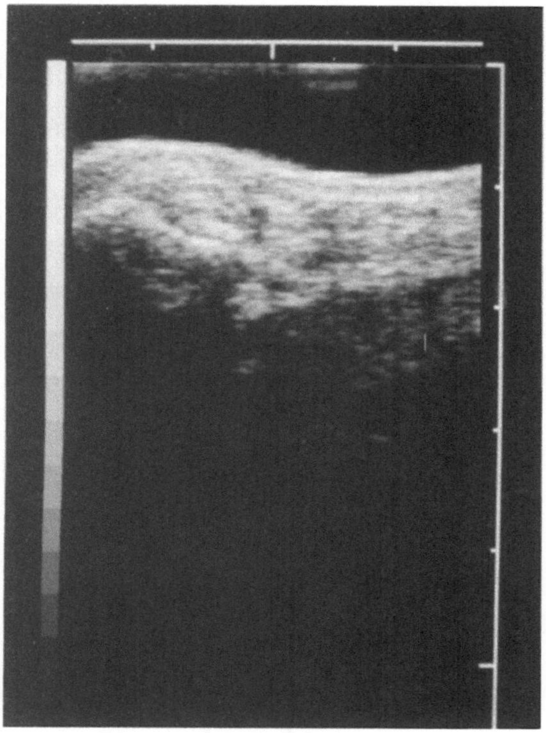

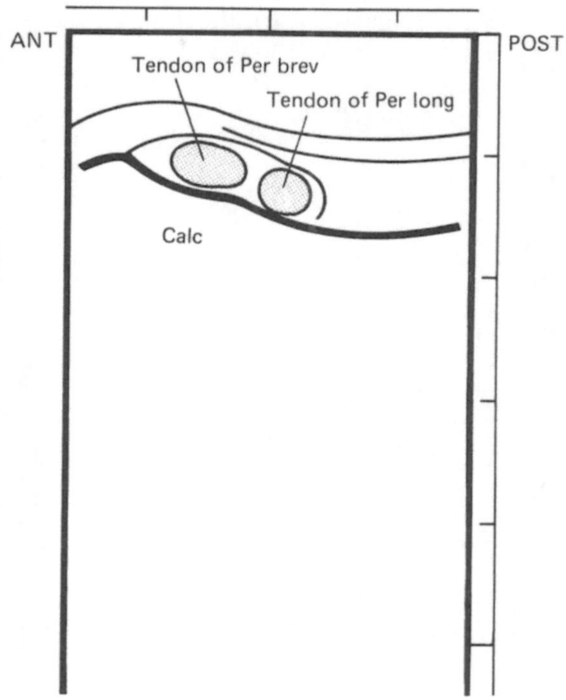
ANT
POST
Tendon of Per brev
Tendon of Per long
Calc

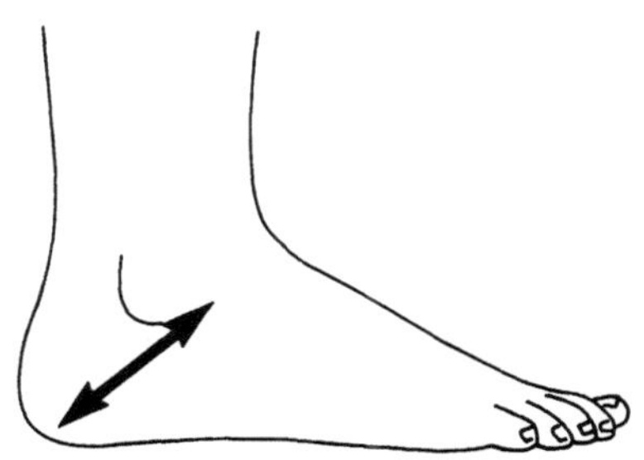

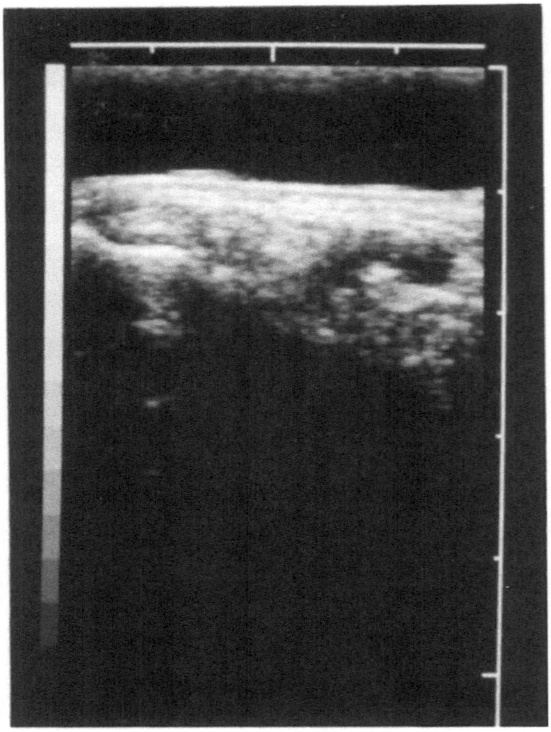

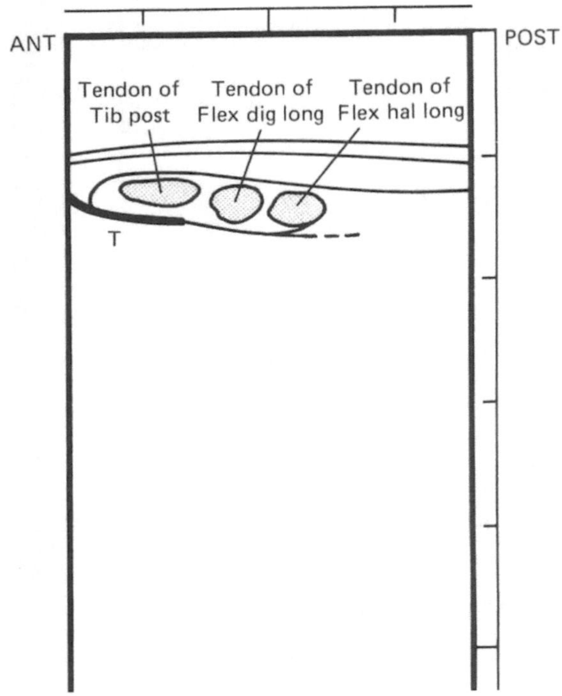
ANT
POST
Tendon of
Tib post
Tendon of
Flex dig long
Tendon of
Flex hal long
T

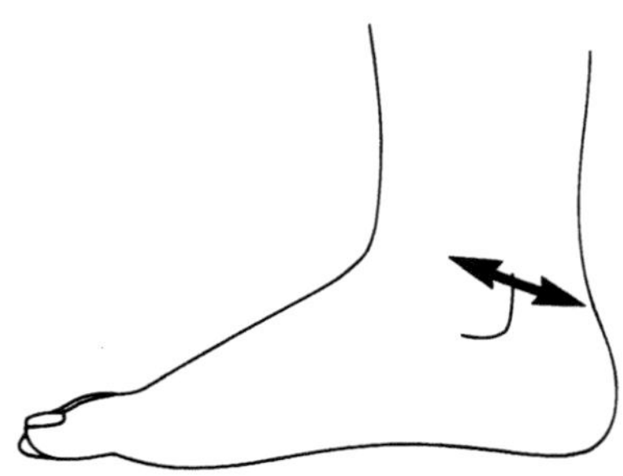

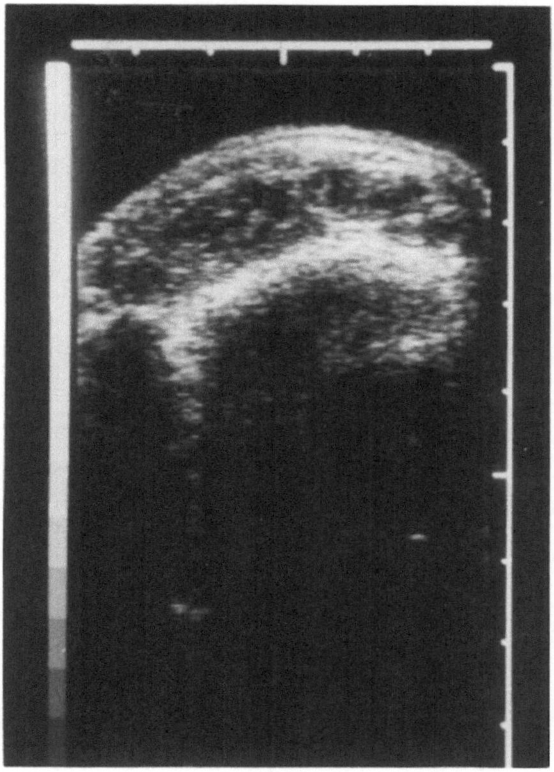

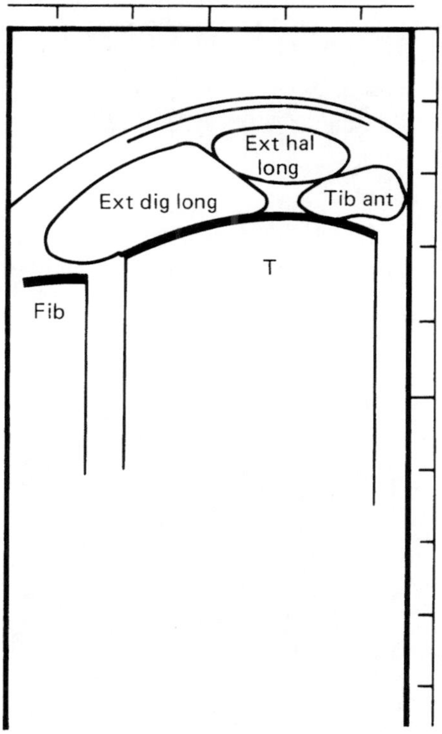
Ext hal
long
Ext dig long
Tib ant
T
Fib

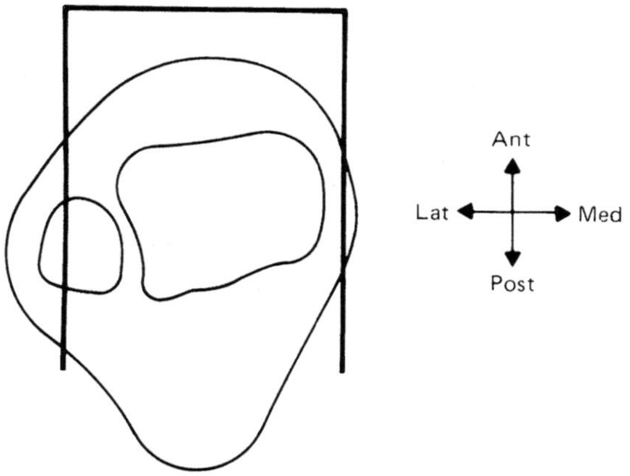
Ant
Lat
Med
Post

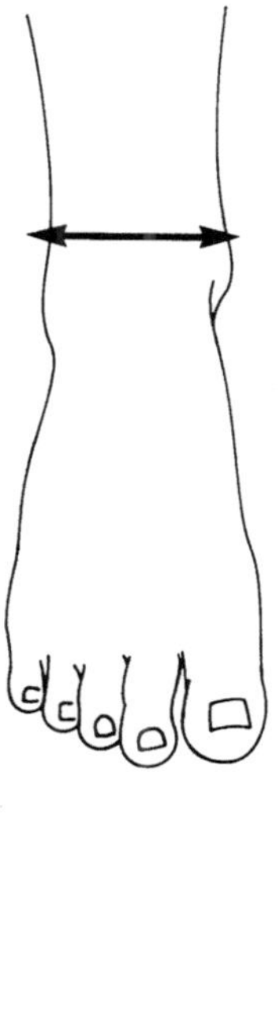

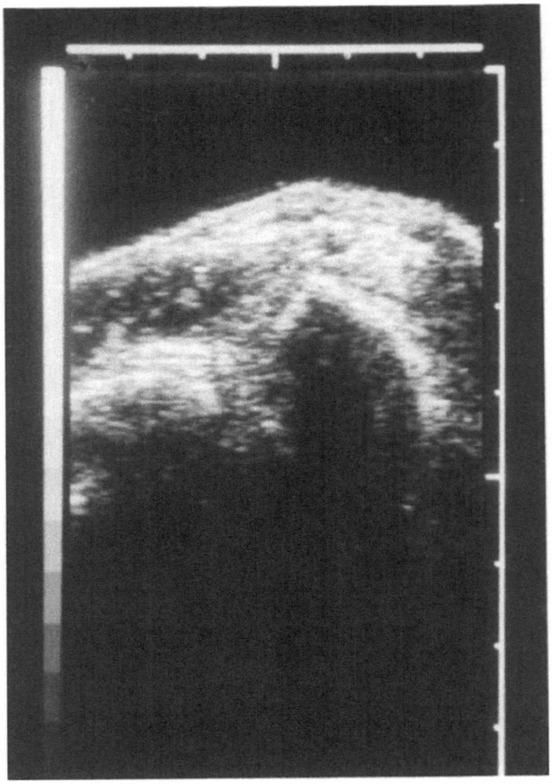

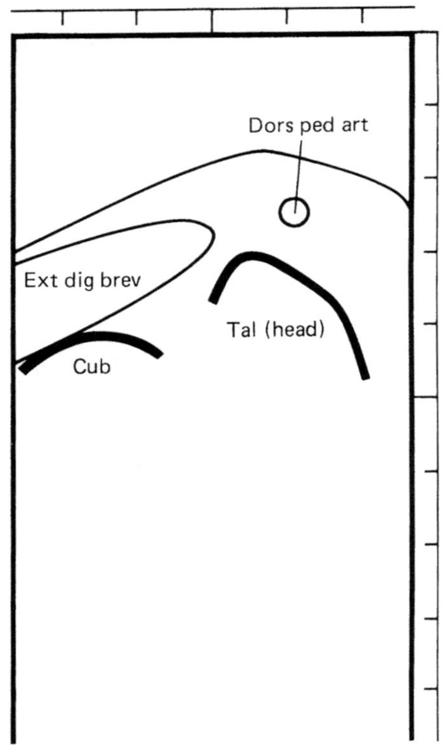
Dors ped art
Ext dig brev
Tal (head)
Cub

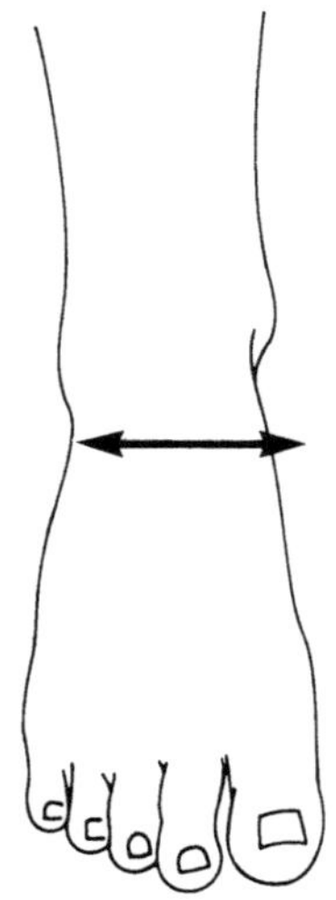

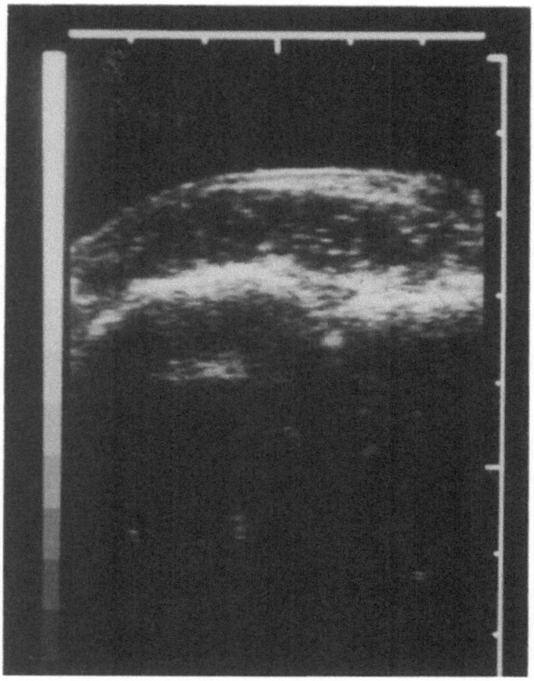

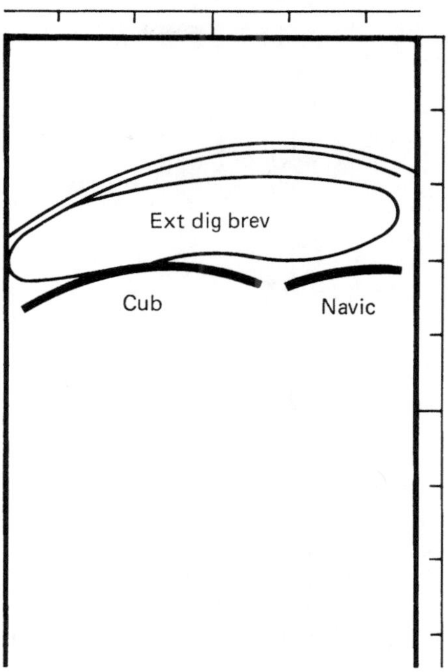
Ext dig brev
Cub
Navic

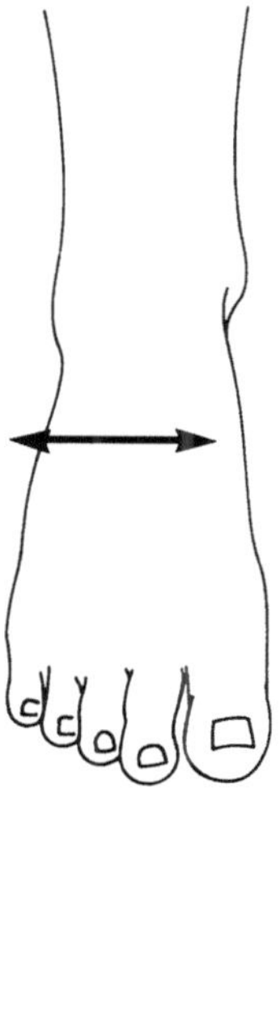

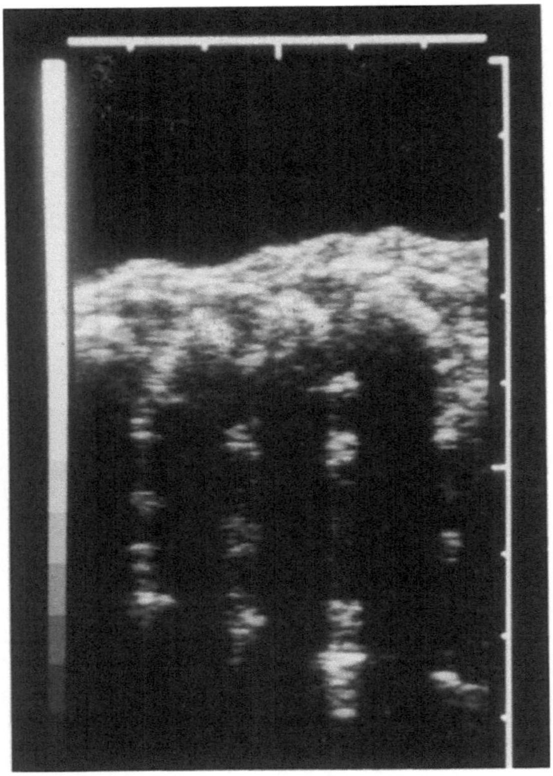

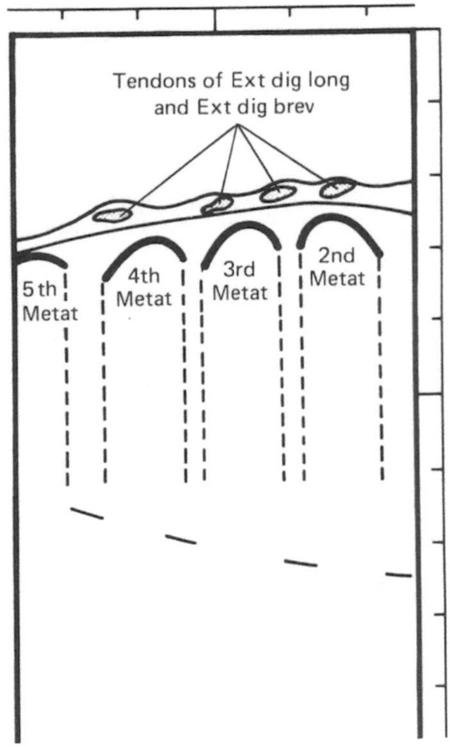
Tendons of Ext dig long
and Ext dig brev
5th
Metat
4th
Metat
3rd
Metat
2nd
Metat

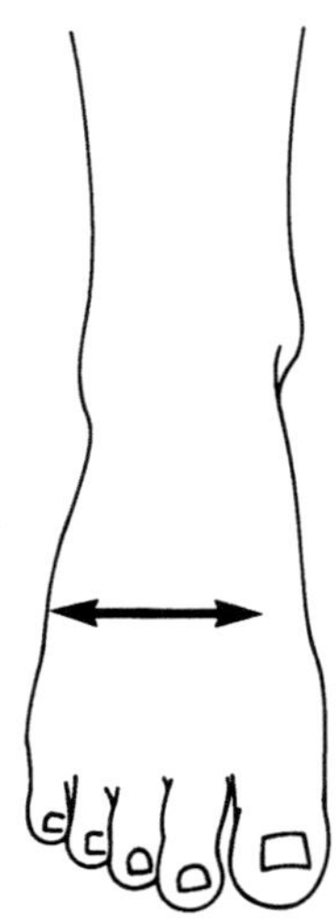

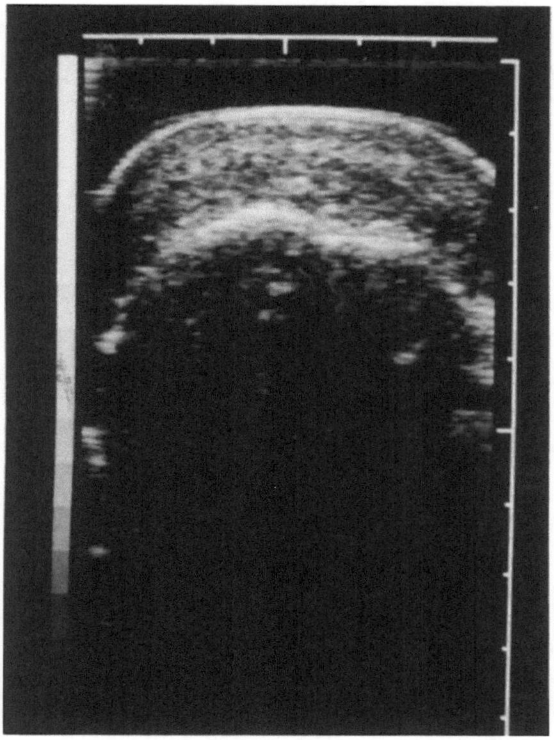

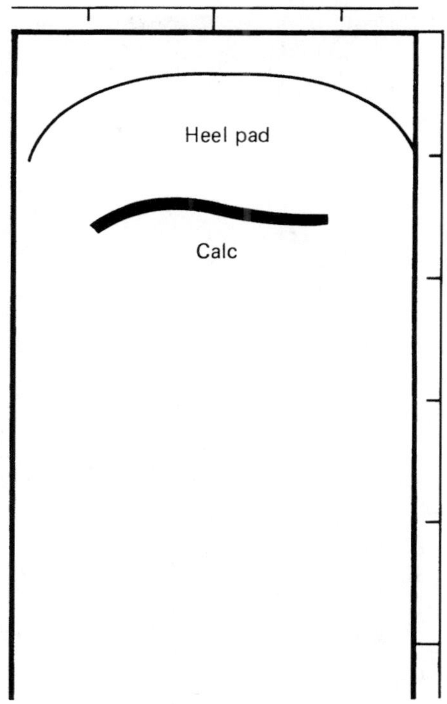
Heel pad
Calc

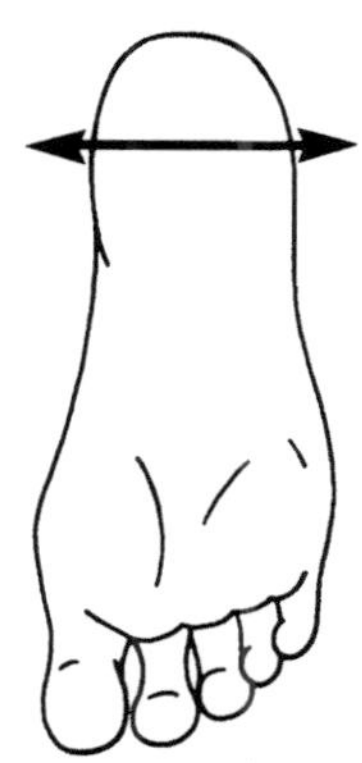

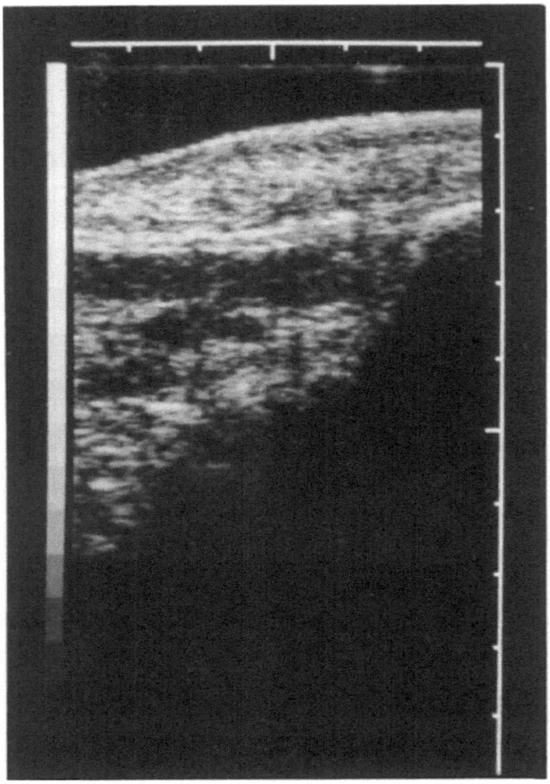

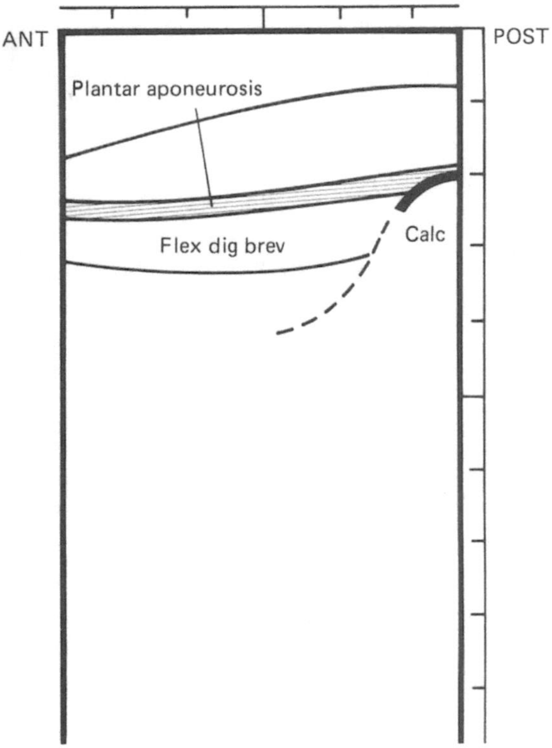
ANT
POST
Plantar aponeurosis
Flex dig brev
Calc

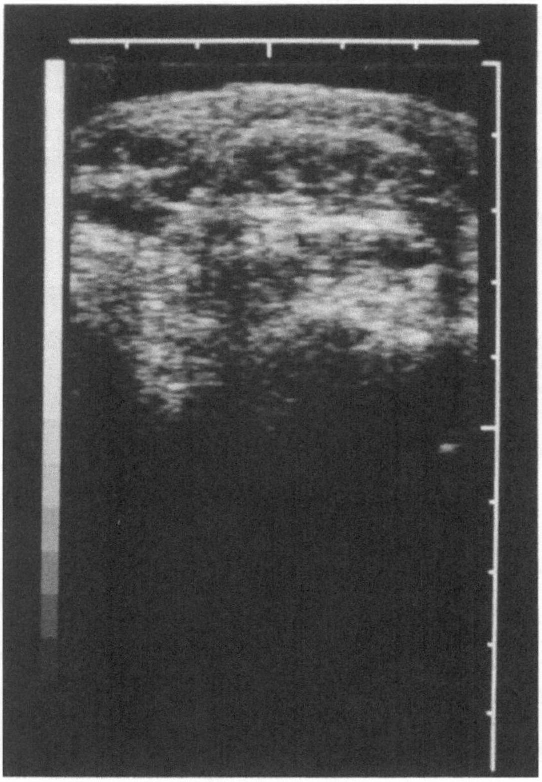

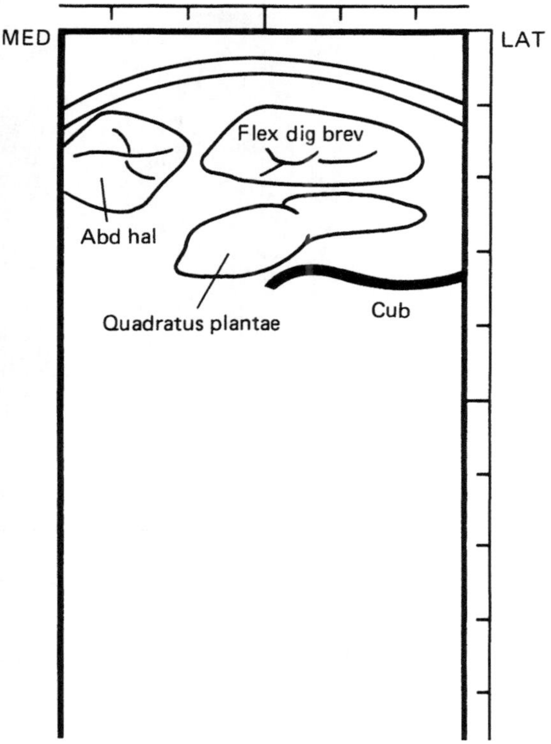
MED
LAT
Flex dig brev
Abd hal
Quadratus plantae
Cub

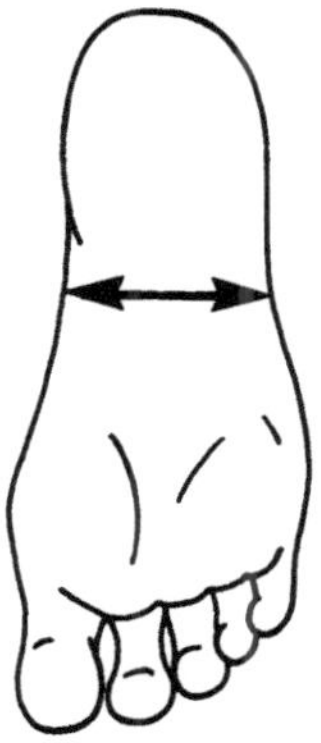

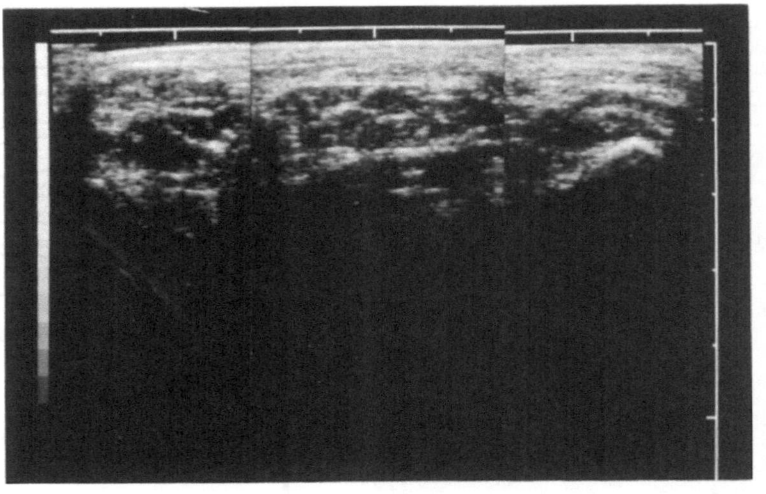

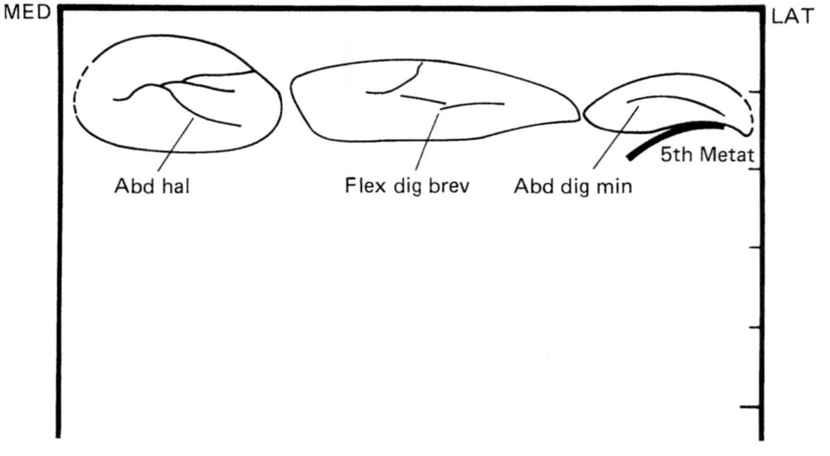
MED
LAT
Abd hal
Flex dig brev
Abd dig min
5th Metat

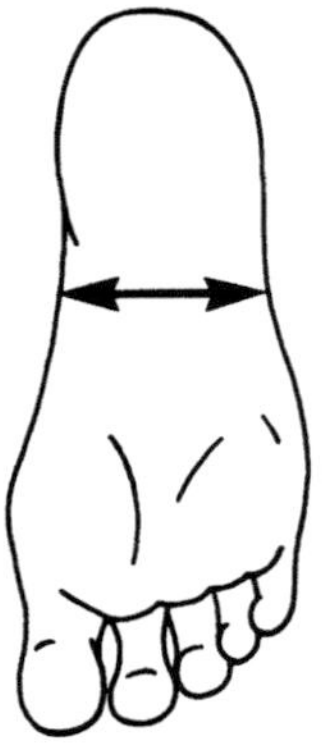

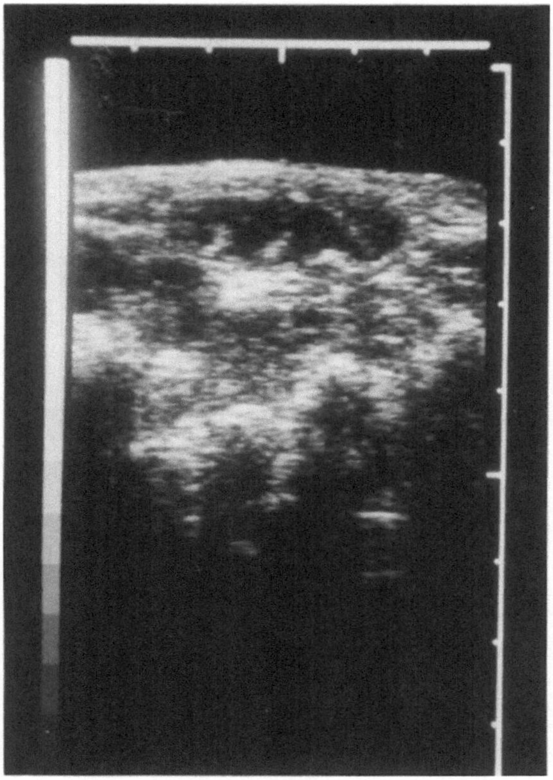

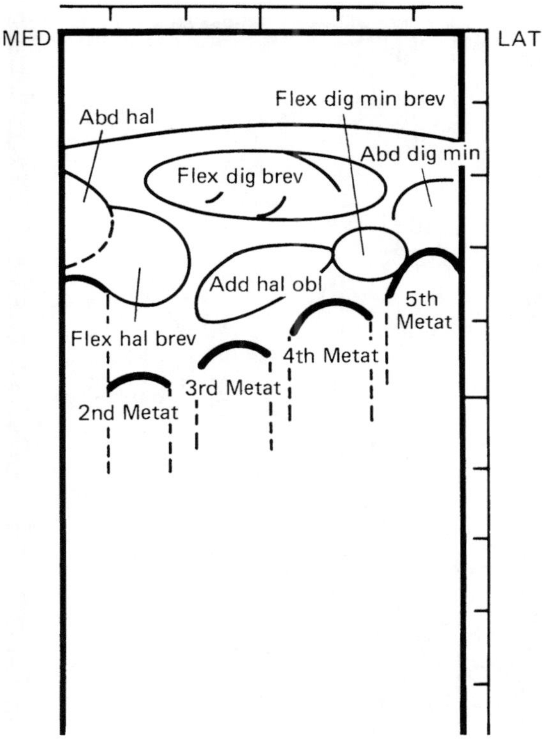
MED
LAT
Abd hal
Flex dig min brev
Abd dig min
Flex dig brev
Add hal obl
Flex hal brev
5th Metat
4th Metat
3rd Metat
2nd Metat

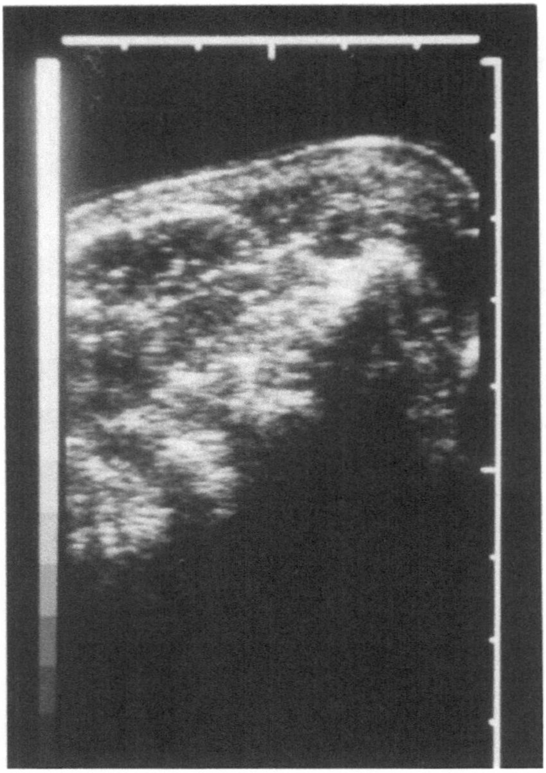

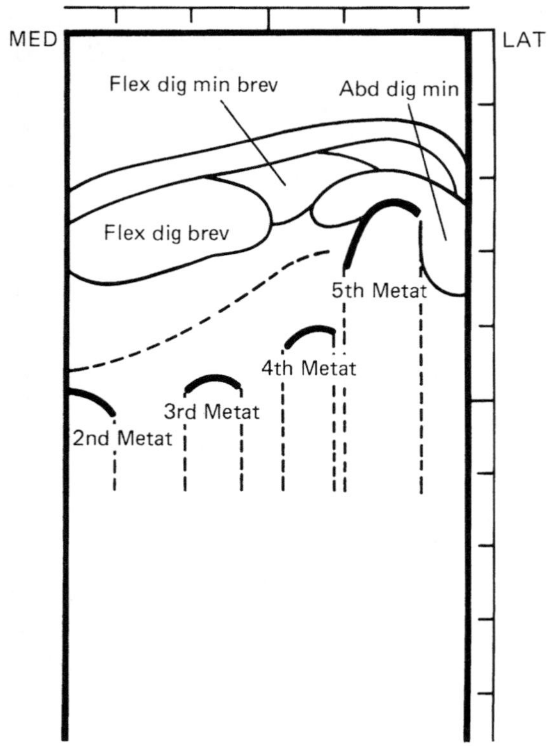
MED
LAT
Flex dig min brev
Abd dig min
Flex dig brev
5th Metat
4th Metat
3rd Metat
2nd Metat

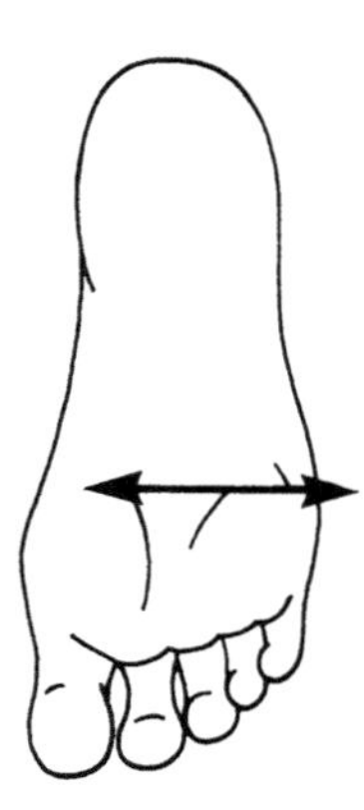

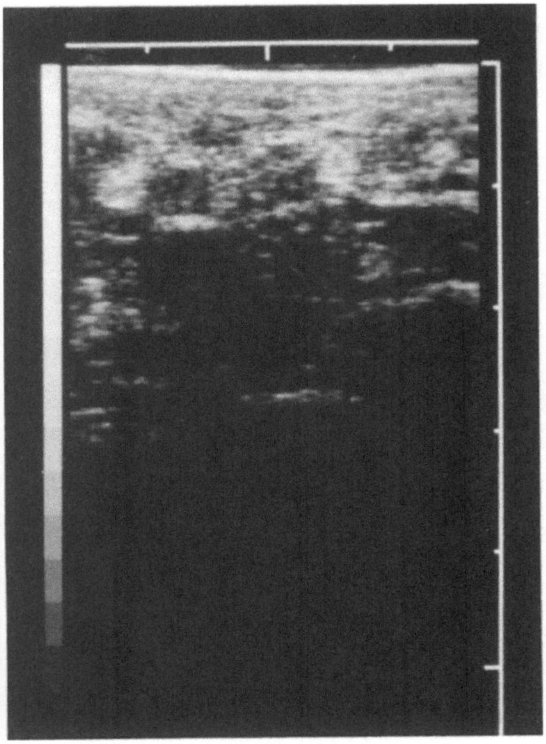

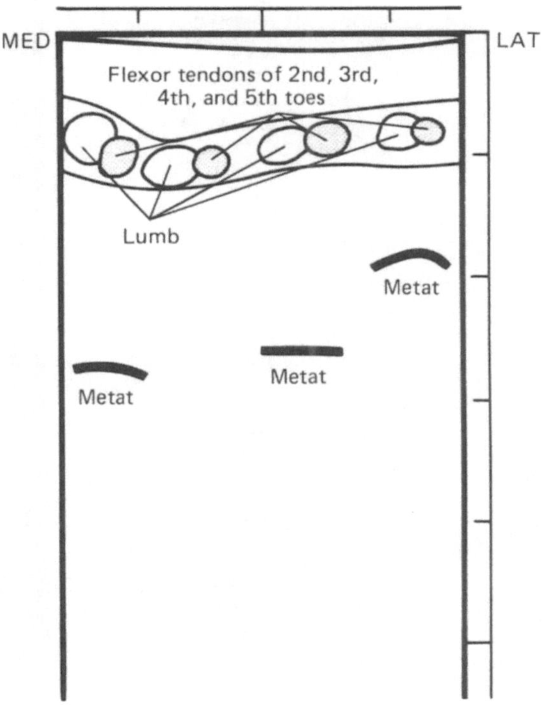
MED
LAT
Flexor tendons of 2nd, 3rd, 4th, and 5th toes
Lumb
Metat
Metat
Metat

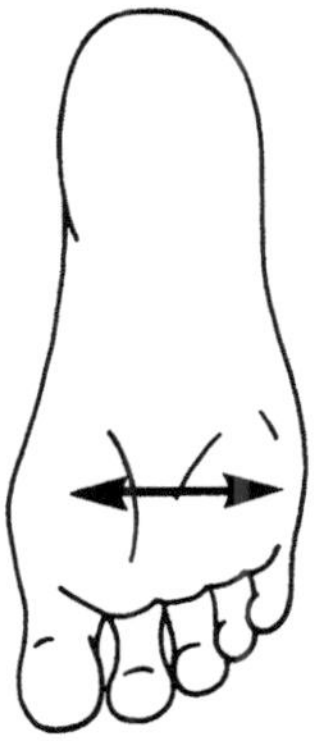

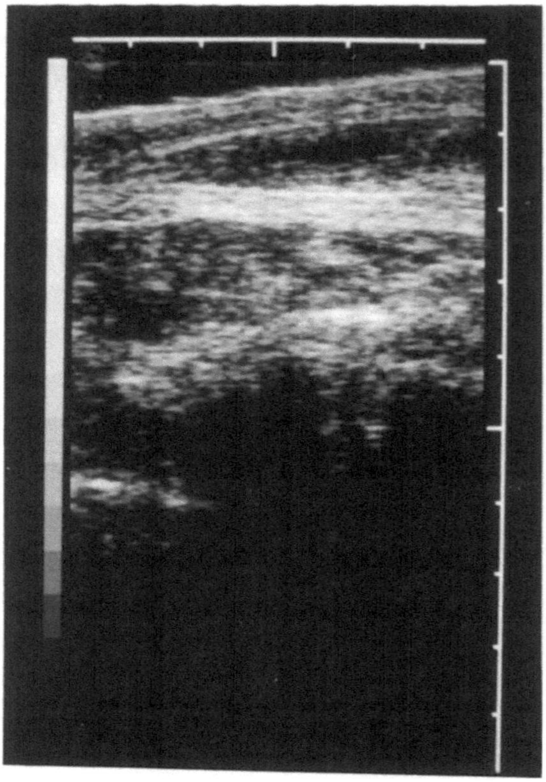

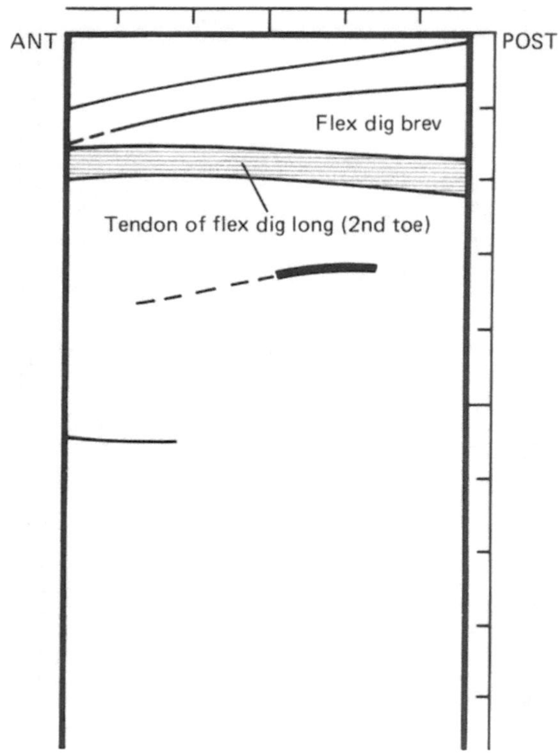
ANT
POST
Flex dig brev
Tendon of flex dig long (2nd toe)

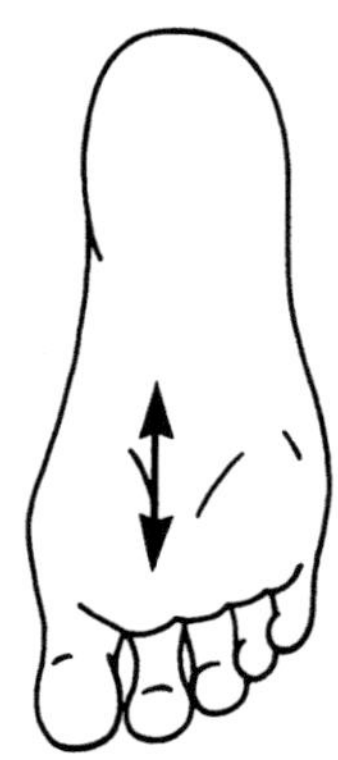

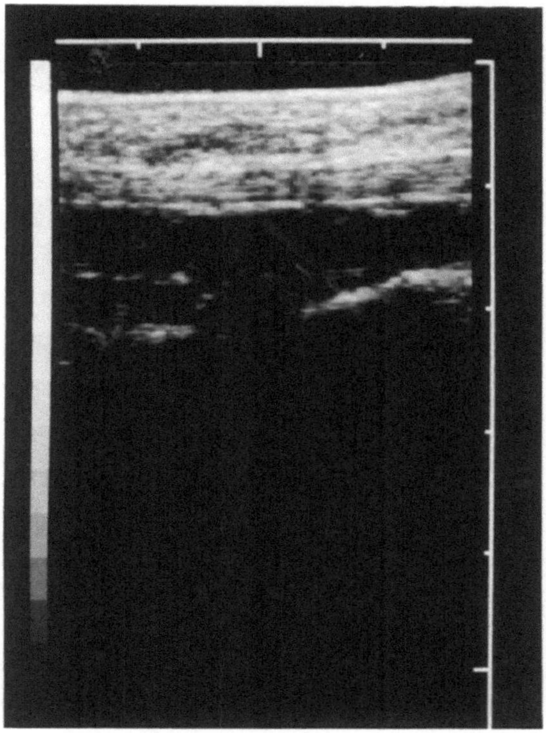

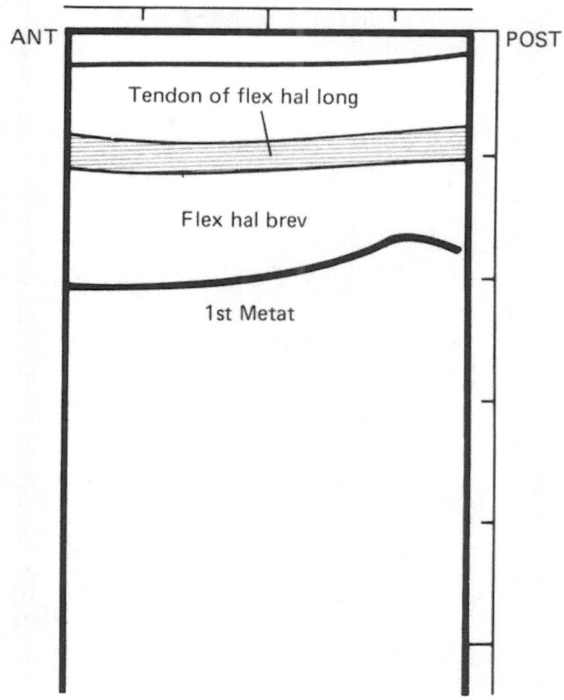
ANT
POST
Tendon of flex hal long
Flex hal brev
1st Metat

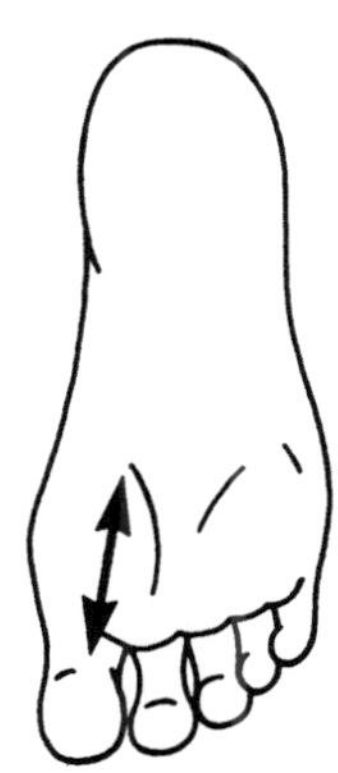

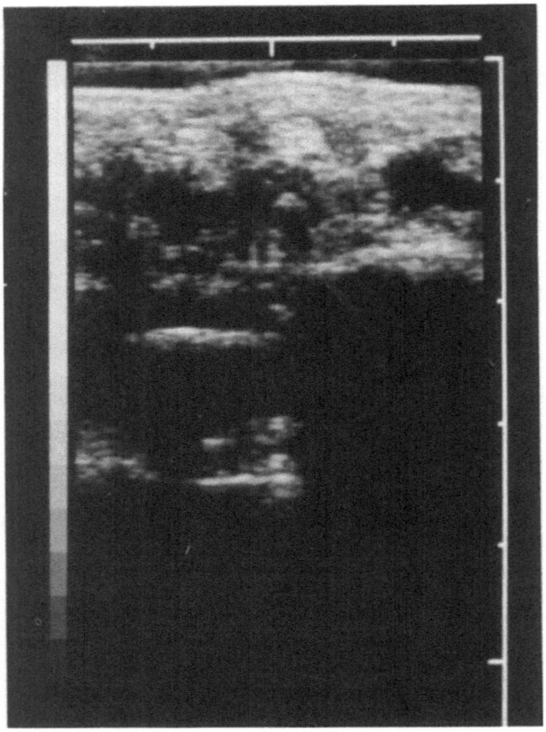

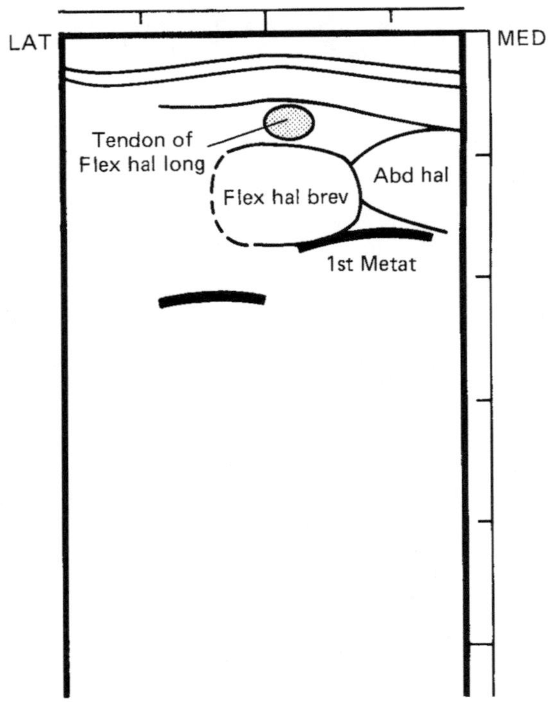
LAT
MED
Tendon of
Flex hal long
Abd hal
Flex hal brev
1st Metat

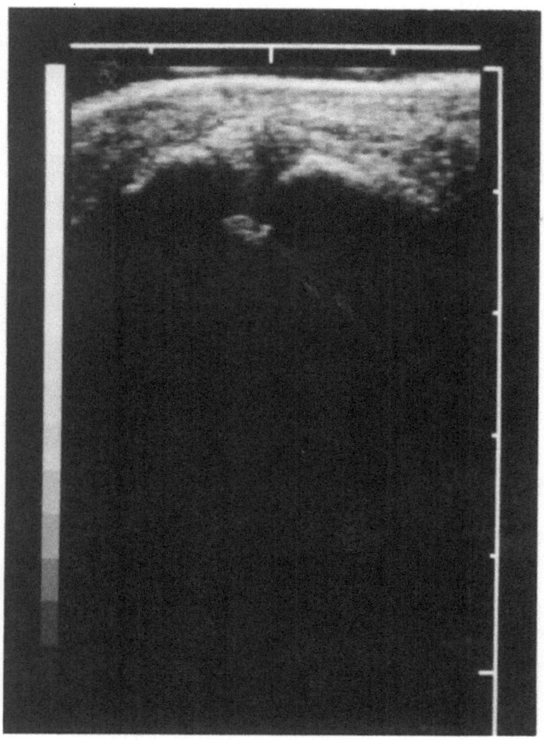

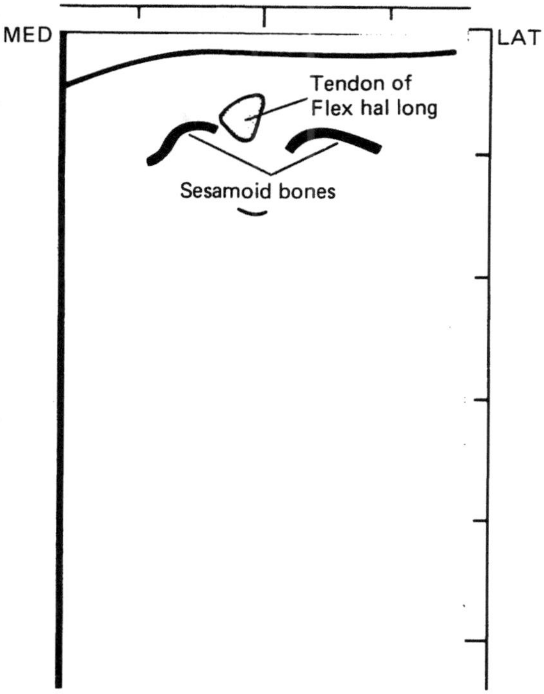
MED
LAT
Tendon of
Flex hal long
Sesamoid bones

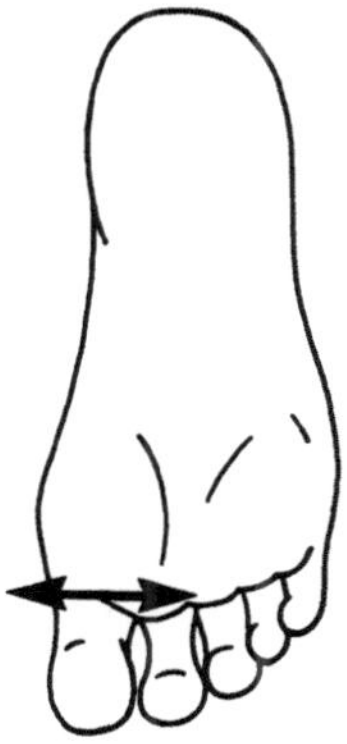

Index

MIX
Papier aus verantwortungsvollen Quellen
Paper from responsible sources
FSC® C105338

If you have any concerns about our products,
you can contact us on
ProductSafety@springernature.com

In case Publisher is established outside the EU,
the EU authorized representative is:
Springer Nature Customer Service Center GmbH
Europaplatz 3, 69115 Heidelberg, Germany

Printed by Libri Plureos GmbH
in Hamburg, Germany